INVISIBLE UNIVERSE INTERNATIONAL CONFERENCE

INVISIBLE UNIVERSE INTERNATIONAL CONFERENCE

Paris, France   29 June – 3 July 2009

EDITORS

Jean-Michel Alimi
Laboratory Universe and Theories (LUTh)
Observatory of Paris, France

André Füzfa
University of Namur
Namur, Belgium

All papers have been peer reviewed

SPONSORING ORGANIZATIONS
CNRS-INSU
CNRS-IN2P3
Observatory of Paris
GEMALTO Enterprise
GIS "Physique des 2 Infinis"
CEA
Cantine Lavorata Enterprises

AIP
American Institute of Physics

Melville, New York, 2010
AIP CONFERENCE PROCEEDINGS ■ VOLUME 1241

Editors:

Jean-Michel Alimi
Laboratory Universe and Theories
Place Jules Janssen
Observatory of Meudon
F-92190 Meudon
France

E-mail: jean-michel.alimi@obspm.fr

André Füzfa
GAMASCO, Unité de Systèmes Dynamiques
Facultés Universitaires Notre-Dame de la Paix
Rempart de la Vierge 8
B-5000 Namur
Belgium

E-mail: andre.fuzfa@fundp.ac.be

L.C. Catalog Card No. 2010903831
ISBN 978-0-7354-0789-3
ISSN 0094-243X
Printed in the United States of America

CONTENTS

CONCLUSION

POSTERS

* **The complete papers are available on the accompanying CD-ROM and online on the AIP website at http://proceedings.aip.org/proceedings/confproceed/1241.jsp**

PAPERS ON CD-ROM AND ONLINE ONLY

OBSERVATIONAL ASTROPHYSICAL ASPECT OF DARK ENERGY AND DARK MATTER

EXPERIMENTAL PARTICLE PHYSICS ASPECTS OF DARK MATTER AND DARK MATTER CANDIDATES

GRAVITATION AND COSMOLOGY

DYNAMICS AND PROPERTIES OF THE INVISIBLE UNIVERSE

POSTERS

Welcome

Welcome to the event "Invisible Universe". So here we are. We did it. The team entrusted with the heady task of organising this event is absolutely delighted that the inauguration is taking place today and that, to mark that event, the conference entitled "Invisible Universe International Conference", which is the lynchpin of the entire project, is to be held in parallel.

Without each and every one of you, there would be no such conference and I would like to thank you wholeheartedly for bringing the project to life.

Our thanks go first and foremost of course to UNESCO who will be our host for the two weeks from today onwards. The support of our true partners within the exact sciences division, and in particular Mr Moustafa El Tayeb and Mrs Diana Malpède, as well as all of the UNESCO exhibition and conference logistics services, has made the entire event possible.

The events "Invisible Universe" commemorates about thirty years of intense research on dark matter and about ten years dedicated to the survey of dark energy. Cosmology has arrived at a crossroads. According to the best data available, from large ground-based telescopes and space observatories: almost 95% of the universe irretrievably escape observational detection. This missing part or invisible part of the cosmos is constituted for 25% by a mysterious form of dark matter and 70%, by a dark energy whose nature is even more exotic and unknown! But what are exactly these new physical entities? What is the nature of dark matter and dark energy ? How can we learn more about these mysterious energies, what they imply, and the questions they raise about our fundamental understanding of the Universe ? The conference brings together scientists working on the various aspects of dark energy, dark matter and gravity modifications and their connection to cosmological observations, particle physics experiments and fundamental physics. We are more than 400 experts gathered in Paris to evaluate the situation, and draw future perspectives. The basic principles of physics appear sometimes to be put into question. Modern cosmology is perhaps at the beginning of a major renewal.

The events "Invisible Universe" comprises an exhibition and two major conferences.

The exhibition "Exploring the Invisible Universe" is a French-English bilingual exhibition especially dedicated to a large audience, is structured around four related themes

- The modern emergence of an Invisible Universe : Along the traces of Galilei. toward a new cosmological vision...
- The obscure and invisible in the art, with original works by the photographers (Elisabeth Alimi, Denis Baudier, Cecile Bortoletti, Marcella Lorenzi and the writer and graphist Frederick Tristan, by the composer Chloe Nicholson and the Sculptor Georges Rasera. The musical piece specifically composed by Chloe Nicholson for the event, The Invisible Universe, was recorded by members of the Royal Philharmonic Orchestra of London.
- The visible and Invisible in Philosophy about these concepts in several philosophical traditions.
- A multimedia area where the visitor is invited to develop his own model of the universe.

which then he can test against the most recent observations.

Special thanks to André Füzfa and Pier-Stefano Corasaniti for their collaboration in the realization of this exhibition.

From 6 to 10 July, or in other words next week, a French-language conference on "Dark Energy, Dark Matter Beyond a Dark Universe" will hold. Exceptionally, it will open up to the general public and looked at the same topics as our own conference.

Four round-table discussions – on

- Color and perception of the Invisible
- Critical discussions about Invisible, Inexplicable or Unintelligible
- Possibility and necessity to develop a new cosmological paradigm ?
- Finally the possible Coherent understanding of reality from infinitely small to infinitely large

should led to interesting exchanges between the public and the scientists, philosophers and artists.

Our conference now, "Invisible Universe International Conference" is divided into nine sessions, the plenary session with twenty-three Invited Speakers, and eight parallel sessions with about 250 oral presentations and fifty posters. The posters are presented in front of the Room 12 of this building, I invite you to take in that poster display, which is sure to provoke fascinating scientific exchanges all week long.

The eight parallel sessions are:

- Observational evidence of dark matter and dark energy
- Dark matter and particle physics experiments and dark matter candidates
- Fundamental theories of physics and cosmology
- Inflation and dark energy
- Inhomogeneous universe and backreaction
- Dark energy as a new energy component
- Dark matter and dark energy as modified theories of gravity
- Theory and simulations of cosmic structures and their evolution

I am tremendously grateful to the international support committee and international scientific committee working for the event "Invisible Universe". The expertise and commitment demonstrated by each and every member of these committees are what enabled us to devise and put into practice the scientific programme for this week. I would also especially like to thank the following for their encouragement: Professors Abbay Ashtekar, Edmund Bertschinger, Vitaly Melnikov, Jim Peebles, Ruben Thiberger and John Barrow. I would, of course, also like to thank the local organizing committee, namely Nabila Aghanim, Gianfranco Bertone, Philipe Brax, Michel Cassé, Cedric Deffayet, Gilles Esposito-Farèse, Joseph Kouneiher, Yann Rasera, Alexandre Refregier and especially Thomas Buchert, Pier-Stefano Corasaniti and André Füzfa and whose invaluable advice and scientific contributions permitted the creation and implementation of the programme for our conference.

But what will truly breathe life into this conference are the many presentations scheduled. I therefore set particular store by thanking the Invited Speakers, all of whom took up our invitation despite their many other commitments, enabling us to draft what I believe to be a quite exceptional agenda. Thank you very much indeed. I look forward to hearing from you and to learning a lot throughout the conference. I have always been fascinated by the formidable ability of the human mind to "understand the world". A world which is, as Einstein himself said, so "incomprehensible".

No matter how good an idea is, it cannot exist without resources and active support. The event "Invisible Universe" has received both scientific and financial assistance from the CNRS, Gemalto Enterprise, the Observatoire de Paris and CEA and we are indebted to them for this. I would like to thank especially again Gemalto Enterprise and his General Director Olivier Piou for the funding they provided for our exhibition and our conferences. Such a support is still exceptional in France, especially for activities in fundamental science as cosmology.

I would like too to thank the Institute for the Science of the Universe of the CNRS and its directors Dominique Lequeau and Jean-Marie Hameury, the Paris Observatory and its president Daniel Egret and its scientific council, and the Institute for Particle and Nuclear Physics of the CNRS and its director Michel Spiro and the interdisciplinarity program in Astroparticle from CNRS as well as its Scientific Council. All of whom have provided us with encouragement and support from the very outset and throughout the preparations for our project, and who, thanks to their great belief in us, made it all possible.

I cannot omit, and especially I do not want omit to thank very much Virginie Hababou and Stéphane Thomas very close collaborators at my Laboratory Universe and Theories for the hard work that they put into the practical organisation of "Invisible Universe". Their dedication, Virginie and Stephane was decisive.

Lastly, on a more personal note, I am proud to thank my all family and most particularly my wife Elisabeth and my son Vincent-Adam who have put up with me over the last two years. My obsession with the preparations for "Invisible Universe" all too often took me far from their sides.

Thank you once again for being here. I feel confident that the event will be a great success and look forward to discovering the sublime mysteries of the Invisible Universe and I hope at least that we will success to clear it up.

Jean-Michel ALIMI
Paris, June 29, 2009

Committees

International Standing Committee

A. Ashtekar (USA), J. Audouze (France), J. Barrow (UK), E. Bertschinger (USA), F. Bouchet (France), B. Carter (France), F. Combes (France), A. Connes (France), E. Copeland (UK), D. Egret (France), G. Ellis (South Africa), M. El Tayeb (UNESCO, Fance), W. Erdelen (UNESCO, France), B. Greene (USA), E. Kolb (USA), D. Lequeau (France), A.-C. Levasseur-Regourd (France), J.C. Mather (USA), V. Melnikov (Russia), M. Milgrom (Israël), V. Mukhanov (Germany), J. Ostriker (USA), T. Padmanabhan (India), J. Peebles (USA), J.-L. Puget (France), A. Riess (USA), M. Rowan-Robinson (UK), J. Schwarz (USA), G. Smoot (USA), M.Spiro (France), D.Spergel (USA), P. Steinhardt (USA), L. Susskind (USA), S. Weinberg (USA), C. Wetterich (Germany), F. Wilczek (USA), E. Witten (USA), E. Wright (USA).

International Advisory Committee

J.-M. Alimi (France), L. Amendola (Italy), A. Ashtekar (USA), P. Astier (France), J. Audouze (France), J. Barrow (UK), J. Bartlett (France), Y. Berenguer (UNESCO, France), P. Binetruy (France), A. Blanchard (France), S. Bonometto (Italy), F. Bouchet (France), B. Carter (France), J.-P. Chieze (France), F. Combes (France), E. Copeland (UK), A. Dekel (Israel), J. Ehlers (Germany), G. Ellis (South Africa), P. Fayet (France), J.-M. Gérard (Belgium), E. Gourgoulhon (France), E. Guendelman (Israel), F. Hammer (France), E. Kolb (USA), D. Lambert (Belgium), O. Le Fevre (France), J.-P. Luminet (France), R. Maartens (UK), D. Malpede (UNESCO, France), V. Melnikov (Russia), M. Milgrom (Israel), J. Ostriker (USA), T. Padmanabhan (India), J. Peacock (UK), J. Peebles (USA), A. Riess (USA), J. Schwarz (USA), A. Serna (Spain), R. Teyssier (France), R. Thieberger (Israel), C. Wetterich (Germany).

Local Organization Committee

J.-M. Alimi (Chairman), N. Aghanim, G. Bertone, P. Brax, T. Buchert, M. Cassé, P.-S. Corasaniti, C. Deffayet, G. Esposito-Farese, A. Füzfa, J. -Y. Giot, V. Hababou, E. Klein, J. Kouneiher, D. Malpede, Y. Rasera, A. Refregier, S. Thomas.

Invisible Universe : Toward a new cosmological paradigm

The essential is invisible to our eye... Astronomy and astrophysics and more in particular modern cosmology have shown that most of the universe is invisible to our eyes and to the most advanced observational devices.

With the formulation of the theory of general relativity by Albert Einstein, the universe in the 20th century has become a physical object like many others, and cosmology has finally become a truly scientific discipline. Nonetheless a puzzling picture of the universe has arisen. It is invisible for more than 95%. In fact a consistent analysis of the observations of the farthest and most obscure regions of the universe, when interpreted in the light of modern physical theories suggests that 25% of the total energy budget of the universe is in a form of a mysterious matter component, dubbed dark matter, and 70% is in a even more exotic form, dubbed dark energy. Suddenly, cosmology becomes a precious and unique experimental laboratory where fundamental theories of physics can be tested, from general relativity to the physics of elementary particles in energy ranges inaccessible to our Earth laboratories.

Among the several questions to which this science has to find an answer, one of the most important concerns the energetic content of the Universe. The exact knowledge of what the Universe is made of and in which amounts, would allow not only to determine its age, its past and future evolution, but also to address from a scientific point of view the question about its finiteness or its final end. This is because the evolutionary history of a universe for a given geometry admits only a particularly fixed energetic content. It is this equivalence between the energetic content and the space-time properties of the Universe that lead cosmologists to make some of the most mysterious and promising discoveries in the history of modern science: the existence of dark matter and dark energy.

Whereas the dark matter is unavoidable to explain the angular fluctuations of the cosmic microwave background radiation, the formation and properties of galaxies, dark energy has been invoked originally to account for the observed acceleration of the cosmic expansion of the universe which took place in the last billions years of its history. The "concordance model" of cosmology assumes that this dark energy is the effect of the cosmological constant introduced by Einstein as an attempt to incorporate Mach's principle in general relativity. However, the usual interpretation of this term as the weight of quantum vacuum fluctuations leads to a dramatic disagreement between theory and observations by several orders of magnitude! Besides, vacuum energy is supposed to be constant everywhere at all times. Therefore one should ask the question as to why it has becomes observable precisely today. This would make us think that we live in one very particular time, even privileged, of the cosmic history. Is this an incredible coincidence or there is something else ? What is dark matter and dark energy ? These are the two big questions, which for their complexity and depth, broad scientific impact and beyond (as we will see later several of these questions in modern astronomy strongly influence artists and philosophers as well) are the subject of the events that we organized.

Dark matter and dark energy are a challenge for astronomy both from an observational and theoretical point of view. In fact how to unveil the invisible? This is often made by combining direct observations of visible properties with a theoretical interpretation. As an example the flatness of the rotation curve of stars inside spiral galaxies is interpreted as the imprint of a halo of invisible matter, which dominates the mass of the galaxy and which allows the stars to keep a sufficient rotation speed far from the center of the galaxy. Similarly the distortions by lensing effects (gravitational and non optical) on larger scales, and the size of galaxy clusters is from this point of view interpreted as to be of the same nature. The analyses of the stability of the cosmic structures, such the clusters, the analyses of the growth of primordial matter fluctuations necessary to explain the formation of galaxies today also proceed from this way to explore our universe. The interpretation of the visible is only possible if founded in a theoretical framework which also allows us to explain more sensitive physical phenomena. Therefore the indirect observation of these mysterious dark matter and dark energy components poses new fundamental questions as one aims to determine their origin. Again, the observational and theoretical constraints seem to impose that the dark matter is made of heavy elementary particles. Thus it is necessary for us to interrogate on understanding the infinitely small to know what particle candidate could be. Concerning dark energy and its repulsive effect on the dynamics of the universe, the question is even more complicated. This time, it seems that even the most adventurous extensions of the standard model and of the infinitely small are not capable to give a satisfactory candidate in terms of elementary constituents that can support its most peculiar property, negative pressure. This is why, dark energy, but also dark matter are interpreted in some scientific literature, not as fundamental constituents but as fundamental modifications of our laws of physics. Again the comparison with observations is determinant, since one has to find the observational methods necessary to disentangle between these different interpretations. After a quarter of century of intense research on dark matter and about ten years dedicated to the survey of dark energy, we organized, in partnership with the **UNESCO** and in the framework of the activities of the World Year of Astronomy 2009 a set of scientific events that have been at the crossroad in the worldwide debate which involves these two open problems, dark matter and dark energy and that can be considered as the premises for a renewal of revolutionary nature of our understanding of the physical world similar to that experienced at the beginning of the twentieth century.

These two questions about the origin and nature of the dark matter and the dark energy illustrate also a significant evolution, it seems to us, on the status of scientific knowledge on the one hand and on the intuitions that science generates in artists whose work has been particularly inspired by the notion of obscure, invisible, dark. For this very reason it was our intention not only to have to organize an international research symposium, but also a symposium opened to the general public, accompanied by an exhibition where works of art and multimedia interactivity were numerous. Many primary figures who play today a determinant role in the development of our understanding of our Universe have been presented. The exchange between science and art was particularly open in the cultural events that constituted the general public symposium and the general public exhibition.

The scientific colloquium includes 8 different topical sessions which cover the whole set of research fields related to the subject of the conference:

I. Observational Astrophysical Aspect of dark energy and dark matter
II. Experimental particle physics aspects of dark matter and dark matter candidates

III. Gravitation and cosmology
IV. Inflation and Dark Energy
V. Inhomogeneous universes and backreaction
VI. Dark energy as a new energy component
VII. Dark energy and Dark Matter as modified gravity
VIII. Theoretical and numerical aspects of cosmic structures and evolution

The series of events have favored the arrival in Paris of a large number of top-notch scientific personalities. Several Nobel laureates in physics and Fields Medal were present. The symposium general public hosted some of the most renowned French-speaking world specialists. They presented to a large public the evolution of our knowledge of the universe through conferences and round tables. The interaction of scientists with the general public. has been encouraged. The exhibition which took place in parallel to these events facilitate also the exchanges between public and experts. All events were hold in Paris, of Monday June 29 to Monday July 10, 2009 at the Palace of **UNESCO**.

We are specially grateful to the CNRS (www.cnrs.fr) and its two institutes Institut National des Sciences de l'Univers (www.insu.fr) and the Institut National de Physique Nucléaire et de Physique des Particules (www.in2p3.fr), the Gemalto Enterprise (www.gemalto.org), the Observatoire de Paris (www.obspm.fr), the GIS "la physique des 2 infinis" (http://events.lal.in2p3.fr/P2I/), the CEA (www.cea.fr) and the Cantine Lavorata Enterprise (www.cantinelavorata.it) for their active support and their financial assistance to the organization of these events

Following the same spirit as the Conference, we decided to construct the outline of the present book as follows. We distinguished six chapters. The first chapter, entitled "Observational astrophysical aspects of dark energy and dark matter" provides an overview of some of the most promising results about the nature of the invisible universe, namely with the presentation by G. Smoot, A. Riess, Ed. Wright, F. Hammer and O. Lefevre.

In the second chapter, entitled "Experimental particle physics and dark matter candidates", the reader will find with the contributions by L. Bergstrom, G. Chardin and P. Fayet, an experimental and theoretical overview about the candidates of the dark matter.

The fundamental and theoretical developments (T. Padmanabhan, A. Ashtekar) in gravitational theory are presented in the third Chapter.

The phenomenological aspects, the cosmological and astrophysical consequences of dark matter and dark energy are presented in the fourth chapter (Ed. Copeland, L. Amendola, M. Milgrom, F. Combes, E. Kolb).

The conclusion is given by Prof P.J.E. Peebles in the fifth chapter.

A word should now be given on the way parallel sessions have been included in this volume. Considering both the impressive number and high scientific quality of these contributions given during the Conference, we decided to allow a consequent number of pages for each of the authors from the parallel sessions who would like to contribute and to include their contribution on a CD-ROM accompanying this volume. Therefore, the book gathers only the abstracts of the parallel contributions that were received and accepted by the Advisory Committee of the Conference for their scientific relevance. Each of these contributions has then been associated to a specific chapter by the Editors in function of their relevancy for the considered topic. The accompanying CD-ROM contains all the contributions to the present book under PDF format. Web pages (HTML format) allow the reader to easily browse amongst the different contributions and to consult them separately in a much more comfortable way than a large book could offer. The CD-ROM also contains other information about the conference as well as a non-exhaustive media gallery. Some Posters are presented in the last chapter.

Finally, we end with our most warm acknowledgements to all the contributors for the high-quality of their contribution to this volume. We wish the reader a pleasant reading of all these fascinating aspects of Invisible Universe. We hope he will find here as many fruitful inspirations for his future work as deep motivations to pursue the still on-going enrichment of problematic of the Invisible Universe.

Jean-Michel Alimi and André Füzfa,

Paris, January 2010

OBSERVATIONAL ASTROPHYSICAL ASPECTS OF DARK ENERGY AND DARK MATTER

Cosmology with the CMB

Professor George F. Smoot

Lawrence Berkeley National Laboratory and Physics Department
University of California, Berkeley, California, USA
Ewha Womans University and Ewha Academy of Advanced Study
Chaire Blaise Pascale Université de Paris

Abstract. The discovery of the CMB radiation in 1965 and of its fluctuations in the CMB by COBE in 1992 have provides us with a powerful probe of the properties of the universe, its contents, and the validity of our models. A myriad of ground-based, airborne, and the WMAP satellite observations have gone far in the utilization of the CMB anisotropies and more recently polarization. A new generation of such experiments are underway and promise to continue and extend this program. A major milestone in this are the launch (May 14, 2009) of the Planck mission and the observations which have begun on the smaller angular scale regions. This talk reviews the overall CMB science program and its progress.

Keywords: CMB, Cosmology, .
PACS: 98.80.-k, 98.80.Bp

THE STANDARD BIG BANG MODEL: THE BASIC FRAMEWORK

The current 'Standard Model' of cosmology contains around 10 free parameters and actually describes the observations well. The basic framework is the Friedmann-Robertson-Walker (FRW) metric (i.e., a universe that is approximately homogeneous and isotropic on large scales), with density perturbations laid down at early times and evolving into today's structures.

The key ingredients of the extended Hot Big Bang Model are the Dark Energy, Dark Matter & other Constituents Along with their equations of state (Lagrangian) plus inflationary perturbations and baryo/leptogenesis.

TABLE a). The Constituents of the Universe

Constituent	Percentage at Present	Eqn of State / Lagrangian
Dark Energy	72%	W = -1
Dark Matter	23%	W = 0
Light Sector		
Baryons	4.4%	W = 0
Neutrinos	0.17%	
Photons (light)	0.005%	W = 1/3

Particle Physicsits and Astronomers are both studying the various key constituents and phases of the universe and its associated cosmology. The particle physicists are generally looking at the forces and particle production in the very early Universe and the astronomers at the large-scale structure left as a result of long-term evolution of the constituents under the forces of Nature – primarily gravitation.

CP1241, *Invisible Universe: Proceedings of the Conference,* edited by J.-M. Alimi and A. Füzfa

THE COSMIC MICROWAVE BACKGROUND (CMB)

The Cosmic Microwave Background (CMB) is a wonderful and critical tool of modern cosmology. A very significant fraction of all the information in cosmology over the last 10 to 15 years has come from CMB observations. It's use has brought us into an era of precision cosmology but also brought out deep mysteries) [2].
The CMB discovery was announced in 1965 by Penzias & Wilson[1], who found what was a 2.7 K blackbody, isotropic (<1%), a relic of hot "big bang". In the 1970's and 1980's we discovered the 3 mK dipole (local Doppler) and that the variations on arcminute angular scales $\Delta T/T < 10^{-5}$. In 1992 COBE satellite mission showed that the CMB was a blackbody 2.728 K [7,8] and found anisotropies on large angular scales ($\ell < 30$) $\Delta T/T \approx 10^{-5}$. It is these anisotropies which we cxploit in the present epoch to learn so much about the early universe and its subsequent development.

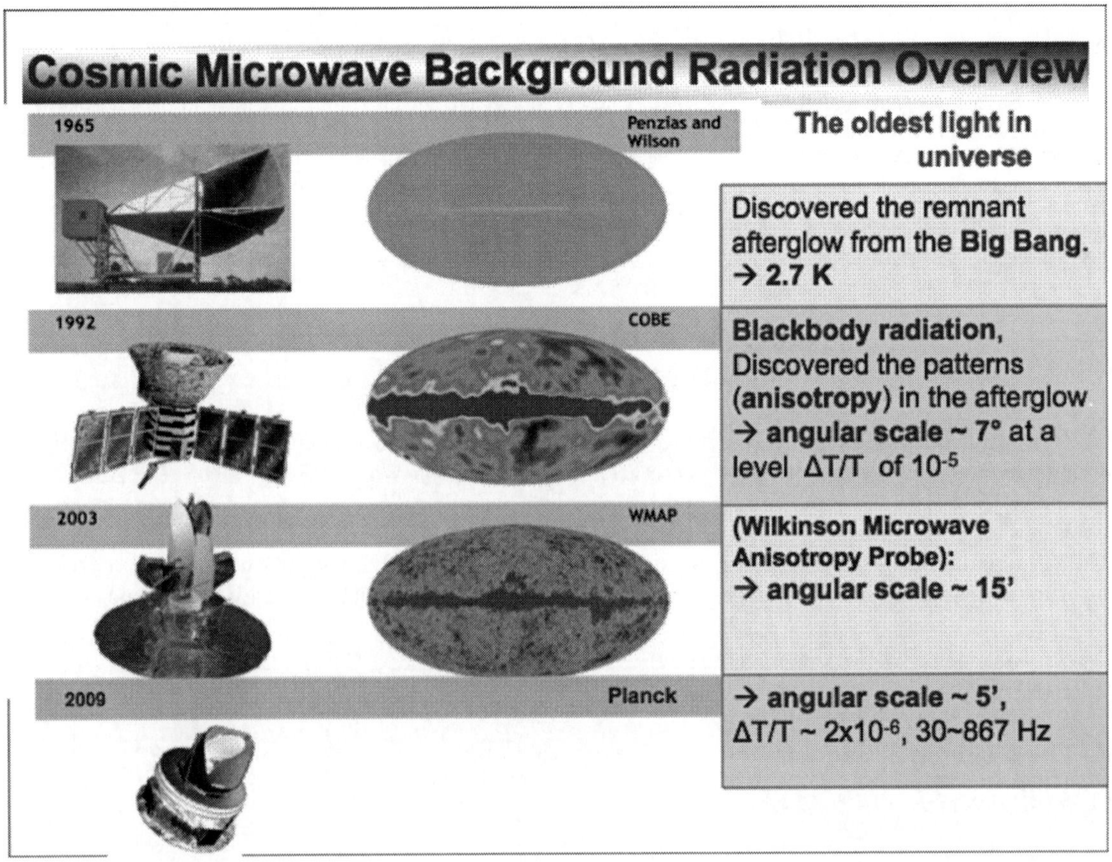

Current Status of CMB Anisotropy Observations:

Since the first detection of CMB anisotropies by the COBE satellite [3], there has been intense activity to map the sky at increasing levels of sensitivity and angular resolution by ground-based and balloon-borne measurements. These were joined in 2003 by the first results from NASA's Wilkinson Microwave Anisotropy Probe (WMAP) [4], which were improved upon by analysis of the 3 and 5 year WMAP data [5,6]. Together these CMB observations have led to a stunning confirmation of the 'Standard Model of Cosmology'. In combination with other astrophysical data, the CMB anisotropy measurements place quite precise constraints on a number of cosmological parameters, and have launched us into an era of precision cosmology. This progress is expected to continue with the improved capabilities of the Planck satellite and new generations of experiments and observational programs.
Observations show that the CMB contains anisotropies at the 10^{-5} level, over a wide range of angular scales. These anisotropies are usually expressed by using a spherical harmonic expansion of the CMB sky:

$$T(\theta,\phi) = \sum_{lm} a_{lm} Y_{lm}(\theta,\phi) \tag{1}$$

A statistically isotropic sky means that all m's are equivalent, i.e., there is no preferred axis. Together With the assumption of Gaussian statistics (tested currently to the 10^{-4} level), the variance of the temperature field (or equivalently the power spectrum in l) then fully characterizes the anisotropies. The power summed over all ms at each l is $(2l + 1)C_l/(4\pi)$, where $C_l \equiv \Sigma|a_{lm}|^2$.

Thus averages of a_{lm}s over m can be used as estimators of the C_ls to constrain their expectation values, which are the quantities predicted by theoretical models. For an idealized full-sky observation, the variance of each measured C_l (i.e., the variance of the variance) is $[2/(2l + 1)]C^2_l$. This sampling uncertainty (known as 'cosmic variance') comes about because each C_l is χ^2 distributed with $(2l + 1)$ degrees of freedom for our observable volume of the Universe. For fractional sky coverage, f_{sky}, this variance is increased by $1/f_{sky}$ and the modes become partially correlated.
It is important to understand that theories predict the expectation value of the power spectrum, whereas our sky is a single realization. Hence the cosmic variance is an unavoidable source of uncertainty when constraining models; it dominates the scatter at lower ls, while the effects of instrumental noise and resolution dominate at higher ls.

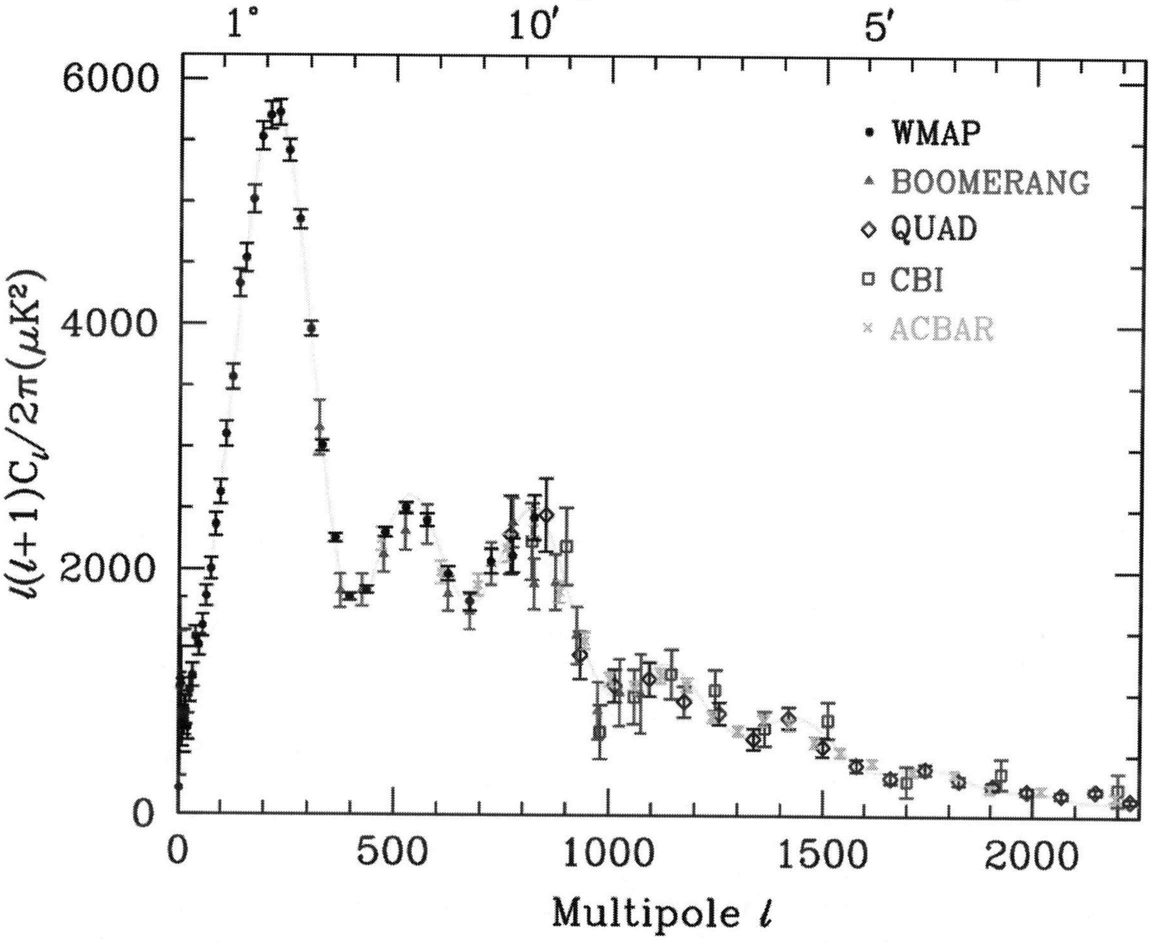

FIGURE 1. Band power estimates of the CMB angular power spectrum from ACBA, CBI, QUAD, BOOMERANG and WMAP observations. Some of the low and high frequency band powers with large error bars are omitted for presentation. The plot is linear so that the Sachs-Wolfe primordial plateau is difficult to see but the acoustic peaks and dampling region are clear. The light curve plotted is the best fit model.

There has been a steady improvement in the quality of CMB data that has led to the development of the present-day cosmological model. At this point we have the observations of CMB anisotropies have given us great confidence by steadily filling our check list of items anticipated by this model (see Table I).

TABLE I CMB Check List

Checklist for CMB Anisotropies

Acoustic oscillations below horizon scale

✓ nearly harmonic series in sound horizon scale

✓ signature of super-horizon fluctuations (horizon crossing starts clock

✓ even-odd peak heights baryon density controlled

✓ a high third peak signature of dark matter at recombination

Nearly flat geometry

✓ peak scales given by comoving distance to last scattering

Primordial plateau above horizon scale

✓ signature of super-horizon potential fluctuations (Sachs-Wolfe)

✓ nearly scale invariant with slight red tilt ($n\approx0.96$) and small running

Damping of small-scale fluctuations

✓ baryon-photon coupling plus delayed recombination (& reionization)

CMB Polarizaton:

Since Thomson scattering of an anisotropic radiation field generates linear polarization, the CMB is predicted to be polarized at the roughly 5% level of the temperature anisotropies [42]. Polarization is a spin-2 field on the sky, and the algebra of the modes in *l*-space is strongly analogous to spin-orbit coupling in quantum mechanics [43]. The linear polarization pattern can be decomposed in a number of ways, with two quantities required for each pixel in a map, often given as the Q and U Stokes parameters. However, the most intuitive and physical decomposition is a geometrical one, splitting the polarization pattern into a part that comes from a divergence (often referred to as the 'E-mode') and a part with a curl (called the 'B-mode') [44]. More explicitly, the modes are defined in terms of second derivatives of the polarization amplitude, with the Hessian for the E-modes having principle axes in the same sense as the polarization, while the B-mode pattern can be thought of simply as a 45° rotation of the E-mode pattern. Globally one sees that the E-modes have $(-1)^l$ parity (like the spherical harmonics), while the B-modes have $(-1)^{l+1}$ parity.

The most general possible set of density perturbations is a linear combination of an adiabatic density perturbation and some isocurvature perturbations. Adiabatic means that there is no change to the entropy per particle for each species, i.e., $\delta\rho/\rho$ for matter is $(3/4)\delta\rho/\rho$ for radiation. Isocurvature perturbations mean that the set of individual density perturbations adds to zero, for example, matter perturbations compensate radiation perturbations so that the total energy density remains unperturbed, i.e., $\delta\rho$ for matter is $-\delta\rho$ for radiation. These different modes give rise to distinct (temporal) phases during growth, with those of the adiabatic scenario being strongly preferred by the data. Models that generate mainly isocurvature type perturbations (such as most topological defect scenarios) are no longer considered to be viable. However, an admixture of the adiabatic mode with up to about 10% isocurvature contribution is still allowed [18].

Looking at Figure 2 the anti-correlation at $l \simeq 150$ and the peak at $l \simeq 300$ are now quite distinct. The measured shape of the cross-correlation power spectrum provides supporting evidence for the adiabatic nature of the perturbations, as well as directly constraining the thickness of the last scattering surface. Since the polarization anisotropies are generated in this scattering surface, the existence of correlations at angles above about a degree demonstrates that there were super-Hubble fluctuations at the recombination epoch. The sign of this correlation also confirms the adiabatic paradigm.

FIGURE 2. The cross power spectrum of the temperature anisotropies and the E-mode polarization anisotropies for the experiments listed in the figure. Note that the ordinates (y-axis) hire is not multiplied by an additional ℓ, which shows the large and small angular scale features more clearly.

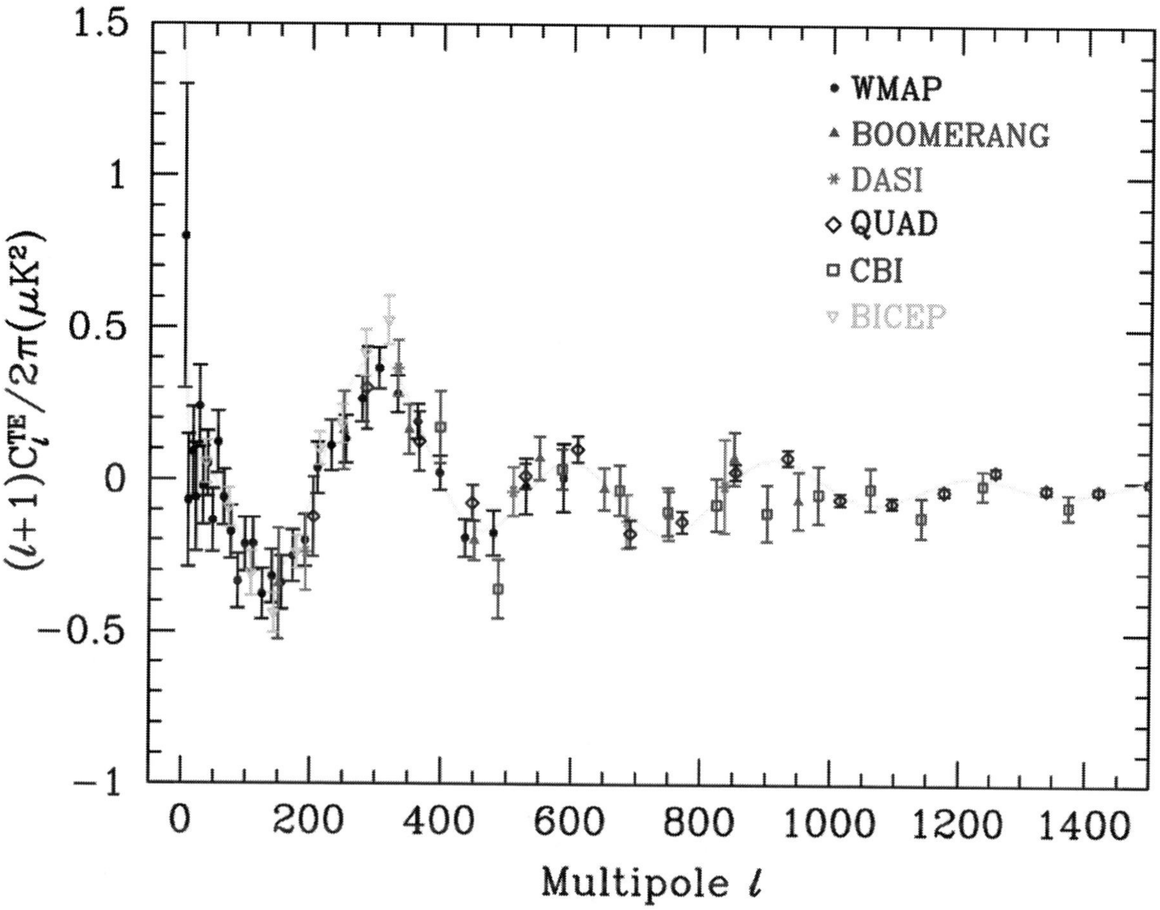

The oscillating photon-baryon fluid also results in a series of acoustic peaks in the polarization C_l s. The main 'EE' power spectrum has peaks that are out of phase with those in the 'TT' spectrum, because the polarization anisotropies are sourced by the fluid velocity. (In a harmonic oscillator the turning points (maximum compression and rarefaction) are 90° out of phase with the maximum velocity) The 'TE' part of the polarization and temperature patterns comes from correlations between density and velocity perturbations on the last scattering surface, which can be both positive and negative, and is of larger amplitude than the EE signal. There is no polarization 'Sachs-Wolfe' effect, and hence no large-angle plateau. However, scattering during a recent period of reionization can create a polarization 'bump' at large angular scales.

Because the polarization anisotropies have only a fraction of the amplitude of the temperature anisotropies they took longer to detect. The first measurement of a polarization signal came in 2002 from the DASI experiment [45], which provided a convincing detection, confirming the general paradigm, but of low enough significance that it lent little constraint to models. As well as the E-mode signal, DASI also made a statistical detection of the TE correlation.

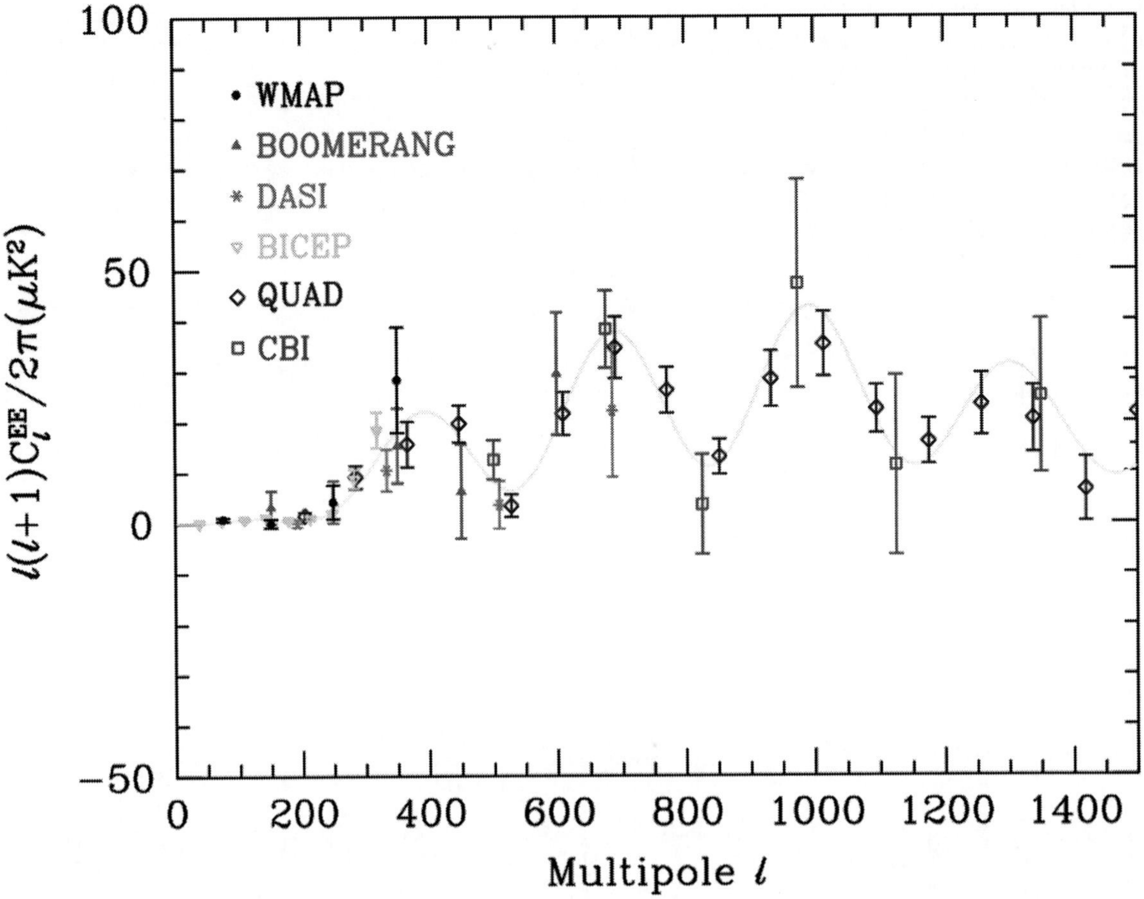

FIGURE 3. The angular power spectrum of the E-mode polarization as observed by several different experiments are plotte along with a theoretical model which matches the CMB anisotropy model of Figure 1.

B-mode Polarization

In principle, the mechanism that produces primordial perturbations could generate scalar, vector, and tensor modes. However, the vector (vorticity) modes decay with the expansion of the Universe. The tensors (transverse trace-free perturbations to the metric) generate temperature anisotropies through the integrated effect of the locally anisotropic expansion of space. Since the tensor modes also redshift away as gravitational waves after they enter the horizon, they contribute only to angular scales above about 1°. Hence some fraction of the low-l CMB anisotropy signal could be due to a gravitational wave contribution, although small amounts of tensors are essentially impossible to discriminate from other effects that might raise the level of the plateau. However, the tensors can be distinguished using polarization information.
Because scalar (e.g. density) perturbations have no handedness, the B-mode power spectrum can only be sourced by vectors or tensors. Moreover, since inflationary scalar perturbations give only E-modes, while tensors generate roughly equal amounts of E- and B-modes, then the determination of a non-zero B-mode signal is a way to measure the gravitational wave contribution (and thus potentially derive the energy scale of inflation), even if it is rather weak. However, one must first eliminate the foreground contributions and other systematic effects down to very low levels.

Several experiments have reported upper limits on C^{BB}_{l}, but they are currently not very constraining. In general the current upper limits are roughly an order of magnitude above the highest anticipated level of C^{BB}_{l} predicted by theories give the slope of the primodial fluctuation power spectrum and anticipated energy scales for Inflation. This situation should change as increasingly ambitious experiments report results.

TABLE II. CMB Check List Continued

Checklist for CMB Polarization

CMB is polarized

✓ acoustic peaks in E-mode spectrum from velocity perturbations

✓ E-mode peaks 90° out-of-phase for adiabatic perturbations

✓ vanishing small-scale B-modes

✓ ? reionization enhanced low ℓ polarization

Gravitational Waves from Inflation

— B-MODES FROM GRAVITY WAVE TENSOR FLUCTUATIONS

— very nearly scale invariant with extremely small red tilt (n≈0.98)

— decay within horizon (ℓ≈100)

— tensor/scalar ratio r from energy scale of inflation ~ $(E_{inf}/10^{16}\ \mathrm{GeV})^4$

Our inflationary hot Big-Bang theory is standing up well to the observations so far! Now for those gravity waves… C^{BB}_l

Current and pending CMB experiments have two main goals: i) measuring small angular scale temperature fluctuations (below a few arc minutes), and ii) measuring the CMB polarisation power spectrum. These goals are fundamental for our understanding of the universe. The small-scale anisotropies are directly related to the presence of structures in the universe whereas the two types of polarisation (E and B-modes) probe both the reionisation of the universe, i.e. the formation of the first emitting objects, and the inflationary potential.

There have been rapid and important advances in the recent past. We already have on the one hand measurements, by ACBAR, CBI, BIMA, VSA, of the temperature power spectrum for $2000 < \ell < 4000$ with CBI and BIMA data showing an excess of power as compared with the predicted damping tail of the CMB (Figure 1). On the other hand, DASI, Archeops, Boomerang, Maxipol, CBI, QUaD and WMAP have direct measurements of the E-mode polarisation. The situation will change even more in the near future with anticipated results from experiments currently taking data or in preparation (QUaD, BICEP, EBEX, CLOVER, QUIET, SPIDER, Planck).

Secondary Anisotropies – The CMB After Recombination

The Cosmic Microwave Background fluctuations provide a powerful probe of the dark ages of the universe through the imprint of the secondary anisotropies associated with the reionisation of the universe and the growth of structure. There is a strong relationship between the secondary anisotropies and and the primary anisotropies that are directly generated by quantum fluctuations in the very early universe. The physics of secondary fluctuations is strongly dependent on the ionisation history and the evolution of structure. The different signatures arising from the secondary effects in terms of their induced temperature fluctuations, polarisation and statistics reveal information about the reionization and the evolution of structure in a strong way. The theory and observations of the secondary anisotropies are being actively pursued at present and this talk presents a short review the future and current observational status.

With increasingly precise measurements of the primary anisotropies, there is growing theoretical and experimental interest in 'secondary anisotropies', pushing experiments to higher angular resolution and sensitivity. These secondary effects arise from the processing of the CMB due to ionization history and the evolution of structure, including gravitational lensing and patchy reionization effects [57]. Additional information can thus be extracted about the Universe at $z \ll 1000$. This tends to be most effectively done through correlating CMB maps with other cosmological probes of structure. However, there are several observations such as the Sunyaev-Zeldovich effect which first come from the CMB alone. Secondary signals are also typically non-Gaussian, unlike the primary CMB anisotropies.

Sunyaev-Zel'dovich Effect:

A secondary signal of great current interest is the Sunyaev-Zel'dovich (SZ) effect [58], which is Compton scattering ($\gamma\, e \rightarrow \gamma'\, e'$) of the CMB photons by hot electron gas, which creates spectral distortions by transferring energy from the electrons to the photons. It is particularly important for clusters of galaxies, through which one observes a partially Comptonized spectrum, resulting in a decrement at radio wavelengths and an increment in the submillimeter. The imprint on the CMB sky is of the form $\Delta T/T = y\, f(x)$, with the y-parameter being the integral of Thomson optical depth times kTe /me c2 through the cluster, and f (x) describing the frequency dependence. This is simply $x \coth(x/2) - 4$ for a non-relativistic gas (the electron temperature in a cluster is typically a few keV), where the dimensionless frequency $x \equiv h\nu / kT\gamma$. As well as this 'thermal' SZ effect there is also a smaller 'kinetic' effect due to the bulk motion of the cluster gas, this being $\Delta T/T \sim \tau\, (v/c)$, with either sign, but having the same spectrum as the primary CMB anisotropies.

A significant advantage in finding galaxy clusters this way is that the SZ effect is largely independent of redshift, so in principle clusters can be found to arbitrarily large distances. The SZ effect can be used to find and study individual clusters, and to obtain of the Hubble constant. There is also the potential to constrain the equation of state of the Dark Energy through counts of detected clusters as a function of redshift [59]. Many experiments (including the Planck satellite) are currently in operation which will probe clusters in this way. The promise of the method has been realized through the first detections of clusters purely through the SZ effect, by the SPT [60] and ACT [61] experiments.

TABLE III. CMB Check List

Checklist for CMB Secondary Anisotropy Predictions from inflation-inspired models:

Late-Time Dark Energy Domination

✓ low ℓ ISW bump correlated with large scale structure (potentials)

Late-Time Non-Linear Structure Formation

✓ gravitational lensing of CMB

✓ Sunyaev-Zeldovich effect from deep potential wells (clusters)

Late-Time Reionization

— overall supression and tilt of primary CMB spectrum

— doppler and ionization modulation produces small-scale anisotropies

Higher-order Statistics:

Although most of the CMB anisotropy information is contained in the power spectra, there will also be weak signals present in higher-order statistics. These statistics can measure any primordial non-Gaussianity in the perturbations, as well as non-linear growth of the fluctuations on small scales and other secondary effects (plus residual foreground contamination of course). Although there are an infinite variety of ways in which the CMB could be non-Gaussian, there is a generic form to consider for the initial conditions, where a quadratic contribution to the curvature perturbations is parameterized through a dimensionless number f_{NL} . This weakly non-linear component can be constrained through measurements of the bispectrum or Minkowski functionals, for example. The constraints depend on the shape of the triangles in harmonic space, and it has become common to distinguish the 'local' or 'squeezed' configuration (in which one side is much smaller than the other two) from the 'equilateral' configuration. The results from the WMAP team are $-9 < f_{NL} < 111$ (95% confidence region), for the local mode and $-151 < f_{NL} < 253$ for the equilateral mode [14]. Different estimators used by other authors give results of similar magnitude [62]. The level of f_{NL} expected is small, so that a detection of $f_{NL} > 10$ would rule out all single-field, slow-roll inflationary models. However, with the capabilities of Planck and other future experiments, it seems that a measurement of primordial non-Gaussianity may be feasible for a wide class of models, and therefore much effort is expected to be devoted to predictions and measurements in the coming years.

Constraints on Cosmologies

The most striking outcome of the newer experimental results is that the standard cosmological paradigm is in excellent shape. A large amount of high precision data on the power spectrum is adequately fit with fewer than 10 free parameters. The framework is that of FRW models, which have nearly flat geometry, containing Dark Matter and Dark Energy, and with adiabatic perturbations having close to scale invariant initial perturbations.

Within this framework, bounds can be placed on the values of the cosmological parameters. Of course, much more stringent constraints can be placed on models which cover a restricted parameter space, e.g., assuming that $\Omega t_{ot} = 1$, $n = 1$ or $r = 0$. More generally, the constraints depend upon the adopted prior probability distributions, even if they are implicit, for example by restricting the parameter freedom or their ranges (particularly where likelihoods peak near the boundaries), or by using different choicesof other data in combination with the CMB. When the data become even more precise, these considerations will be less important, but for now we caution that restrictions on model space and choice of priors need to be kept in mind when adopting specific parameter values and uncertainties.

TABLE ii. CMB Check List Continued

Checklist for CMB Final

Structure predictions from inflation-inspired models:

Late-Time Non-Linear Structure Formation (revisited)

✓ gravitational lensing of CMB

✓ Sunyaev-Zeldovich effect from deep potential wells (clusters)

Growth of Matter Power Spectrum

✓ primordial power-law above current sound horizon

✓ CMB acoustic peaks as baryon oscillations

Dark Energy Domination at Late Times

✓ correlation of galaxies with Late ISW in CMB

— cluster counts (SZ) reflect LCDM growth and volume factors

It appears our current Universe is dominated in energy density by a Dark Energy (Lambda) component!

Fundamental Lessons

More important than the precise values of parameters is what we have learned about the general features which describe our observable Universe. Beyond the basic hot Big Bang picture, the CMB has taught us that:

• The Universe recombined at $z \simeq 1100$ and started to become ionized again at $z \simeq 10$.
• The geometry of the Universe is close to flat.
• Both Dark Matter and Dark Energy are required.
• Gravitational instability is sufficient to grow all of the observed large structures in the Universe.
• Topological defects were not important for structure formation.
• There are 'synchronized' super-Hubble modes generated in the early Universe.
• The initial perturbations were adiabatic in nature.
• The perturbations had close to Gaussian (i.e., maximally random) initial conditions.

It is very tempting to make an analogy between the status of the cosmological 'Standard Model' and that of particle physics (see earlier Sections of this Review). In cosmology there are about 10 free parameters, each of which is becoming well determined, and with a great deal of consistency between different measurements. However, none of these parameters can be calculated from a fundamental theory, and so hints of the bigger picture, 'physics beyond the Standard Model', are being searched for with ever more ambitious experiments.

CMB observations present a beautiful success for the standard cosmological model. The acoustic oscillations in the CMB angular power spectrum, a major prediction of themodel, have now been detected at the 5σ C.L. for the first peak and the 3σ C.L. for the second and third peaks and about 2σ C.L beyond. Furthermore, when constraints on cosmological parameters are determined under the assumption of adiabatic primordial perturbations their values are in excellent agreement with the predictions of theory and/or with independent observations.

All in all one finds this an exciting and rewarding time in cosmology as we are making tremendous strides in our understanding and surety and as the results give us the promise of more exciting science to follow. CMB anisotropy and polarization measurements constrain the cosmological parameters and test new theoretical scenarios.

ACKNOWLEDGMENTS

I would like to acknowledge the support from Chaire Blaise Pascal, from Ewha Institute of Advanced Studies and the Institute for the Early Universe at Ewha Univesity and the Berkeley Center for Cosmological Physics at the University of California at Berkeley and specifically the DOE HEP Contract to LBNL.

I would like to thank the organizers of the Invisible Universe Conference for the excellent program and venues and to my fellow participants and attendees for making it such an interesting and fun meeting.

REFERENCES

1. A.A. Penzias and R. Wilson, Astrophys. J. 142, 419 (1965); R.H. Dicke et al., Astrophys. J. 142, 414 (1965).
2. M. White, D. Scott, and J. Silk, Ann. Rev. Astron. & Astrophys. 32, 329 (1994);
W. Hu and S. Dodelson, Ann. Rev. Astron. & Astrophys. 40, 171 (2002).
3. G.F. Smoot et al., Astrophys. J. 396, L1 (1992).
4. C.L. Bennett et al., Astrophys. J. Supp. 148, 1 (2003).
5. N. Jarosik et al., Astrophys. J. Supp. 170, 263 (2007).
6. G. Hinshaw et al., Astrophys. J. Supp. 180, 225 (2009).
7. J.C. Mather et al., Astrophys. J. 512, 511 (1999).
42. W. Hu and M. White, New Astron. 2, 323 (1997).
43. W. Hu and M. White, Phys. Rev. D56, 596 (1997).
44. M. Zaldarriaga and U. Seljak, Phys. Rev. D55, 1830 (1997);
M. Kamionkowski, A. Kosowsky, and A. Stebbins, Phys. Rev. D55, 7368 (1997).

Seeing Dark Energy

A. G. Riess

Space Telescope Science Institute, Baltimore, MD USA

Abstract. This talk is largely based on the work now published in Riess et al. (2007).

The accelerating cosmic expansion first inferred from observations of distant type Ia supernovae (SNe Ia; Riess et al. 1998; Perlmutter et al. 1999) indicates unexpected gravitational physics, frequently attributed to the dominating presence of a "dark energy" with negative pressure.

(For students of the history of science, a copy of my lab notebook page from 1997 showing where I recorded the first indication that the data pointed to present acceleration and dark energy can be found in Symmetry Magazine,Fall 2007, http://www.symmetrymagazine.org/cms/?pid=1000557)

Increasingly incisive samples of SNe Ia at $z < 1$ have reinforced the significance of this result (Tonry et al. 2003; Knop et al. 2003; Barris et al. 2004; Conley et al. 2006; Astier et al. 2006). Using the new Advanced Camera for Surveys (ACS) and refurbished NICMOS camera on the *Hubble Space Telescope (HST)*, our collaboration secured observations of a sample of the most-distant known SNe Ia. These half-dozen SNe Ia, all at $z > 1.25$, helped confirm the reality of cosmic acceleration by delineating the transition from preceding cosmic deceleration during the matter-dominated phase and by ruling out simple sources of astrophysical dimming (Riess et al. 2004b, hereafter R04). The expanded sample of 23 SNe Ia at $z \geq 1$ presented here are now used to begin characterizing the early behavior of dark energy.

Other studies independent of SNe Ia now strongly favor something like dark energy as the dominant component in the mass-energy budget of the Universe. Perhaps most convincingly, observations of large-scale structure and the cosmic microwave background radiation provide indirect evidence for a dark-energy component (e.g., Spergel et al. 2006). Measurements of the integrated Sachs-Wolfe effect (e.g., Afshordi, Loh, & Strauss 2004; Boughn & Crittenden 2004; Fosalba et al. 2003; Nolta et al. 2004; Scranton et al. 2004) more directly suggest the presence of dark energy with a negative pressure. Additional, albeit more tentative, evidence is provided by observations of X-ray clusters (Allen et al. 2004) and baryon oscillations (e.g., Eisenstein et al. 2005).

The unexplained existence of a dominant, dark-energy-like phenomenon presents a stiff challenge to the standard model of cosmology and particle physics. The apparent acceleration may result from exotic physics such as the repulsive gravity predicted for a medium with negative pressure or from entirely new physics. The explanation of strongest pedigree is Einstein's famous "cosmological constant" Λ (i.e., vacuum energy; Einstein 1917), followed by a decaying scalar field similar to that already invoked for many inflation models (i.e., quintessence –Wetterich 1995, Caldwell, Davé, & Steinhardt 1998; Peebles & Ratra 2003). Competitors include the Chaplygin gas (Bento, Bertolami, & Sen 2002), topological defects, and a massless scalar field at low temperature. Alternatively, alterations to General Relativity may be required as occurs from the higher-dimensional transport of gravitons in string theory models (Deffayet et al. 2002) and braneworlds, or by finely-tuned, long-range modifications (e.g., Cardassian type, Freese 2005; or Carroll et al. 2004; see Szydlowski, Kurek, & Krawiec 2006 for a review). Empirical clues are critical for testing hypotheses and narrowing the allowed range of possible models.

SNe Ia remain one of our best tools for unraveling the properties of dark energy because their individual measurement precision is unparalleled and they are readily attainable in sample sizes of order 10^2, statistically sufficient to measure dark-energy-induced changes to the expansion rate of $\sim$1%. Specifically, the equation-of-state parameter of dark energy, w (where $P = w\rho c^2$) determines both the evolution of the density of dark energy,

$$\rho_{DE} = \rho_{DE,0} \exp\{3 \int_a^1 \frac{da}{a}(1+w(a))\},$$

and its gravitational effect on expansion,

$$\frac{\ddot{a}}{a} = -\frac{4\pi G}{3}(1+3w(a)),$$

CP1241, *Invisible Universe: Proceedings of the Conference,* edited by J.-M. Alimi and A. Füzfa

where $\rho_{DE,0}$ is the present dark-energy density. Measuring changes in the scale factor, a, with time from the distance and redshift measurements of SNe Ia,

$$\frac{d_l(z)}{c(1+z)} = \int_t^{t0} \frac{dt'}{a(t')} = \int_0^z \frac{dz'}{H(z')},$$

constrains the behavior of $w(a)$ or $w(z)$ and is most easily accomplished at $z < 2$ during the epoch of dark-energy dominance.

Ideally, we seek to extract the function $w(z)$ for dark energy or its mean value at a wide range of epochs. Alternatively, we might constrain its recent value $w_0 \equiv w(z=0)$ and a derivative, $dw/dz \equiv w'$, which are exactly specified for a cosmological constant to be $(-1,0)$. Most other models make less precise predictions. For example, the presence of a "tracker" dark-energy field whose evolution is coupled to the (decreasing) dark matter or radiation density may be detected by a measured value of $w' > 0$. In truth, we know almost nothing of what to expect for $w(z)$, so the safest approach is to assume nothing and measure $w(z)$ across the redshift range of interest. SN Ia at $z > 1$ are crucial to constrain variations of w with redshift. These measurements can only be made from space, and we report here on that endeavor. We have discovered and measured 21 new SN Ia with HST and used them to constrain the properties of the dark energy.

HIGHER-Z SNE IA

In Figure 1 we show the Hubble diagram of distance moduli and redshifts for all of the *HST*-discovered SNe Ia in the Gold and Silver sets from our program. The new SNe Ia span a wide range of redshift ($0.21 < z < 1.55$), but their most valuable contribution to the SN Ia Hubble diagram remains in the highest-redshift region where they now well delineate the range at $z \geq 1$ with 23 SNe Ia, 13 new objects since R04. This territory remains uniquely accessible to *HST*, which has discovered the dozen highest-redshift SNe Ia known, and its exploration is the focus of the rest of this paper.

In the inset to Figure 1 we show the residual Hubble diagram (from an empty Universe) with the Gold data uniformly binned. Here and elsewhere, we will utilize uniform, unbiased binning achieved with a fixed value of $n\Delta z$, where Δz is the bin width in redshift and n is the number of SNe in the bin.[1] In Figure 1 we use $n\Delta z = 6$ which yields seven bins for our sample. Although binning is for illustrative purposes in the Hubble diagram, there are some specific advantages of binning such as the removal of lensing-induced asymmetrical residuals by flux averaging (Wang 2005) and the ease of accounting for systematic uncertainties introduced by zeropoint errors in sets of photometric passbands used at similar redshifts.

The distance-redshift relation of SNe Ia is one of few powerful tools available in observational cosmology. A number of different hypotheses and models can be tested with it, including kinematic descriptions of the expansion history of the Universe, the existence of mass-energy terms on the right-hand side of the Friedman equation, and the presence of astrophysical sources of contamination. Testing all interesting hypotheses is well beyond the scope of this paper and is best left for future work. Instead, we now undertake a few narrowly posed investigations.

ALTERNATIVES TO DARK ENERGY

After the detection of the apparent acceleration of cosmic expansion (and dark energy) by Riess et al. (1998) and Perlmutter et al. (1999), alternative hypotheses for the apparent faintness of high-redshift SNe Ia were posed. These included extragalactic gray dust with negligible tell-tale reddening or additional dispersion (Aguirre 1999a,b; Rana 1979, 1980), and pure luminosity evolution (Drell, Loredo, & Wasserman 2000).

In R04 we found that the first significant sample of SNe Ia at $z > 1$ from *HST* rejected with high confidence the simplest model of gray dust by Goobar, Bergstrom, & Mörtsell (2002), in which a smooth background of dust is present (presumably ejected from galaxies) at a redshift greater than the SN sample (i.e., $z > 2$) and diluted as the Universe expands. This model and its opacity was invented to match the 1998 evidence for dimming of supernovae at z 0.5 without invoking dark energy in a universe with $\Omega_m = 1$. This model is shown in the inset of Figure 1. The

---

[1] The last bin ends abruptly with the highest-redshift SN; thus, its $n\Delta z \leq$ value is smaller than the rest.

present Gold sample (at the best fitting value of H_0) rejects this model at even higher confidence ($\Delta\chi^2 = 194$, i.e., 14 σ), beyond a level worthy of further consideration.

Dark Energy

Strong evidence suggests that high-redshift SNe Ia provide accurate distance measurements and that the source of the apparent acceleration they reveal lies in the negative pressure of a "dark energy" component. Proceding from this conclusion, our hard-earned sample of SNe Ia at $z > 1.0$ can provide unique constraints on its properties. Strong motivation for this investigation comes from thorough studies of high-redshift and low-redshift SNe Ia, yielding a consensus that there is no evidence for evolution or intergalactic gray dust at or below the current statistical constraints on the average high-redshift apparent brightness of SN Ia (see Filippenko 2004, 2005 for recent reviews). We summarize the key findings here. (1) Empirically, analyses of SN Ia distances versus host stellar age, chemical abundance, morphology, and dust content indicate that SN Ia distances are relatively indifferent to the evolution of the Universe (e.g., Sullivan et al. 2003; Astier et al. 2006; Wang et al. 2006; Riess et al. 1999). (2) Detailed examinations of the distance-independent properties of SNe Ia (including the far-UV flux, e.g., as presented in the last section) provide strong evidence for uniformity across redshift and no indication (thus far) of redshift-dependent differences (e.g., Sullivan et al. 2005; Howell et al. 2005; Blondin et al. 2006). (3) SNe Ia are uniquely qualified as standard candles because a well understood, physical limit (the Chandrasekhar limit) provides the source of their homogeneity. Based on these studies, we adopt a limit on redshift-dependent systematics is to be 5% per $\Delta z = 1$ at $z > 0.1$ and make quantitative use of this here.

Many have studied the constraint placed by the redshift-magnitude relation of SNe Ia on the parameter combination Ω_M-w, where w (assumed to be constant) is the dark energy equation-of-state parameter. There are few models for dark energy that predict an equation of state that is *constant*, different from the cosmological constant, and not already ruled out by the data. On the other hand, a prominent class of models does exist whose defining feature is a time-dependent dark energy (i.e., quintessence). While the rejection of $w = -1$ for an assumed constant value of w would invalidate a cosmological constant, it is also possible that apparent consistency with $w = -1$ in such an analysis would incorrectly imply a cosmological constant. For example, if $w(z)$ is rising, declining, or even sinusoidal, a measured derivative could be inconsistent with zero while the average value remains near -1. Therefore, when using $w(z)$ to discriminate between dark-energy models, it is important to allow for time-varying behavior, or else valuable information may be lost. Here, we seek to constrain the value of $w(z > 1)$ and bound its derivative across the range $0.2 < z < 1.3$. This is unique information afforded by the *HST*-discovered SN Ia sample.

Finally, we may consider whether three additional parameters to describe $w(z)$ are actually needed to improve upon a flat, Λ-cold-dark-matter (ΛCDM) model fit to the data. To determine this we can calculate the improvement to the fit,

$$\chi^2_{eff} \equiv -\Delta(2\ln\mathscr{L}) = 2\ln\mathscr{L}(w=-1) - 2\ln\mathscr{L}(w_i = \mathscr{W}_i), \tag{1}$$

with i additional free parameters. For the weak, strong, and strongest priors we find an improvement of $\chi^2_{eff} = 4$, 5.5, and 5.5, respectively, for the three additional degrees of freedom, in no case requiring the additional complexity in dark energy (improvements of > 14 would be noteworthy). Likewise, there is no improvement at all for the Akaike Information Criterion (i.e., ΔAIC=$\Delta\chi^2 - 2i$; Liddle et al. 2004) with changes of -2, -0.5, and -0.5, respectively which fail to overcome the penalty of increased complexity in the model.

CONCLUSIONS

(1) We present 21 new *HST*-discovered SNe Ia and an improved calibration of the previous sample from R04. Together this sample contains 23 SNe Ia at $z \geq 1$, extending the Hubble diagram over 10 Gyr.

(2) We derive uncorrelated, model-independent estimates of $H(z)$ which well-delineate current acceleration and preceding deceleration. The *HST*-discovered SNe Ia measure $H(z > 1)$ to slightly better than 20% precision.

(3) The full *HST*-discovered SN Ia sample, presented here, provides a factor of two improvement over our present ability to constrain simple parameterizations of the equation-of-state parameter of dark energy (w) and its evolution.

(4) Stronger priors and tighter constraints on the preferred cosmological model can be extracted from independent measurements tied to the surface of last scattering, but the use of these requires assumptions about the behavior of dark energy across a wide range of redshift ($1.8 < z < 1089$). The strongest of these priors, like the simplest dark

energy parameterizations, appears unjustified in the presence of our current ignorance about dark energy. Assuming the effect of dark energy at $z > 1.8$ is minimal, we derive meaningful constraints on the early properties of dark energy: $w(z > 1) = -0.8^{+0.6}_{-1.0}$ and $w(z > 1) < 0$, i.e., negative pressure, at 98% confidence.

(5) At present, we find that the use of additional parameters to describe $w(z)$ does not provide a statistically significant improvement to the fit of the redshift-magnitude relation over the use of a simple cosmological constant.

(6) An analysis of the $z > 1$ sample-averaged spectrum shows it to be consistent with the mean spectrum of SNe Ia over the last 10 Gyr, failing to reveal direct evidence for SN Ia evolution.

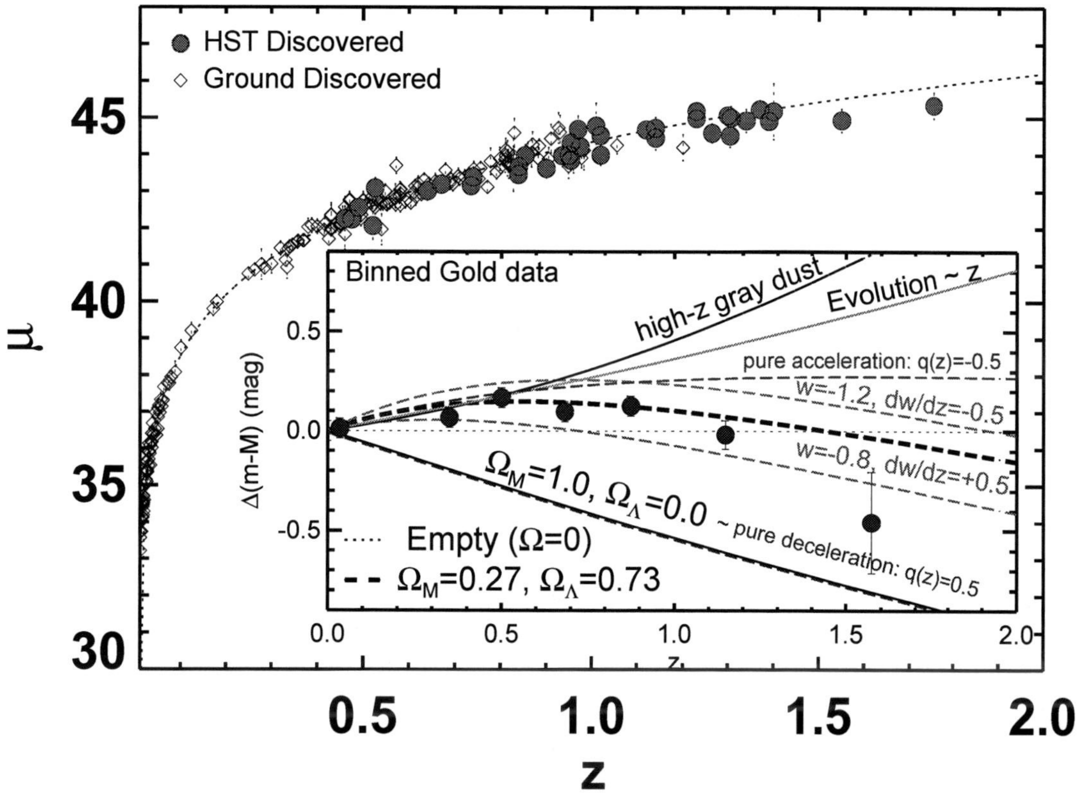

HST Discovered
Ground Discovered
45
40
35
30
μ
0.5
1.0
1.5
2.0
z
Binned Gold data
high-z gray dust
Evolution ~ z
pure acceleration: q(z)=-0.5
w=-1.2, dw/dz=-0.5
w=-0.8, dw/dz=+0.5
Ω_M=1.0, Ω_Λ=0.0
~ pure deceleration: q(z)=0.5
Empty (Ω=0)
Ω_M=0.27, Ω_Λ=0.73
Δ(m-M) (mag)

FIGURE 1.

Cosmological Observations of Dark Matter and Dark Energy

Edward L. Wright

UCLA Physics & Astronomy, Los Angeles CA 90095-1547, USA

Abstract. Cosmology is an observational science, since we are unable to make different Universes with varying conditions or initial conditions. Even so, by finding relatively simple initial conditions that evolve under the actions of known physical laws into a system very much like our observations, we can have fairly secure knowledge of the current contents and conditions in our Universe.

Keywords: Observational Cosmology
PACS: 98.80.Es

INTRODUCTION

When Isaac Newton observed the fall of an apple, he was not discovering gravity. Rather he was able to show that the same law of gravity that caused the fall of an apple on Earth also caused the orbital motion of the Moon around the Earth, and guided all the planets in their orbits around the Sun. This eliminated the need for a special material, the quintessence or fifth element, that followed different physical laws than the earth, water, fire and air of ancient chemistry. Following along Newton's path, astronomers have been able to use the Universal Law of Gravitation to determine the masses of distant objects like stars that can never be placed on the pans of a laboratory balance. That is why the Cavendish experiment [1], which measures the minute gravitational force between masses in the laboratory, has always been correctly regarded as "weighing the Earth" (and the stars). These difficult experiments to measure Newton's constant G provide the bridge between laboratory mass measurements defined by comparison to the standard kilogram in Paris, and the astronomical scale of masses where one only measures the combination GM.

But when astronomical measurements were extended to larger and larger scales, two discrepancies became apparent. The first discrepancy involves dark matter. In the outer reaches of galaxies, the orbital motion of stars is fairly rapid, and requires large gravitational forces to maintain equilibrium. These forces could be supplied by gravity from dark matter that was more extended than the light from stars we see when looking at a galaxy. And on longer scales, in clusters of galaxies, we also saw higher than expected orbital speeds. In clusters of galaxies the dark matter has to have the same extent as the normal matter. Historically unexplained perturbations on the motion of Uranus led to a hypothetical 8th planet which was discovered - Neptune. So dark matter found gravitationally can sometimes be seen in other ways. On the other hand, unexplained perturbations on the motion of Mercury were credited to a hypothetical planet Vulcan, but shown to be due to a modification of Newton's Laws - Einstein's General Relativity. Some scientists (and speakers at this conference) believe that the evidence for dark matter in the outer reaches of galaxies and comingled with the galaxies and hot gas in clusters of galaxies can be better explained by another modification of Newton's Laws: the MOND hypotheses. But I think that the various circumstances in which dark matter has been seen by its gravitational effects cannot all be explained by a universal modification to gravity, but can be explained by dark matter which need not have a universal distribution with respect to normal matter.

The second discrepancy observed in cosmological settings is the unusual faintness of distant Type Ia supernovae. This can be explained by a cosmological constant, a modification of General Relativity introduced by Einstein [2] to achieve a homogeneous and isotropic but static model of the Universe. The cosmological constant can be viewed as an energy density of the vacuum. In order for the energy density to be constant as the Universe expands, it is necessary that the vacuum also have a negative pressure, making the stress-energy tensor of the vacuum proportional to the metric and thus Lorentz invariant. These properties of the vacuum have been called "dark energy" by Mike Turner and "smooth tension" by Sean Carroll. The energy density and pressure can be attributed to a field or fields instead of the vacuum, but the fields must have very small gradients, both in space and time, in order to conform to the observations. And it is also possible to develop different modifications to General Relativity to explain the observations, although to my mind, a cosmological constant is already modified gravity.

These dark components of cosmology, both matter and energy, would not have been accepted as well as they have

CP1241, *Invisible Universe: Proceedings of the Conference,* edited by J.-M. Alimi and A. Füzfa

been without the presence of theories of high energy particle physics that contained natural niches for dark matter and dark energy. Supersymmetry proposes that all fermions have boson partners like the selectron and the squarks; and that all bosons have fermion partners, the "inos" like the photino, the gravitino, the zino and wino (but not the neutrino). Since the supersymmetric partner particles have not been seen they are presumed to be quite heavy, and if "R-parity" is conserved, supersymmetric partner particles could not decay into normal particles, making the lightest supersymmetric partner particle absolutely stable, very weakly interacting, and an excellent candidate for a dark matter particle.

In naive quantum field theory a sea of virtual pairs of mass m particles produces a vacuum energy density of $O(mc^2(mc/\hbar)^3)$ which is much larger than the 4 keV/cc allowed by observations. In fact, for the Planck mass this value is 10^{120} times larger than the observed value. The old "cosmological constant problem" was to explain now this large value got reset to zero. The new "dark energy problem" is to explain why this large value is reset to a small specific value.

Dark energy fits in very well with the Higgs mechanism for giving masses to particles like the $W^{\pm}$ and the Z^0. A set of fields has a potential energy density that is symmetric on the internal space of the fields, but the value is large for zero field, near zero for a preferred magnitude of the fields, and high for large fields. The "Mexican hat" potential has a high frequency (high mass) radial oscillation, and a zero frequency (massless) mode for transverse oscillations. In order to explain the cosmological constant, it is only necessary for the "near zero" of the potential be not quite zero.

SUPERNOVA OBSERVATIONS

Eleven years ago two independent groups found that distant supernovae were fainter than expected[3, 4]. Much better datasets are being presented at this meeting. Since these data depend on fluxes of standardizable candles, they give the luminosity distance. The observed luminosity distance is defined as

$$D_L = \sqrt{F/4\pi L} \qquad (1)$$

where F is the observed flux in erg/cm^2/sec while L is the luminosity in erg/sec. This formula also works for νF_ν and νL_ν if the flux is measured at $\lambda_{obs} = (1+z)\lambda_{em}$.

The theoretical calculation of the luminosity distance goes as follow: the metric is given by

$$ds^2 = c^2dt^2 - a(t)^2R_\circ^2(d\psi^2 + S^2(\psi)[d\theta^2 + \sin^2\theta d\phi^2]) \qquad (2)$$

where $S(x)$ is $\sinh(x)$ if $\Omega_{tot} < 1$ and $\sin(x)$ for $\Omega_{tot} > 1$. $R_\circ = (c/H_\circ)/\sqrt{|1-\Omega_{tot}|}$. The past light cone is given by $cdt = a(t)R_\circ d\psi$ so the current radial distance is

$$D_{now} = R_\circ\psi = \int \frac{cdt}{a} = \int_{1/(1+z)}^{1} \frac{cda}{a\dot{a}}. \qquad (3)$$

We can write $\dot{a}$ as $H_\circ\sqrt{X}$ with

$$X(a) = \Omega_{m\circ}/a + \Omega_{r\circ}/a^2 + \Omega_{v\circ}a^2 + (1-\Omega_{tot}) \qquad (4)$$

where $H = \dot{a}/a$, and $\Omega_{m\circ}$ is the current matter density divided by the critical density $\rho_c(0) = 3H_\circ^2/8\pi G$, and $\Omega_{r\circ}$ or $\Omega_{v\circ}$ are the same ratio for radiation an vacuum. The above assumes that the vacuum energy density is a constant, or that the ratio of pressure to energy density is given $w = P_v/\rho_v c^2 = -1$. If w is not -1 then the vacuum energy density changes with time, and then a more general formula

$$X(a) = \Omega_{m\circ}/a + \Omega_{r\circ}/a^2 + \frac{\rho_v(z)}{\rho_c(0)}a^2 + (1-\Omega_{tot}) \qquad (5)$$

is needed.

Let us define

$$Z = \int_{1/(1+z)}^{1} \frac{da}{a\sqrt{X}} \qquad (6)$$

so $D_{now} = (cZ/H_\circ)$. Then the angular size distance is

$$\begin{aligned} D_A &= \frac{c}{H_\circ}\frac{S(\sqrt{|1-\Omega_{tot}|}Z)}{(1+z)\sqrt{|1-\Omega_{tot}|}} \\ &= \frac{D_{now}}{(1+z)}\left(1+\frac{1}{6}(1-\Omega_{tot})Z^2+\frac{1}{120}(1-\Omega_{tot})^2Z^4+\ldots\right) \end{aligned} \tag{7}$$

We can define a function $J(x)$ given by

$$J(x) = \begin{cases} \frac{\sin\sqrt{-x}}{\sqrt{-x}}, & x<0; \\ \frac{\sinh\sqrt{x}}{\sqrt{x}}, & x>0; \\ 1+x/6+x^2/120+\ldots+x^n/(2n+1)!+\ldots, & x\approx 0. \end{cases} \tag{8}$$

Then since $D_L = (1+z)^2 D_A$ we can write

$$D_L = (1+z)^2\frac{cZ(z)}{H_\circ}\frac{J([1-\Omega_{tot}]Z^2)}{1+z} \tag{9}$$

An interesting aspect of this formula is that the second order term of the expansion of the luminosity distance *vs.* distance is given by

$$D_L = \frac{cz}{H_\circ}(1+0.5(1-q_\circ)z+\ldots) \tag{10}$$

so it only depends on the deceleration parameter

$$q_\circ = -\frac{a\ddot{a}}{\dot{a}^2}. \tag{11}$$

The effects of $w \neq -1$ and $(1-\Omega_{tot}) \neq 0$ therefore change D_L only in the 3rd and higher orders in an expansion in powers of z. Since both effects enter in the same order it is very important to fit for both simultaneously. This rule has often been neglected however.

A flat ΛCDM model with $w = -0.9$ can be matched by a non-flat ΛCDM model with $w = -1$ to an accuracy of 2 millimagnitudes [5]. The effects of w' are even smaller, entering only in the 4th order of z in the expansion. As a result the differences in supernova brightness induced by a non-zero w' are only ± 1 millimagnitude.

Observations of distant supernovae cannot uniquely determine D_L if the luminosity can evolve as a function of redshift. Consider two models with $D_L(z)$ and $D'_L(z)$. If we change the luminosity by a factor

$$f(z) = [D'_L(z)/D_L(z)]^2 \tag{12}$$

then we can change the consistent model from the unprimed to the primed model. For example, let the consistent model without luminosity evolution be the flat ΛCDM model, and let the primed model be a standard CDM Einstein-de Sitter model [6]. Then $f(z)$ is very well approximated by an exponential function of cosmic time[7]. This means that a “steady state dust” model with grey dust having a constant *physical* density in an EdS Universe is consistent with the supernova data. The common claims that high redshift supernova data rule out grey dust only apply to models with a constant *comoving* dust density. This exponential function of cosmic time is also predicted by axion-photon mixing[8], so it is important to obtain independent confirmation of the existence of dark energy.

ACOUSTIC OSCILLATION OBSERVATIONS

The most precise confirmations of dark energy come from using the the baryon acoustic oscillations as a standard ruler. The peaks seen in the angular power spectrum of CMB anisotropies have a spacing that corresponds to the distance sound can travel between the Big Bang and the time of last scattering. This spacing is the acoustic scale [9]:

$$\ell_a = \frac{\pi(1+z_{LS})D_A(z_{LS})}{\int_0^{1/(1+z_{LS})} c_s da/(a\dot{a})} \tag{13}$$

where z_{LS} is last scattering, c_s is the sound speed, a is the scale factor, $\dot{a}$ is its time derivative, and D_A is the angular size distance. The shift parameter [10, 11] $R = \Omega_M^{1/2} H_\circ (1+z_{LS}) D_A(z_{LS})/c$ is similar to ℓ_a but makes some approximations about the sound speed and expansion rate in the early Universe that are not as precise as the current accuracy (better than $\pm 0.3\%$) on the acoustic peak spacing and certainly much worse than the accuracy that will be provided by the Planck mission.

The acoustic scale can also be seen in the distribution of galaxies [12]. Here the comoving length scale

$$\lambda_a = \int_0^{1/(1+z_{LS})} c_s da/(a\dot{a}) \tag{14}$$

shows up as a secondary peak in the galaxy two-point correlation function. The angular scale corresponding to this standard ruler can be measured at different redshifts using galaxy redshift surveys to different depths. The CMB acoustic peak spacing is determined by this angular scale at $z = 1089$.

These angular measurements do not depend on flux measurements so grey dust or luminosity evolution do not affect them. The acoustic scale at very low redshift depends only on Ω_m, while at very high redshifts like the CMB it depends on both Ω_m and Ω_v. A combination of low redshift and CMB acoustic scale data gives a well constrained intersection that is very consistent with a flat ΛCDM Universe.

STRUCTURE GROWTH OBSERVATIONS

A different kind of data involves the growth of structure as a function of redshift. The growth function is given by

$$D(t) = \frac{\sqrt{X}}{a} \int X^{-3/2} da \tag{15}$$

and it helps to determine the level of gravitational potential perturbations as a function of redshift. These perturbations can be seen in the late-integrated Sachs-Wolfe effect [13], which depends on the time derivative of the potential, and in gravitational lensing, which depends on spatial derivatives of the potential. Measurements of the LISW definitely show the correlation of the CMB temperature with large scale structure expected in ΛCDM models, while an EdS model would give no correlation, and a super-Sandage closed CDM model would give a negative correlation.

Gravitational lensing has also been used to confirm the existence of dark energy. Observations of multiple strongly lensed arcs in the cluster A2218 show that an EdS model is inconsistent with the data [14].

DISCUSSION

Multiple types of data have converged on a concordance model with flat spatial sections ($\Omega_{tot} = 1$), a matter density only a quarter to a third of the critical density, and hence a large vacuum energy density. We now have enough data to simultaneously show that $\Omega_{tot} = 1$ and that the equation of state is $w = -1$ with fairly small errorbars. Thus of the three models seriously considered after the COBE DMR results [15] – open CDM, mixed hot and cold DM, and ΛCDM – the cosmolgical observations have decisively favored ΛCDM.

ACKNOWLEDGMENTS

The *WMAP* mission is made possible by support from the Science Mission Directorate at NASA Headquarters and by the hard and capable work of scores of scientists, engineers, technicians, machinists, data analysts, budget analysts, managers, administrative staff, and reviewers.

REFERENCES

1. H. Cavendish, *Annalen der Physik* **2**, 1–62 (1799).
2. A. Einstein, *Sitzungsberichte der Königlich Preußischen Akademie der Wissenschaften (Berlin), Seite 142-152.* pp. 142–152 (1917).

3. A. G. Riess, A. V. Filippenko, P. Challis, A. Clocchiatti, A. Diercks, P. M. Garnavich, R. L. Gilliland, C. J. Hogan, S. Jha, R. P. Kirshner, B. Leibundgut, M. M. Phillips, D. Reiss, B. P. Schmidt, R. A. Schommer, R. C. Smith, J. Spyromilio, C. Stubbs, N. B. Suntzeff, and J. Tonry, *AJ* **116**, 1009–1038 (1998), URL `http://adsabs.harvard.edu/cgi-bin/nph-bib_query?bibcode=1998AJ....116.1009R&db_key=AST`.
4. S. Perlmutter, G. Aldering, G. Goldhaber, R. A. Knop, P. Nugent, P. G. Castro, S. Deustua, S. Fabbro, A. Goobar, D. E. Groom, I. M. Hook, A. G. Kim, M. Y. Kim, J. C. Lee, N. J. Nunes, R. Pain, C. R. Pennypacker, R. Quimby, C. Lidman, R. S. Ellis, M. Irwin, R. G. McMahon, P. Ruiz-Lapuente, N. Walton, B. Schaefer, B. J. Boyle, A. V. Filippenko, T. Matheson, A. S. Fruchter, N. Panagia, H. J. M. Newberg, W. J. Couch, and The Supernova Cosmology Project, *ApJ* **517**, 565–586 (1999), URL `http://adsabs.harvard.edu/cgi-bin/nph-bib_query?bibcode=1999ApJ...517..565P&db_key=AST`.
5. E. L. Wright, *New Astronomy Review* **49**, 407–412 (2005), `arXiv:astro-ph/0409094`.
6. A. Einstein, and W. de Sitter, "On the Relation between the Expansion and the Mean Density of the Universe," in *Proceedings of the National Academy of Sciences*, 1932, vol. 18, p. 213.
7. E. L. Wright, *ArXiv Astrophysics e-prints* (2002), `arXiv:astro-ph/0201196`.
8. C. Csáki, N. Kaloper, and J. Terning, *Physical Review Letters* **88**, 161302–+ (2002), `arXiv:hep-ph/0111311`.
9. L. Page, C. Jackson, C. Barnes, C. Bennett, M. Halpern, G. Hinshaw, N. Jarosik, A. Kogut, M. Limon, S. S. Meyer, D. N. Spergel, G. S. Tucker, D. T. Wilkinson, E. Wollack, and E. L. Wright, *ApJ* **585**, 566–586 (2003).
10. J. R. Bond, G. Efstathiou, and M. Tegmark, *MNRAS* **291**, L33–L41 (1997), URL `http://adsabs.harvard.edu/cgi-bin/nph-bib_query?bibcode=1997MNRAS.291L..33B&db_key=AST`.
11. Y. Wang, and P. Mukherjee, *Astrophys. J.* **606**, 654–663 (2004), `astro-ph/0312192`.
12. D. J. Eisenstein, I. Zehavi, D. W. Hogg, R. Scoccimarro, M. R. Blanton, R. C. Nichol, R. Scranton, H.-J. Seo, M. Tegmark, Z. Zheng, S. F. Anderson, J. Annis, N. Bahcall, J. Brinkmann, S. Burles, F. J. Castander, A. Connolly, I. Csabai, M. Doi, M. Fukugita, J. A. Frieman, K. Glazebrook, J. E. Gunn, J. S. Hendry, G. Hennessy, Z. Ivezić, S. Kent, G. R. Knapp, H. Lin, Y.-S. Loh, R. H. Lupton, B. Margon, T. A. McKay, A. Meiksin, J. A. Munn, A. Pope, M. W. Richmond, D. Schlegel, D. P. Schneider, K. Shimasaku, C. Stoughton, M. A. Strauss, M. SubbaRao, A. S. Szalay, I. Szapudi, D. L. Tucker, B. Yanny, and D. G. York, *ApJ* **633**, 560–574 (2005).
13. R. K. Sachs, and A. M. Wolfe, *ApJ* **147**, 73 (1967), URL `http://adsabs.harvard.edu/cgi-bin/nph-bib_query?bibcode=1967ApJ...147...73S&db_key=AST`.
14. G. Soucail, J.-P. Kneib, and G. Golse, *A&A* **417**, L33–L37 (2004).
15. E. L. Wright, S. S. Meyer, C. L. Bennett, N. W. Boggess, E. S. Cheng, M. G. Hauser, A. Kogut, C. Lineweaver, J. C. Mather, G. F. Smoot, R. Weiss, S. Gulkis, G. Hinshaw, M. Janssen, T. Kelsall, P. M. Lubin, S. H. Moseley, T. L. Murdock, R. A. Shafer, R. F. Silverberg, and D. T. Wilkinson, *ApJL* **396**, L13–LL18 (1992), URL `http://adsabs.harvard.edu/cgi-bin/nph-bib_query?bibcode=1992ApJ...396L..13W&db_key=AST`.

The formation of disks in massive spiral galaxies

François Hammer[a] with Mathieu Puech[a] & Hector Flores[a]

[a] *Laboratoire Galaxies Etoiles Physique et Instrumentation, Observatoire de Paris, UMR8111 du CNRS, 5 place Jules Janssen, 92195 Meudon France*

Abstract. The flatness of the rotation curve inside spiral galaxies is interpreted as the imprint of a halo of invisible matter. Using the deepest observations of distant galaxies, we have investigated how large disks could have been formed. Observations include spatially resolved kinematics, detailed morphologies and photometry from UV to mid-IR. Six Giga-years ago, half of the present-day spirals had anomalous kinematics and morphologies that considerably affect the scatter of the Tully Fisher relation. All anomalous galaxies can be modelled through gas-rich, major mergers that lead to a rebuilt of a new disk. The spiral-rebuilding scenario is proposed as a new channel to form large disks in present-day spirals and it accounts for all the observed evolutions since the last 6 Giga-years. A large fraction of the star formation is linked to merging events during their whole durations.

Keywords: Galaxies: formation --Galaxies: spiral -- Galaxies: kinematics and dynamics
PACS: 98.35.Ac, 98.35.Gi, 98.52.Cf, 98.52.Nr, 98.52.Sw, 98.56, 98.62.Ai, 98.62.Dm, 98.62.Gq, 98.62.Hr , 98.62.Lv

INTRODUCTION

A considerable fraction, 72%, of massive galaxies is made of spiral galaxies, including the Milky Way and M31 in the Local Group. By massive galaxies, I arbitrarily consider those with stellar masses larger than $10^{10}M_{sun}$, the Milky Way being five times more massive than this value. Since the early 20th century it is well known that the baryonic matter is not massive enough to equilibrate the observed rotation of stars in the giant disks, requiring the presence of a massive unseen component [1]. Besides the important question of the nature of the dark matter, astrophysicists have debated a lot the following question: how large disks formed in massive spirals? Disks are supported by their angular momentum that may be acquired by early interactions in the framework of the tidal torque theory [2,3]. In this theory, galactic disks are then assumed to evolve without subsequent major mergers, in a secular way.

However this theory faces with at least two major problems. First, galaxy simulations demonstrate that such disks can be easily destroyed by collisions of relatively big satellites[1] [4], and such collisions might be too frequent to let disks survive; Second, the disks produced by simulations in such a way are too small or have a too small angular momentum when compared to the observed ones, and this is so-called the "angular momentum catastrophe". Besides this, an early and gradual formation of the disk of the Milky Way is well supported by observations.

THE MILKY WAY IS AN EXCEPTIONAL LOCAL SPIRAL

Observations have considerably improved since the early 70s and it is now possible to analyze with much more details the properties of progenitors of present-day galaxies by observing distant galaxies. Moreover our knowledge of the two massive spirals in our neigbourhood, the Milky Way and M31, has considerably improved. It is now well established that the Milky Way experienced very few minor mergers and no major merger during the last 10-11 Gyrs

[1] In the following I will assume that major mergers are those which destroy the disk of the main interloper, which correspond to various mass ratio depending on the orbital parameters; generally major mergers are assumed to have mass ratio larger than 1:5, 1:4 or 1:3.

CP1241, *Invisible Universe: Proceedings of the Conference,* edited by J.-M. Alimi and A. Füzfa

[5,6]. The old stellar content of the thick disk let possible a merger origin at that early epoch. The Milky Way is presently absorbing the Sagittarius dwarf though this is a very tiny event given the fact that the Sagittarius mass is less than 1% of the Milky Way mass [7]. If the Magellanic Clouds are truly bound to the Milky Way, a collision is expected within the next Giga-year, which may have more impact to the Milky Way disk since they carry on a quite significant amount of orbital angular momentum. Milky Way past history strongly contrasts with that of M31 (Andromeda). Numerous works have revealed the very tumultuous history of M31 [8, 9, 10]. Its outskirts is surrounded by gigantic structures at low surface brightness, including the Giant stream, the extended thick disk, rings etc.. Several attempts have been successfully made to relate some of these structures to interactions with smaller neighbouring galaxies (e.g. M33). However we are still lacking of a full model of M31 and its outskirts, possibly because its past history was too rich in merger events. It is still uncertain whether M31 experienced only several minor interactions, or also a major merger.

Observations of galaxies 'outskirts at similar depths are in progress, and superb structures have been identified surrounding NGC 4013, NGC 5907, M 63 and M 81 [11, 12]. They are all reminiscences of past mergers events. It has to be established whether or not most spiral galaxies are showing similar richness of past merger events in their halo. It would be also interesting to simulate how would appear the Milky Way and the Sagittarius stream from an external observer. There is an important clue that many spirals' haloes contain more relics of past merger events than does the Milky Way. In fact the discovery that RGB stars in the halo in very nearby spirals [13] are redder than those of the Milky Way implies that they have been metal enriched by a way or another. The same applies to the important population of local spirals from the SDSS as it has been shown by [14] after stacking more than one thousand spiral halo emission. Figure 1 displays the corresponding metal abundance of stars in the inner halo of various galaxies, including the Milky Way and M31. In all massive galaxies with $V_{flat} > 150$ km/s, halo stars have metal abundances much larger than those of the Milky Way. Most of these spiral galaxies have been observed by [13] and are very nearby. The agreement with the average value from SDSS brings a lot of support to the idea that most spirals have an inner halo much more enriched than that of the Milky Way. NGC 891 has been observed by [17] because it shows a lot of similarities with the Milky Way, although its pristine halo is not discovered yet. Examination of the very outskirts of M31 has revealed [16] the existence of a pristine halo at very large distances, up to 150 kpc. The M31 halo, as well as its globular cluster system, is similar to the sum of a pristine halo, or the Milky Way halo, with numerous metal enriched stars, probably due to merger residuals. Conversely, the Milky Way halo and globular cluster system may require the absence of previous merger of satellites with mass larger or equal to $10^9 M_{sun}$ [18].

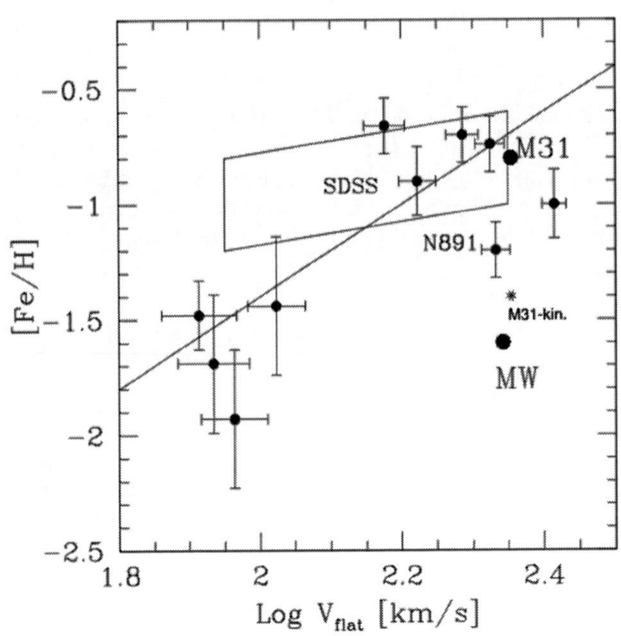

FIGURE 1. Iron abundances of RGB stars estimated in the outskirts (2-10kpc from the minor axis) of edge-spirals galaxies from [13] plotted against log V_{flat}. Large points represent the values for the Milky Way and M31. The solid lines delineate the range of

values from [14], after stacking 1047 edge-on SDSS galaxies. This Figure is similar to Figure 6 of [15], to which we have added the value for the pristine halo of M31 that has been recovered at 60-150 kpc ([16]). NGC 891 measurements from [17] are also added, as this galaxy shows many similarities with the Milky Way. The vertical error bars have been taken from [13].

Thus the Milky Way outskirts appear to be quite exceptional. It is useful to compare the fundamental parameters of its disk to those of other galaxies. The main difficulty is coming from the fact that we are lying in the Milky Way, and then this is not an easy task. By chance, very detailed models of the light distribution of the Galaxy were necessary to remove at best its signal to recover the CMB emission. Hipparcos also provided very useful data to model in detail the Galaxy [18]. Using these data, [15] have shown that the Galactic disk radius, R_d= 2.3 +/- 0.6 kpc, is quite small especially when compared to that of M31 (R_d= 5.8 +/- 0.6 kpc). The whole emission of both the Milky Way and M31 in K-band have been well recovered by Spitzer, and provides M_K=-24.02 and -24.51, for the Milky Way and M31, respectively. The difference between the two values indicates that the stellar mass of M31 is twice that of the Milky Way, after accounting for their respective stellar mass to K-band luminosity ratios. Even if the Milky Way is approximately twice gas rich than M31, the baryonic mass ratio is still 2, because the gas content is rather marginal in both galaxies. Interestingly the two galaxies have the same circular, or flat velocities, implying a different location in the fundamental volume delineated by mass, radius and velocity. On the basis of a very detailed study of the local fundamental relations (mass-velocity or Tully Fisher, radius-velocity) for local spirals, [15] showed that M31 is quite a typical spiral, while the Milky Way is surprisingly exceptional[2], being offset by 1σ in both relationships. To summarize, only 7% of local spirals have similar properties than the Milky Way.

Some interesting lessons can be learned. The Milky Way is lacking approximately half of its stellar mass, radius and angular momentum when compared to an average spiral, or to M31. Its halo and globular system is pristine, almost free of merger residuals in contrast to the vast majority of other galaxies that systematically show redder halo stars, or more metal-enriched RGB stars. This is quite suggestive that the vast majority of spirals experienced a quite rich merger history, as M31 certainly did, contrasting to the well-established quiescent history of the Milky Way. This would explain why the Milky Way is lacking of angular momentum when compared to other galaxies. Indeed by accounting for the additional effect of the orbital angular momentum provided by major mergers, [19] showed that the angular momentum catastrophe could be solved.

Future observations of halo stars in a larger volume will bring a lot of new constraints to the above, as well as new measurements of the detailed structure of the Galaxy and of M31. Within our present knowledge the Galaxy is not an ordinary spiral, and understanding the formation of large disks in massive spirals requires observing in great details other spirals. Consequently, a gradual elaboration of large disks might not be the rule for most disks, and we may have to identify different channels for their formation. Another method to investigate the disk formation is to verify directly whether or not many spirals have experienced merger events. The present generation of very large telescopes allows us to study in detail galaxies at large look-back times, which have to be in a statistical sense the progenitors of the present-day spirals.

WHAT IS THE PAST HISTORY OF PRESENT-DAY SPIRALS?

There have been many studies of distant galaxies, and reviewing them is out of the scope of this paper. I would simply mention the first significant survey of distant galaxies up to z=1, the Canada France Redshift Survey, that evidenced for the first time the strong decrease of the star formation density since the last 8 Giga-years [20]. This has been confirmed by infrared photometry [21], using ISO and more recently Spitzer. Integration of star formation density over look-back time can be compared to evolution of the stellar mass density, which is obtained from near-IR photometry [22, 23]. Both approaches give the same result, which could be considered as robust: 50% of the present-day stellar mass has been formed at $z < 1$ or approximately since the last 8 Gyrs. Analyses by [24, 25] of representative samples selected in near-IR and observed in mid-IR have revealed that the star formation below z=1 is dominated by luminous IR galaxies (LIRGs) with star formation rates from 19 to 190 M_{sun}/year. Because the fraction of red E/S0 does not change from z=1 to z=0, the stars formed during the last 8 Gyrs are mostly assembled in spiral galaxies [24, 26], and this is confirmed by analyses of LIRG morphologies that range from peculiar or merger, to massive spirals.

---

[2] I have to mention here that both John Kormendy and Joe Silk told me that Allan Sandage firstly mentioned the hypothesis of an exceptional Milky Way.

A Strategy to Observe Distant Galaxies and to Relate Them to Local Galaxies

Observational Requirements

A remarkable suite of telescopes and instruments is now available to study the distant Universe. As a European scientist I may focus -in a quite biased manner- on the Very Large Telescopes operated by ESO. These 4 telescopes offer a suite of 12 instruments allowing us to cover very large wavelength, spectral and spatial resolution ranges and give us the possibility to observe several sources at the same time. FLAMES/GIRAFFE is successfully operating since July 2002 at the UT2 of the VLT. It is still the only instrument offering to observe, simultaneously up to 15 galaxies using integral field units (IFUs). Using such an instrument, our team has been able to gather the largest existing sample of velocity fields of distant galaxies (see Figure 2).

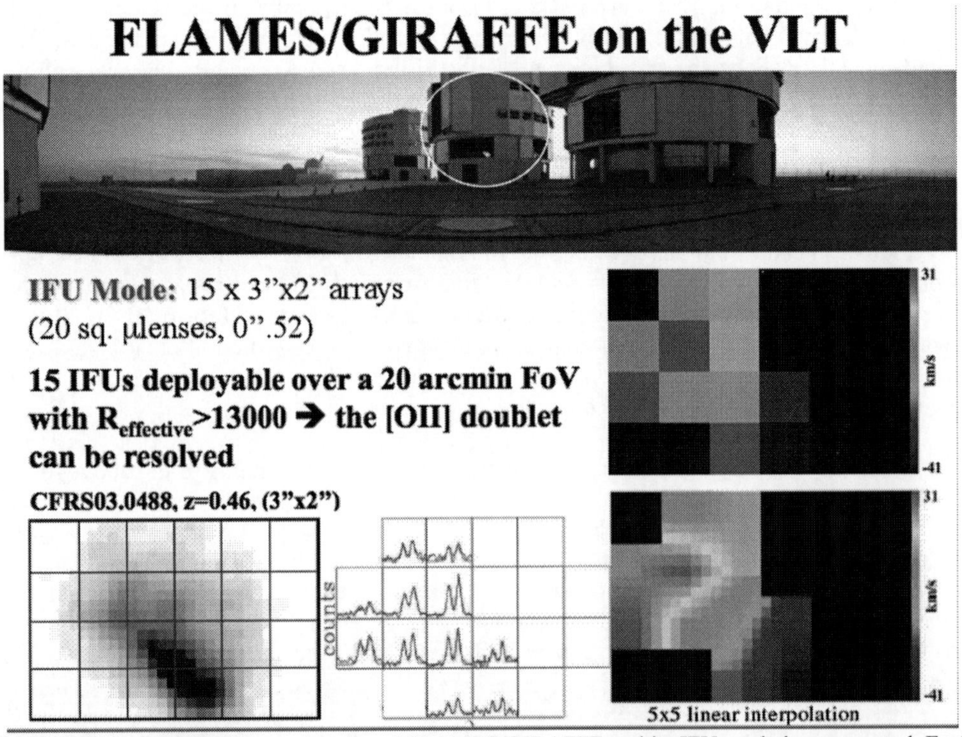

FIGURE 2. A sketchy representation of FLAMES/GIRAFFE and its IFU mode is represented. Each galaxy is dissected in up to 20 small apertures or micro-lenses, each of them providing a spectrum at intermediate resolution. In the illustration the OII doublet is well resolved, and each line can be used to trace the velocity and the dispersion of an individual galaxy region. In the right panels the corresponding distribution of the velocity field is shown, that has been interpolated for practical reason in the bottom panel.

However kinematics is not all and other important information need to be collected by other facilities. The best strategy is to perform kinematic surveys within existing cosmological fields that have been surveyed by HST, Spitzer and other telescopes. Because distant galaxies may present various evolutionary stages, it is mandatory to compare kinematical maps to morphological details. For example, the dynamical axis has to be compared to the disk optical axis: if they are severely misaligned, the galaxy is likely not at equilibrium, or not relaxed. The Hubble Space Telescope equipped with the Advanced Survey Camera is able to provide details of few hundred of parsecs at $z < 1$. However it is only a 2.5 meters telescope and depth is an important issue for distant galaxies whose emissions are severely affected by cosmological dimming. For example, the optical radius (3.2 times the disk scale length) of a Milky Way like galaxy at z=0.5 (z=1.3) requires 3 (100) hours of HST/ACS observations to be properly recovered, and this is without accounting for extinction effects. To be predictive enough, studies of kinematics at $z > 1$ would have to be undertaken in extremely deep fields, and up to now, there is only the small Ultra Deep Field.

In the following I will focus on the comparison between z ~ 0.65 and local galaxies. Recall that z= 0.65 represents a 6 Gyrs look-back time, which is not negligible. Within this redshift range, we may indeed compare galaxies in a quite homogeneous manner: most quantities can be evaluated in a similar way, independently of the redshift[3]. Figure 2 illustrates that kinematics of distant galaxies implies to dissect each of them in several parts. Given the faintness of these galaxies, large amounts of time are needed to reach useful S/N. Using GIRAFFE at VLT, [27, 28] showed that from 8 to 24 hours of exposure time are requested to recover the kinematics of almost all emission line galaxies with S/N $\geq$ 4 in at least 5 pixels or micro lenses. A special care is needed to select samples of galaxies for which we need to examine together their velocity fields, their morphologies and their masses.

Representative Samples of Local and Distant Galaxies

First, we need to determine the simplest method in selecting galaxies. At first glance, the absolute luminosity in near-IR is a relatively good proxy of the stellar mass. It indeed limits the evolutionary effects of the $M_{stellar}/L$ that would be severe in bluer broad bands. By applying $M_J < -20.3$ to both samples of local and z=0.65 galaxies, [26, see their Appendix A] demonstrated that they essentially select galaxies within the same baryonic mass range, because the evolutionary effect of $M_{stellar}/L_J$ (~ 0.15 dex) is almost exactly compensated by the gas fraction increase with redshift (~ -0.16 dex). Second, we need to ensure that our selection is representative of the luminosity function of galaxies at each redshift. This has been verified by [26, 28] after comparison with various luminosity functions derived from large surveys. Third, we have to account for more subtle effects linked with observational techniques. For example, derivation of kinematics for $M_J < -20.3$ distant galaxies is only feasible for sources with W_0(OII) $\geq$ 15A [27, 28]. We should complement the sample of emission line galaxies by a sample of "quiescent" galaxies with W_0(OII) < 15A [26]. The only residual difficulty would be the absence of kinematic information for the distant, quiescent galaxies, out of reach of 8 meters telescopes. This is not a stringent difficulty because 60% of z=0.65 galaxies have W_0(OII) $\geq$ 15A [29]. Fourth, we have to ensure that the resulting selections of galaxies are not affected by cosmological variance. This has been done by [27, 28] by selecting 63 galaxies in 4 different fields of view, including the GOODS-S, the CFRS 03 and 22h and the HDF-S. However some results discussed in the following are based on the sole GOODS-S field (33 galaxies). Besides being the most surveyed field at high Galactic latitude, it shows two large-scale structures at z= 0.668 and z=0.735, which may affect the luminosity density evolution between z=0.25-0.5 and z=0.5-0.75 by up to 0.4-0.5 dex [30]. This should be kept in mind during the discussion of the results based on the sole GOODS-S field.

The Strong Evolution of Massive Galaxies at z=0.65: Morphologies, Kinematics, Star Formation and Gas & Metal Fraction

Morphological Changes of Massive Galaxies since the last 6 Giga-years

The HST/ACS provides details of 200 pc in z=0.65 galaxies, whose z-band observations can be directly compared to r-band of SDSS galaxies at z=0. Inspections of the morphologies evidence that galaxies have strongly evolved. Peculiar galaxies represent half of z=0.65 galaxies [31, 32, 26], while the fraction of early type galaxies is similar to that at z=0 (see Figure 3). It means that more than half of the progenitors of present-day spirals are peculiar galaxies at z=0.65 [26, 32]. Here I need to make two important statements. First, morphologies should be examined in detail, either from examination by a well-experimented observer [32] or by using a complete and full analysis of the light distribution in several colours (bulge-disk decomposition using GALFIT, see [33]) and of a semi-automatic "decision tree" (see [26] and their Figure 4). Galaxies are too complex systems to have their morphologies classified by a set of only 2 parameters such as compactness or asymmetry: those classifications often overestimate the number of spiral morphologies [34]. Second, it should be pointed out that peculiarities in distant galaxies are generally observed at rest-frame UV to near-IR, meaning that all the galaxy light distribution is

---

[3] K-correction effects can be simply ignored when observing z=0.65 galaxies in filters with central wavelengths that are 1.65 times that of filters used to observe z=0 galaxies. For spectral energy distribution and star-formation rates this is feasible using Spitzer IRAC & MIPS as well as HST/ACS broad-band filters that could be compared to SDSS AND 2MASS filters (see [26]). Morphological analyses of SDSS can be easily compared to those of distant galaxies in the GOODS-S field.

affected[4]. Some peculiarities could be due to complex networks of star forming regions as it could be observed in local irregular dwarves. The large variety of peculiar morphologies at such high surface brightness (see Figure 3) could be also attributed to mergers [34]. Solving out this important question requires investigating the internal motions in distant galaxies.

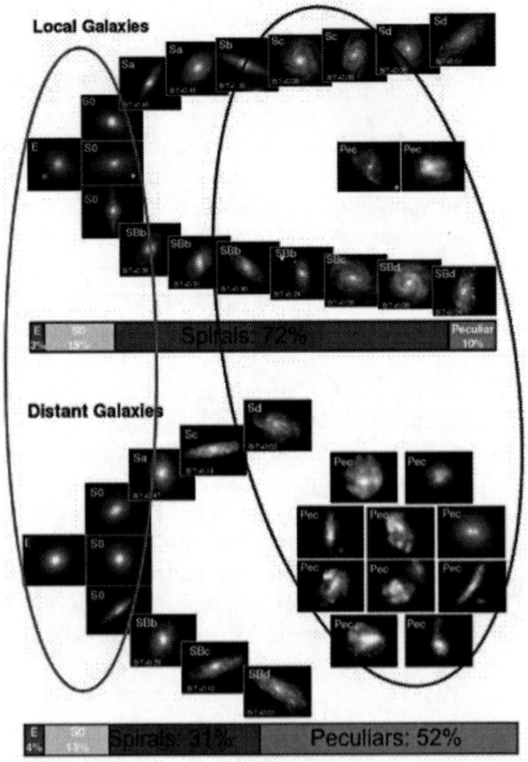

FIGURE 3. Morphologies of galaxies at z=0 (top) and z=0.65 (bottom) from [26]. Each stamps represents 5% of the $M_J \leq -20.3$ galaxy population and are 3-colours composite of u, g and r or of v, i and z bands, respectively. The fraction of regular spirals has significantly increased since 6 Gyrs, while for early-type (E/S0) it appears unchanged. At least half of the present-day spirals have a progenitor which is peculiar at z=0.65.

Strong Evolution of the Kinematics since the last 6 Giga-years

Kinematical study of distant galaxies is a recent field of investigations. At present all studies are limited by the spatial resolution, i.e. to gas motions at scales from 1 to 3 kilo-parsecs. However, such motions are crucial to disentangle disk rotation from other mechanisms affecting galaxy dynamics at large scales (Figure 4). To reach higher spatial resolution requires combining both adaptive optics techniques and extremely large exposure time, because of the apparent faintness of distant galaxies. GIRAFFE observations are limited by the pixel size (0.52 arc second or 3.5 kpc at z=0.65), although the *precision* in identifying the location of kinematical features is much better than that. This comes from simulations of GIRAFFE observations using redshifted local templates degraded down to the GIRAFFE IFU resolution, and assuming a similar S/N that that from observations of distant galaxies[5]. It shows that kinematic features are stable within variations in positions of one half of an IFU pixel, meaning that the *precision* in recovering their location is close to 0.25 arcsec [36]. This sub-spatial pixel *precision* largely results from the high spectral resolution of GIRAFFE, which provides us with a very accurate kinematic measurement

[4] It has been claimed by [35] that distant LIRGs have regular, disk morphologies on the basis of an adaptive optic survey conducted at the Keck telescope. However the exposure time was too limited to allow an observation beyond the core of individual galaxies. An examination of much deeper HST/ACS observations reveals their peculiar morphologies with rings, tidal tails, irregularities and even mergers, in rest-frame V-band.
[5] A generalization of this simulator is now used to prepare several phases A of E-ELT instruments, especially the link between spatial resolution provided by different adaptive optics techniques and spatially resolved kinematics (see [37]).

along the spectral axis. Finally we have learnt from our experience with GIRAFFE that the accuracy in comparing galaxy kinematics with morphology is generally dominated by the relative astrometry (0.23 arcsec). The half pixel accuracy has been illustrated in the observation of a satellite (1:18) in falling into a z=0.7 galaxy, responsible of the offset of the dispersion peak [38].

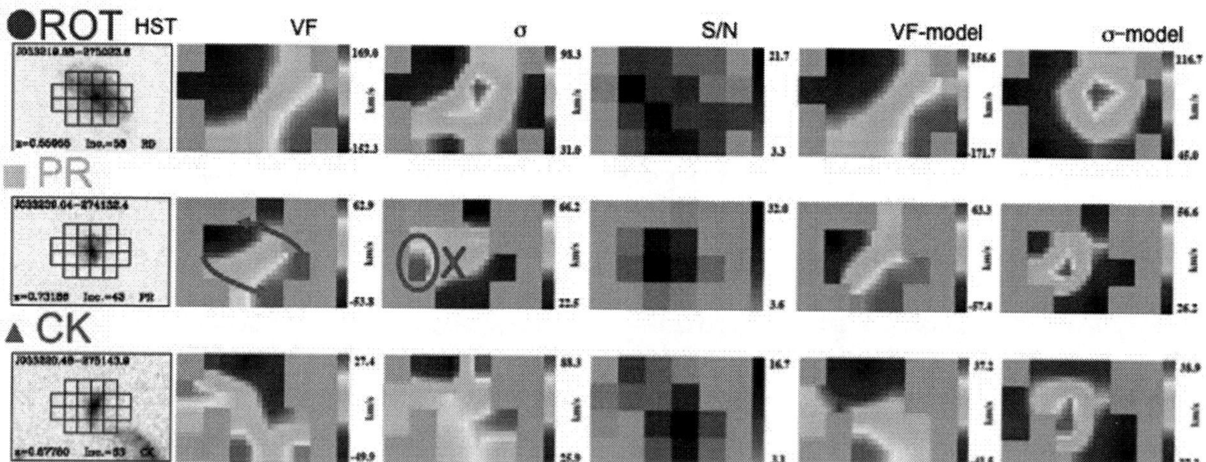

FIGURE 4. It illustrates three different examples (or class) of kinematical properties of distant galaxies. The upper row shows a rotating disk showing a dynamical axis aligned with the optical axis. Because of the low spatial resolution most of the gradient between the two extreme velocities provides a dispersion peak (see 3rd and 6th panels from the left) just at the centre of the velocity field and of the galaxy, i.e. the centre of rotation. Velocity models assume the inclination from the optical image and a velocity gradient from the observations. The middle row shows a galaxy with an almost aligned dynamical axis, thus indicative of a rotation. However the dispersion peak falls far from the centre of rotation, indeed far from any optical emission of the galaxy. This is indicative of a strong disturbance of the gaseous emission, probably due to shocks in gas flows, which perturbs the rotation (called perturbed rotation, PR). In the bottom row, the kinematics is complex (called complex kinematics, CK) from both velocity field and dispersion maps. The galaxy is likely after its first passage during its merger with a more massive spiral [39].

Among the 63 z=0.4-0.75 galaxies observed by GIRAFFE, most of them show anomalous kinematics. Accounting for a representative sample of $M_J \leq -20.3$ galaxies (in emission and in absorption) there is only 19% of rotating disks, 15% of perturbed rotators and 26% of complex kinematics, the resting 40% being made of "quiescent" galaxies that could not been observed due to the faintness of their emission lines. The latest galaxies have been examined by [26] and can be further subdivided into 18% of E/S0, 12% of spirals and 10% of peculiar galaxies. Taking into account the non-evolving fraction of E/S0 let us with 72% of present-day spirals and only 31% of rotating disks at z=0.65. Then at least half of the present-day spirals progenitors at z=0.65 have anomalous kinematics. This implies a strong evolution of spiral kinematics[6] with amplitude similar to what is found from morphological analyses.

Combining Morphological and Kinematical Analyses

A thorough comparison of morphology and kinematics of 52 distant galaxies has been made by [31]. They find that 92% of galaxies with anomalous kinematics (PR or CK) have peculiar morphologies, and that 80% of rotating galaxies have spiral morphologies. This result stands for morphological classifications assuming a strict similarity in classifying local and distant galaxies [26, 32], but it does not for automatic or semi-automatic methods based on asymmetry and compactness. Nevertheless it has profound implications, because kinematics is sampling the ionised gas ([OII] emission lines) while galaxies morphologies are due to bright stars emitting at all optical wavelengths. Then a common mechanism is responsible of both modes of disturbances in half of the present-day spiral

[6] It is assumed here that local spiral galaxies are fully dominated by their rotation, but see [40] for a detailed analysis of kinematics of local galaxies.

progenitors. In fact the combination of detailed morphologies with HST/ACS with large-scale kinematics from VLT/GIRAFFE is well appropriate. For example, most distant galaxies show numerous blue or dusty regions due to recent star formation (SF). The characteristic SF time (~ 100Myrs) is similar to the characteristic dynamical time to affect the galaxy large-scale kinematics (for motions assumed to be 100 to 200 km/s, see [39]). In other words, the turbulent gas motions observed by GIRAFFE give a good approximation of the dynamical state of both gas and star-forming regions, the latter dominating the morphological disturbances of distant galaxies. This is a good start to identify the physical mechanism(s), which reproduce the large variety of both peculiar morphologies and anomalous kinematics of spiral' progenitors.

Gas Fraction and Metal Content

Conversely to local galaxies for which the gas content can be easily estimated from HI measurements [41], at z=0.65, most galaxies are out of the reach of present-day radio telescopes. On the other hand one may rely on the existing correlation between surface densities of the gas and of the star formation rate, known as the Kennicutt-Schmidt relation [42]. It is natural to expect a correlation between star formation efficiency and gas content, although the tightness of the observed relation from normal spirals to powerful ULIRGs is quite impressive as it linked very different processes generating star formation. A correlation between star formation and molecular gas is also observed [43], which is easier to understand from the theory of star formation. Gas content can be estimated from the inversion of the Kennicutt-Schmidt relation [39, 44], whose slope may slightly vary depending on the inclusion of very powerful ULIRGs. The uncertainties related to the exact slope of the KS relation are certainly marginal because the distribution of SFR surface densities is concentrated within a quite small range, $\Sigma(SFR)=0.01$-$0.1\ M_O/yr/kpc^2$ (see Figure 5, bottom-right). Uncertainties in estimating the gas mass are probably dominated by those on the determination of the gas radii. Several distant galaxies are compact from their stellar continuum, although they almost all show relatively large extent of their ionised gas emission, which can be only detected by IFU techniques.

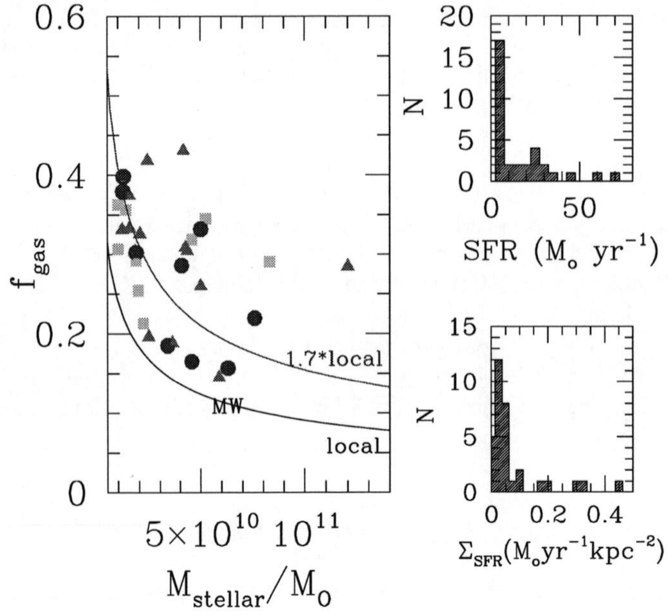

FIGURE 5. Gas fraction versus stellar mass for 33 z=0.4-0.75 galaxies observed in the GOODS-S field with symbols representing their kinematics as shown in Figure 4. The black curve represents the mean value for local galaxies from [41] and the location of the Milky Way is also shown. All distant galaxies lie above the local relation and assuming a similar relation between stellar mass and gas fraction implies that distant galaxies have gas fraction 1.7 times the average values at z=0. On the left the two histograms show the distribution of SFR and SFR surface density of the sample.

Figure 5 shows a median value of 31% for the gas fraction in galaxies at z=0.65. Because those galaxies (with emission lines) represent the majority of present-day spirals' progenitors, we may compare this value to that from local spirals, that is, in a similar mass range, 1.7 times smaller. Considering a sample of 65 similar galaxies at z=0.4-0.75, [45, 46] found that the Oxygen content of the gaseous phases is two times smaller at z=0.65 than at z=0. At the average stellar mass of the distant samples (3.5 10^{10} M_O), [46] found an evolution of the O/H abundance of -0.3dex. Assuming that the average gas fraction of local galaxies is 15% (see Figure 5), one may deduce a fraction of 35-40% for z=0.65 galaxies, on the basis of a close box model (see [45, 46]). This provides an independent support to the estimate from [44] based on the inversion of the KS relation.

Lessons from the Tully Fisher Relation and from Disk Dynamics

Mass and velocity of disk-dominated galaxies define a very tight relationship. When including all the baryonic mass, this relation follows $M_{baryonic}$=61(+60,-30) X V_{flat}^4 over almost five mass decades, from (gas-rich) dwarves to massive spirals [47]. Well-resolved rotational curves of nearby galaxies reveal the predominance of flat velocities in their outskirts. Interestingly most of the uncertainties in the above relationship are related to the estimate of the stellar mass budget. For local and distant galaxies we observe only the brightest stars and not the main sequence stars, which dominates the stellar mass. There are still efforts to be done to provide a better accuracy on $M_{stellar}$ [48], including using information's from the star formation rate and from a better estimate of the red stellar component based on Lick indices. Conversely the gaseous mass, except possibly that from the molecular gas, is much better estimated from HI measurements. The tightness of the "universal" baryonic Tully Fisher and the accuracy of velocity measurements allow [49] predicting the $M_{stellar}/L_K$ behaviour with colour, which is found to be consistent with estimates from [50].

It is not possible to derive accurate rotational curves for most of the distant galaxies because of the coarse spatial resolution. The problem is even more complex because of the predominance of anomalous kinematics. However [51] were able to derive V_{flat} values with a reasonable accuracy, i.e. a median uncertainty of 0.12dex. This has been done by a series of Monte Carlo simulations comparing models from various rotational curves (see Figure 4) and measurements that essentially correct for the limited surface sampled by the IFU. This has been calibrated using data from local galaxies and velocities for anomalous galaxies have been derived assuming that the largest velocity gradient is indeed due to rotation. The resulting stellar-mass Tully Fisher found by IFU measurements [27, 51] show a much larger scatter (0.63dex) than that from local spirals (0.12dex), similarly to what is found from slit measurements [52]. It has been convincingly shown that a tight Tully Fisher relation still hold for rotational disks and then that all the scatter is related to galaxies with anomalous kinematics, principally those with complex kinematics [27, 51].

Adding the gas mass for the sample of 33 z=0.65 galaxies, [44] have been able to derive, for the first time, the evolution of the baryonic Tully Fisher relation. Although the result is based on only 7 distant rotating disks, they fall exactly onto the local relation. If confirmed, this result is quite fundamental to link local disks to their progenitors, 6 Giga-years ago. Rotating disks at z=0.65 could have already assembled all their baryonic mass[7] (gas and stars), and may have evolved into the present-day spirals simply by transforming two-thirds of their gaseous content into stars during the 6 Giga-years elapsed time. Evolution of fundamental relationships involving mass (e.g. Tully Fisher & mass-metallicity) is consistent with a close box model for a significant part of the present-day spirals 'progenitors. This may be extended to progenitors with perturbed kinematics, as those lie on both sides of the local baryonic Tully Fisher relation (see [44] and their Figure 5). Such a result does not exclude external sources of gas but it simply does not require them. It means that the close-box model works quite well for linking the two samples at z=0.65 and at z=0: however [46] demonstrated that this is no more the case when relating z > 2 galaxies to local galaxies, i.e. there should be other sources of gas between z> 2 and z=0.65 or z=0.

Having identified the probable mass evolution of spirals in the Tully Fisher relationship –just gas transformed into stars- I consider again the origin of the dramatic redshift increase of the scatter of the Tully Fisher relation. From the above, it is certainly not due to variation or problem with the mass estimates. Then it should be caused by strong and fundamental changes of both gas kinematics and galaxy morphologies. Another change is that of the

[7] Interestingly the distant rotational disks indicate a zero-point in the stellar-mass Tully Fisher that is slightly different from that of local spirals, implying a smaller stellar mass by a factor 1.6 to 2.1 for a given velocity [51]. However the slope of the stellar mass Tully Fisher is not established on similar robust grounds than that of the baryonic Tully Fisher, and the previous factor could be well affected by small changes of the corresponding slope.

gravitational support to the gaseous disk, which may be evaluated by the ratio V/σ. Giant thin disks have σ in the range of 10-30km/s and then V/σ from 8 to 20. For distant rotating disks and galaxies with perturbed rotation [53] found median values of 4 and 3 for V/σ, respectively. All distant galaxies present large values of σ, generally above 50km/s (see also [54]). It leads to two interesting consequences:

- Rotational gaseous disks at z=0.65 have a rotational support similar to that of the thick disk of the Milky Way (V/σ =3.9), although they appear to be the most relaxed galaxies, 6 Giga-years ago;
- A considerable fraction of the scatter in the Tully Fisher relation is removed by assuming a transfer from bulk motion to random motions (see [55] and references therein) as it can be expected in major mergers from their hydro-dynamical simulations.

In the following I will consider the different possible mechanisms that can explain the evolutionary changes of massive galaxies.

GAS-RICH GALAXIES MERGERS COULD FORM LARGE DISKS

Why the major merger hypothesis is unavoidable in linking z=0.65 to z=0 galaxies?

We have to link distant galaxies with peculiar morphologies and anomalous kinematics with the numerous present-day spiral galaxies (see Figure 3). As I have shown above, such an exercise is far much easier at z=0.65 than at larger redshift, for which the existing observations are difficult to be compared to those of local galaxies. At first order, most of the mass changes since the last 6 Gyrs is due to the gas-to-stars transformation. The strong link between morphological and kinematical disturbances let few physical mechanisms able to account for them. Physical processes related to star formation have been advocated to explain the large dispersions observed in z=2 galaxies [56]. However at z=0.65 no significant outflows have been detected by [39] from the comparison of emission lines and absorption lines systems. This is not surprising because all the galaxies considered here have large baryonic masses, in excess of $10^{10}M_{O}$, and not very large star formation rates: as such the energy developed by massive stars is always negligible when compared to that from gravitational motions. A considerable effort has been made to explain the unusual properties of very distant galaxies at z=2. Cosmological simulations predict large reservoirs of cold gas in the very distant Universe, which prompted [57] to propose cold and clumpy gas flows to explain the high star formation in z=2 galaxies. Given the uncertainties in linking z > 2 galaxies to local ones, it is still unclear whether this scenario reproduce the progenitors of spirals galaxies, or alternatively of more early-type galaxies. All in all, no such large cold gas reservoirs are predicted at z=0.65 and as such this scenario has never been considered for such redshifts.

The coeval evolution of star formation, stellar mass assembly, fraction of peculiar galaxies, interacting pair statistics and gas content have lead [24] to propose the rebuilding disk scenario. To reproduce the observed evolutions requires that 50 to 75% of present-day spirals have been reprocessed through mergers. In parallel a considerable effort has been made in simulating mergers using appropriate hydro-dynamical simulations [58, 59, 60, 61]. Given the fact that distant galaxies have large gas content, all the studies agree on the fact that gas-rich mergers may produce a new disk in their remnant phase, and that such a new disk is mainly supported by the orbital angular momentum provided by the collision. Having in hand the observations, and all the details they brought, as well as the theoretical frame, the rebuilding disk scenario, it has to be investigated whether or not both are matching together.

Modelling the properties of distant galaxies

The wealth of the data accumulated by the IMAGES study allows us to generate a detailed model for each individual galaxy. This is a considerable effort given the fact that we have such observations for 100 galaxies. Up to now this has been conducted only for 5 distant galaxies that have been fully modelled on the basis of hydro-dynamical simulations of mergers ([38, 62, 63, 64, 65]; see Figures 6 and 7). Completion of such a project has the important goal to reproduce the present-day Hubble sequence on the sole basis of the observed galaxies having emitted their light 6 Gyrs ago[8]. A success of this project may possibly solve the question of the origin of the Hubble sequence. It would also generate important consequences on the properties of present-day spirals, including about the origin of their substructures such as bulge, spiral arms, bars and rings. This is challenged by the important question of the model uniqueness. In fact, the further evolution of the merger remnant is not unique and uncertainties on the modelling have to be considered. We might solve this important issue by considering the conservation of important quantities, such as the mass and the angular momentum. For example, remnants of distant galaxies have to produce present-day spirals that delineated the Tully Fisher relation and the angular momentum velocity relationship.

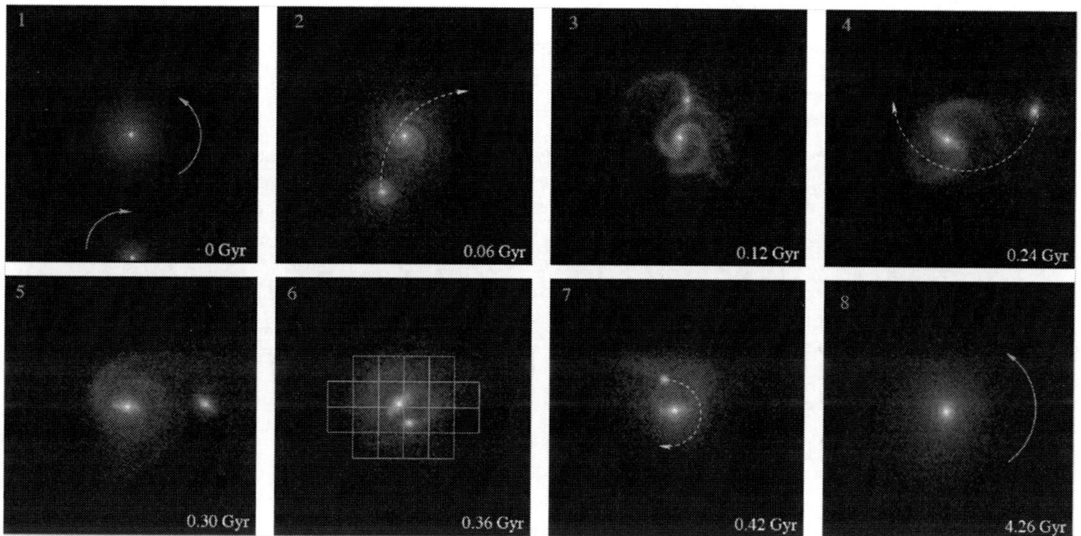

FIGURE 6. It shows the evolution of a 1:3 merger that reproduces both the morphology and kinematics of a distant galaxy (6th panel, [62]). The simulation has been performed using GADGET2 with stellar and gaseous masses provided by the observations. The spin of the smaller component is inverted relatively to that of the main galaxy. The interaction is responsible of the formation of the observed blue-giant bar as well as the dynamical axis that passes through the minor component. Letting the simulation evolving to z=0 (8th panel) provide a galaxy that is an S0_a: this is the object in the sample for which the simulation produces the largest B/T value because of the spin inversion.

Major mergers can reproduce all anomalous distant galaxies

In this section I present results based on the sample of 32 galaxies that have been observed in the GOODS_S field, and including 26 galaxies showing peculiar morphologies and/or anomalous kinematics. For reasons of simplicity, [39] have assumed a limited series of simulations provided by [66], i.e. 48 different merger models. For half of the anomalous galaxies, [39] has been able to distinguish the two components by using the exquisite spatial resolution of the HST/ACS. Among these 13 anomalous galaxies, 10 have both their morphologies and kinematics well reproduced by the above simple model. Interestingly the remaining 3 galaxies are obvious cases of merger, and the failure of the model is certainly linked to its simplicity. All the 13 galaxies but one show a mass ratio larger than 1:4 implying that minor merger is not the dominant process. This is because minor mergers should have a much lower efficiency to distort galaxy morphologies and kinematics [60]. In several cases, including those having been studied at depth, there is no compelling mechanism other than merging to explain the whole properties of the galaxy (see Figure 7).

What could be the nature of the remaining anomalous galaxies showing only a single component? Detailed simulations of galaxy mergers by [59] indicate that for a significant period of 1 to 1.5 Gyr after the fusion, the remnant shows distorted morphology and kinematics. However it is uneasy to evaluate the mass ratio and the orbit of those remnants. It has been a long exercise for me (several months) to examine all these individual galaxies and to

[8] Of course a limitation to such an ambitious project is due to the fact that it neglects possible further interactions for the remnant with other galaxies but this effect should be small as discussed in the Appendix of [26].

try to derive some clues of their past history. The problem is even more complex because in some cases the remnant is dusty because star formation is quite active during that phase (e.g. [59]). Indeed by using different techniques, including a full model of the light distribution of each galaxy, [39] have been able to evidence the presence of features such as bars, rings and other central features, sometimes after removing the adopted disk-plus-bulge model. This has been well illustrated by [63] for a galaxy (see Figure 8) that is probably an archetype of a disk that is rebuilt few 100 Myrs after a major fusion. The dynamical axis is 45° offset from the optical axis of the dusty disk that produces stars at a rapid rate (20 M_{O}/yr). The central features (see Figure 8 top-left panels) are helicoidally structured and are likely the response of the system to the gravitational torque after the fusion (see e.g. [60]) of galaxies with mass ratio larger than 1:3 and an inclined orbit. Accounting for the star formation in the disk, the system will evolve into a late-type spiral, with bulge to total ratio less than 0.2. On the basis of central features within the 13 anomalous galaxies with a single component, [39] have been able to well reproduce 7 of them by a major merger model. This let the major merger hypothesis as unrivalled to reproduce both kinematics and morphology of distant galaxies.

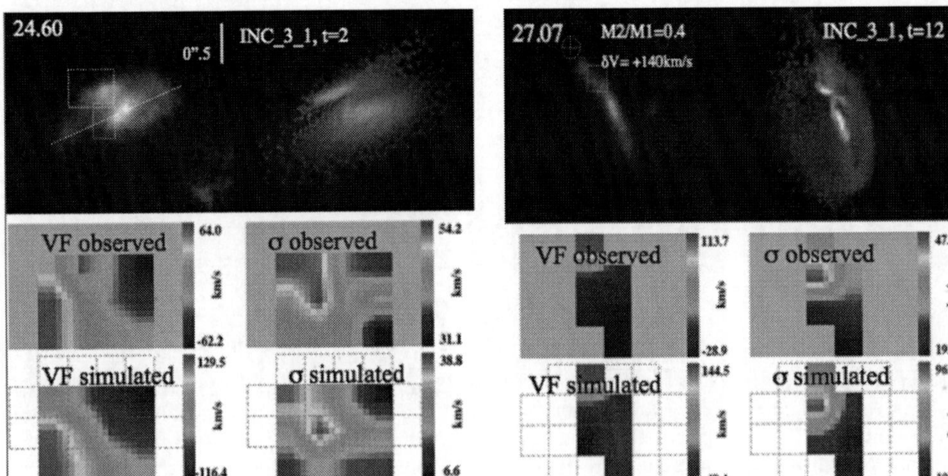

FIGURE 7. It shows two galaxies (J033224.60-274428.1 at z=0.5368 and J033227.07-274404.7 at z=0.7381) for which two components have been identified and which have been simulated. Images of galaxies are composite of b+v, I and z filters of HST/ACS and on the right, simulations show the gas component with the coding of the [66] model. The second galaxy on the right is a good example of an object that could not be explained by any other process than a merger, in a quite subtle way. At first glance it could resemble a giant luminous arc (e.g. [67]), but the absence of any cluster and the moderate redshift of the galaxy rules out this possibility. On the other hand it could resemble an edge on galaxy, but the velocity gradient along 90% of its length is less than 10 km/s, which again rules out this alternative. It let the single possibility that it is a giant bar inserted in a face-on spiral that has been activated by a merger for which the smaller component is identified near the dispersion peak, on the top of the galaxy.

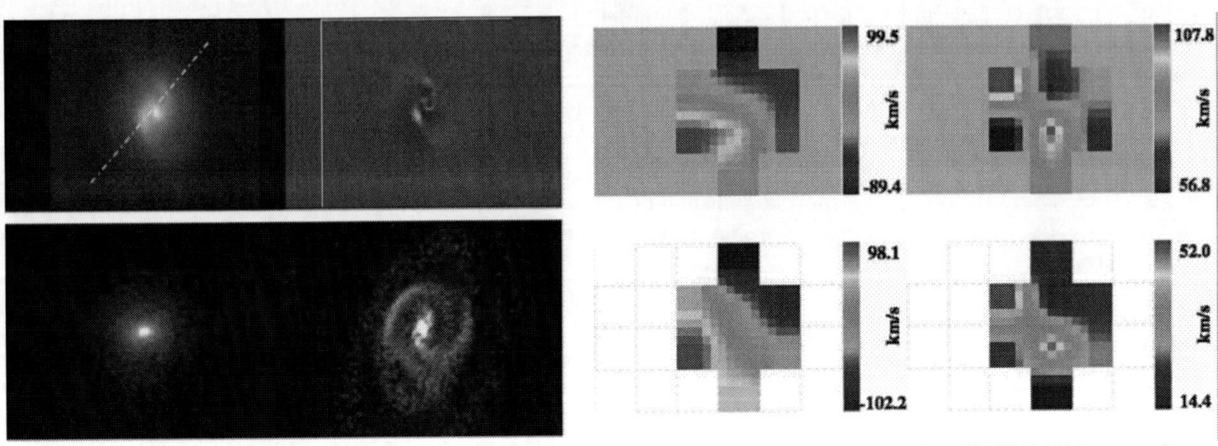

FIGURE 8. J033245.11-274724.0 at z=0.4346: the image in the top-left is a b+v, I and z composite of HST/ACS imagery from the UDF. The disk is much redder than the central region and because of its large dust content it can be hardly detected by

shallower imagery from GOODS. The dynamical axis is indicated by the dotted-dash line and come from the observed velocity field displayed in the third top panel. The second top panel shows the residual of the galaxy after subtracting the best fit of the galaxy light from a bulge-plus-disk model. The bottom panels show modelled stars, gas distribution, velocity field and dispersion map (from left to right).

Properties of the merging progenitors of distant galaxies

Following the assumption that most anomalous galaxies can be major mergers or their remnants, it may be of interest to investigate their progenitors properties. Figure 9 shows the observed gas fraction as well as the derived value for progenitors. To estimate the last value, I have estimated the characteristic time of star formation (T_{SFR}= $M_{stellar}$/SFR) for all galaxies classified in the 4 different merger phases. For each modelled galaxy, we know from the model the time it has spent during the merger. For each merger phase we can derive the median T_{SFR}, and reasonably assume that it is representative of each galaxy of this phase. Comparing the two times for each galaxy we can derive the fraction of gas that has been transformed into stars during the event. It results that most progenitors have gas fraction well above 50%, which is the condition to reform significant disks after the merger [58, 59, 60, 61]. The progenitors of these assumed mergers were individual galaxies at the time they were pairs, such a look-back time can be evaluated for each observed object in each phase. For example the 8 galaxies which are modelled near fusion would have their progenitors in pairs approximately 1.8 Gyrs before z=0.65, which corresponds to z=1. The median stellar mass of their progenitors would be 5 10^9 M_O and their gas fraction 80%. Gas fraction of similar galaxies (same stellar mass) at z=0 average to 30%, and it is not surprising that the gas content of similar galaxies is more than twice at z=1. The numbers displayed in that section slightly differ from those of [39], because they account for a more realistic distribution of the characteristic merger times (see caption in Figure 9). Nevertheless they are even more compelling to support that the disk-rebuilding scenario is a serious alternative to form the present-day massive disks.

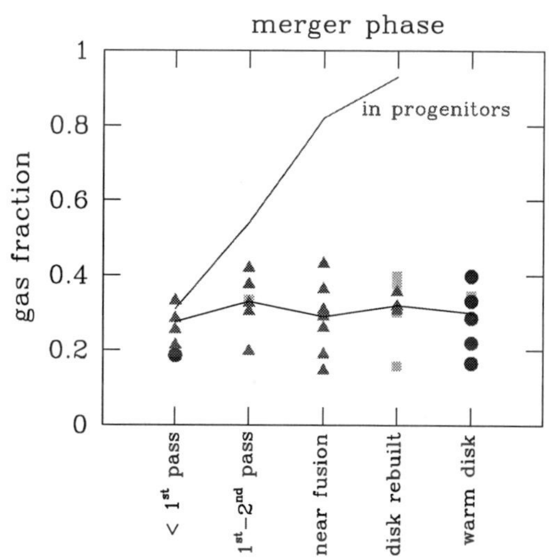

FIGURE 9. Observed gas fraction in distant galaxies from IMAGES (symbols are the same than in Figure 5). This Figure is an up-dated version of Figure 7 of [39] for which I have considered that the characteristic dynamical time scale with mass at the power 1/4 rather than 1, in agreement with [66]. The blue line indicates the gas fraction in the progenitors assuming that some gas has been transformed to stars during the merger (see text).

The fraction of stars that are formed during major mergers since the last 8 Giga-years

A kind of general consensus seems to be emerging in establishing a small contribution of mergers to star formation (about 10%, see [68, 69, 70] and references therein). In fact I think there is still a lot of misunderstanding in estimating such a contribution. For example, [68, 69] have considered samples including a large number of galaxies and have assumed that in most irregular galaxies, mergers do not activate star formation. However, in absence of kinematic data, this conclusion might be premature: [31] found a very good correlation between morphological irregularities (stars) and anomalous kinematics (ionised gas). Either the IMAGES sample is too limited or biased, or the conclusion of [68, 69] requires an unknown process to distort both morphology and kinematics. Another way to minimise the impact of merging in star formation could be to limit the merger-like activity to the only peaks of star formation that are merger induced, e.g. during the first passage and at the fusion, during periods of 10^8 years [70]. This contrasts with simulations of star formation history during mergers by [59] and [71] showing that star formation is enhanced by pressure driven gas during 1 to 3 Gyrs, including for a significant period after the merger[9]. Taking into account these simulations and results from [70], star formation induced by mergers may well account for most of the star formation evolution since the last 8 Gyrs. A more substantial effort should be done to evaluate this, by comparing from simulations how would be the star formation in isolated galaxies assuming they wouldn't have any merger during their evolution. For example, low-surface-brightness galaxies are showing large gas fractions, while they do not reach the threshold for star formation in the Kennicutt-Schmidt relation [72]. If such galaxies were involved in an interaction or a merger, gravitational forces would easily pressure the gas and then activate the star formation.

CONCLUSION: A NEW CHANNEL TO FORM LARGE DISKS

The formation of large galactic disks has been discussed since the early 60s with a general description within massive haloes by [73]. In the frame of the tidal torque theory [2, 3], giant disks have acquired their angular momentum at very high redshift, and are assumed to evolve in a quiescent way by a smooth accretion of gas from the intergalactic medium. In this quiescent mode, substructures in spirals are mostly formed through internal instabilities. Concerning the formation of substructures in disk, a thorough review of secular evolution has been done by [74], and a recent comparison of the different channels of disk formation can be found in [75]. However the observations have considerably improved, especially those of distant galaxies. Before the 80s, they were observed as seeing limited images on either photographic plates or CCDs. Nowadays, it is possible to derive their global physical properties from photometry at UV to mid-IR wavelengths, and moreover HST and very large telescopes allow detailing their morphologies to several hundred of parsecs and their internal kinematics to few kilo-parsecs.

Since the study by [76] and subsequent works, the role of mergers has been re-examined to be an important channel for the formation of elliptical galaxies. In the corresponding simulations, the two progenitors were gas-poor spiral galaxies. Several Giga-years ago, galaxies were much richer in gas than today, and their mergers unavoidably lead to giant disks superposed to a bulge. The disk-rebuilding scenario [24, 39] propose a new channel to form disks that would have acquired their angular momentum much more recently than in the tidal torque theory. The latter theory appears quite efficient in reproducing most of the properties of the Milky Way, although the representativeness of the Milky Way is questioned [15]. It however faces with two major difficulties: the disks formed by an early acquisition of their angular momentum appear too small when compared to observations, and it is at odd with the discovery that a large fraction of spiral' progenitors had peculiar morphologies and anomalous kinematics, 6 Giga-years ago.

The new proposed channel of disk formation accounts by principle to all the evolutionary features that have been observed when comparing distant to nearby spirals [24]. It might solve the angular momentum crisis as illustrated by [19], although this needs to be confirmed on actual data. In such a model, a galaxy like M31 could have been formed after a gas-rich merger at z~ 1 between two galaxies having each a stellar mass of 8 $10^9 M_{\odot}$, i.e. twice the value for M33. Such progenitors have to be gas-rich to rebuild a significant disk after the collision, and the required gas-richness is found to be consistent with high redshift observations [39]. Ages of stars in present-day disks have to be investigated to evaluate whether these disks can be enough young to have been formed more recently than at z=2 to 3. This can be done in a robust way only on galactic disks in our neighbourhood, as is has been shown by [15] for M31 with ages up to 8 Giga-years. For more distant galaxies, derivations of ages from spectral energy distributions cannot discriminating in a robust way stellar ages from 8 to 10-11 Giga-years, i.e. stars formed from z=1 to z=2-3. If true, the spiral-rebuilding scenario could imply that the whole Hubble sequence is mainly a consequence of galaxy mergers, those being predicted by the hierarchical formation of galaxies. To prove or disprove it, one has to compare

[9] To illustrate this point, let us compare it to a medical survey evaluating the impact of the H1N1 flux and which would account for only the sick persons having high fever, discarding all the persons who are affected at other stages of the illness.

it to the merger rate derived from the halo-occupation method [70] and to show how the angular momentum of distant galaxies can be linked to that of present-day spirals. Simulations have also to reproduce the present-day thin disks from the anomalous galaxies and thick disks that dominate the Universe, 6 Giga-years ago.

ACKNOWLEDGMENTS

This review has been written on the basis of the work of numerous collaborators, mostly from the IMAGES team. Such a study would not have been achieved without the important contributions of Yanchun Liang, Yanbin Yang, Xianzhong Zheng, Benoit Neichel, Lia Athanassoula, Sebastien Peirani, Myriam Rodrigues and Rodney Delgado-Serrano.

REFERENCES

1. F. Zwicky, *Astrophys. J.* 1937, 86, 217.
2. P.J. Peebles, 1976 *Astrophys. J.* 1976, 205, 109
3. S. D. M. White, *Astrophys. J.* 1984, 286 ,38
4. G. Toth & J. P. Ostriker, *Astrophys; J.* 1992, 389, 5
5. R. F. G. Wyse, *Proceedings IAU Symposium*, 2009, No. 258
6. G. Gilmore, *ASP Conference Series*, 2001, Vol. 230
7. A. Helmi, *Astronomy and Astrophysics Review*, 2009, Vol. 15, pp. 145-188
8. R. Ibata, M. Irwin, G. Lewis, A. Ferguson & N. Tanvir, *Nature*, 2001, 412, 49
9. R. Ibata, S. Chapman, A. Ferguson, M. Irwin, G. Lewis, & A. McConnachie, 2004, *Monthly Notices of the RAS*, 351, 117
10. T. Brown, R. Beaton, M. Masashi et al., . 2008 *Astrophys. J.*, 685, 121
11. D. Martinez-Delgado, M. Pohlen, R. Gabany et al., 2009 *Astrophys. J.*, 692, 955
12. M. Barker, A. Ferguson, M. Irwin, N. Arimoto & P. Jablonka, 2009 *Astronomical J.*, 138, 1469
13. M. Mouhcine, *Astrophys. J.* 2006, 652, 277
14. S. Zibetti, S. White, & J. Brinkmann, 2004, *Monthly Notices of the RAS*, 347, 556
15. F. Hammer, M. Puech, L. Chemin, H. Flores, & M. Lehnert, 2007, *Astrophys. J.*, 662, 322
16. S. C. Chapman, R. Ibata, G. Lewis et al., 2006 *Astrophys. J.*, 653, 255
17. M. Rejkuba, M. Mouhcine & R. Ibata, 2009, *Monthly Notices of the RAS*, 396, 1231
18. A. Renda, A., B. Gibson, M. Mouhcine et al., 2005 *Monthly Notices of the RAS*, 363, L16
19. A. H. Maller, A. Dekel, & R. Somerville 2002, *Monthly Notices of the RAS*, 329, 423
20. S. Lilly, O. Le Fevre, F. Hammer, & D. Crampton 1996, *Astrophys. J.*, 460, 1L
21. H. Flores, F. Hammer, T.X. Thuan, C. Cesarsky, F. X. Desert, et al. 1999, *Astrophys. J.*, 517, 148
22. M. Dickinson, C. Papovich, H.C. Ferguson, T. Budavari 2003, *Astrophys. J.*, 587, 25
23. P. G. Perez-Gonzalez, G. H. Rieke, V. Villar, G. Barro, M. Blaylock, et al. 2008, *Astrophys. J.*, 675, 234
24. F. Hammer, H. Flores, D. Elbaz, X. Z. Zheng, Y. C. Liang, & C. Cesarsky 2005, *Astron. & Astrophys.*, 430, 115
25. E. Bell, C. Papovich, C. Wolf, E. Le Floc'h, J.A.R. Caldwell, et al. 2005, *Astrophys. J.*, 625, 23
26. R. Delgado-Serrano, F. Hammer, Y. B. Yang, M. Puech, H. Flores, & M. Rodrigues 2009, *Astron. & Astrophys.*, in press, ArXiv:0906.2805
27. H. Flores, F; Hammer, M. Puech, P. Amram, & C; Balkowski 2006, *Astron. & Astrophys.*, 455, 107
28. Y. Yang, H. Flores, F. Hammer, B. Neichel, M. Puech, et al., *Astron. & Astrophys.*, 477, 789
29. F. Hammer, H. Flores, S. J. Lilly, D. Crampton, O. Le Fevre, et al. 1997, *Astrophys. J.*, 481, 49
30. C. D. Ravikumar, M. Puech, H. Flores, D. Proust, F. Hammer, et al. 2007, *Astron. & Astrophys.*, 465, 1099
31. B. Neichel, F. Hammer, M. Puech, H. Flores, M. Lehnert, et al. 2008, *Astron. & Astrophys.*, 484, 159
32. S. Van den Bergh 2002, *Pub. of the Astron. Soc. of the Pacific*, 114, 79
33. C. Y. Peng, L. C. Ho, C. D. Impey, & H. W. Rix 2002, *Astron. J.* 124, 266
34. C. J. Conselice 2006, *Astrophys. J.*, 638, 686
35. J. Melbourne, M. Ammons, S. A. Wright, A. Metevier, E. Steinberg, et al. 2008, *Astron. J.*, 135, 1207
36. M. Puech et al., in preparation
37. M. Puech, M. Lehnert, Y. Yang, J.-G. Cuby, S. Morris, et al. 2009, *Proc. of the AO4ELT meeting* held in Paris, Jun. 2009, ArXiv:0909.1747
38. M. Puech, F. Hammer, H. Flores, B. Neichel, Y. Yang, & M. Rodrigues 2007, *Astron. & Astrophys.*, 476, 21
39. F. Hammer, H. Flores, M. Puech, Y. Yang, E. Athanassoula 2009, *Astron. & Astrophys.*, in press, ArXiv:0903.3962
40. B. Epinat, P. Amram, C; Balkowski, M. Marcelin 2009, *Monthly Notices of the RAS*, in press, ArXiv:0904.3891
41. D. Schiminovich 2008, *AIP Conf. Proc.*, 1035, 180
42. R. Kennicutt 1998, *Astrophys. J.*, 498, 541
43. Y. Gao, P. M. Solomon 2004, *Astrophys. J.*, 606, 271

44. M. Puech, F. Hammer, H. Flores, R; Delgado-Serrano, M. Rodrigues, & Y. Yang 2009, ArXiv:0903.3961
45. Y. C. Liang, F. Hammer, & H. Flores 2006, *Astron. & Astrophys.*, 447, 113
46. M. Rodrigues, F. Hammer, H. Flores, M. Puech, Y. C. Liang, et al. 2008, *Astron. & Astrophys.*, 492, 371
47. S. S. McGaugh 2005, *Astrophys. J.*, 632, 859
48. M. Rodrigues et al., in preparation
49. D. V. Stark, S. S. McGaugh, R. A. Swaters 2006, *Astron. J.*, 138, 392
50. E. F. Bell, D. H. McIntosh, N. Katz, & M. D. Weinberg 2003, *Astrophys. J. Suppl. S.*, 149, 289
51. M. Puech, H. Flores, F. Hammer, Y. Yang, B. Neichel, et al. 2008, *Astron. & Astrophys.*, 484, 173
52. C. J. Conselice, K. Bundy, R. S. Ellis, J. Brinchmann, N. P. Vogt, & A. C. Philips 2005, *Astrophys. J.*, 328, 160
53. M. Puech, F. Hammer, M. D. Lehnert, & H. Flores 2007, *Astron. & Astrophys.*, 466, 83
54. N. M. Forster-Schreiber, R. Genzel, N. Bouche, G. Cresci, R. Davies, et al. 2009, *Astrophys. J.*, in press, ArXiv:0903.1872
55. M. D. Covington, S. A. Kassin, A. A. Dutton, B. J. Weiner, T. J. Cox, et al. 2009, ArXiv:0902.0566
56. M. D. Lehnert, N. P. H. Nesvadba, L. Le Tiran, P. Di Matteo, W; van Driel, et al. 2009, *Astrophys. J.*, 699, 1660
57. A. Dekel & Y. Birboim 2006, *Monthly Notices of the RAS*, 368, 2
58. B. Robertson, J. S. Bullock, T. J. Cox, T. Di Matteo, L. Hernquist 2006, *Astrophys. J.*, 645, 986
59. J. M. Lotz, P. Jonsson, T.J. Cox, & J. R. Primack 2008, *Monthly Notices of the RAS*, 391, 1137
60. P. F. Hopkins, T. J. Cox, J. D. Younger, & L. Hernquist 2009, *Astrophys. J.*, 691, 1168
61. K. Stewart, J. S. Bullock, R. Wechsler & A. Maller, 2009, *Astrophys. J.*, 702, 307
62. S. Peirani, F. Hammer, H. Flores, Y. Yang, & E. Athanassoula 2009, *Astrophys. J.*, 496, 51
63. F. Hammer, H. Flores, Y. B. Yang, E. Athanassoula, M. Puech, et al. 2009, *Astron. & Astrophys.*, 496, 381
64. M. Puech, F; Hammer, H. Flores, B. Neichel, & Y. Yang 2009, *Astron. & Astrophys.*, 493, 899
65. I. Fuentes-Carrera et al., in preparation
66. J. Barnes 2002, *Monthly Notices of the RAS*, 333, 481
67. G. Soucail, Y. Mellier, B. Fort, G. Mathez & F. Hammer, 1987, *Astron. & Astrophys.*, 184, 7
68. A. R. Robaina, E; F. Bell, R. E. Skelton, D. H. McINtosh, R. S. Somerville, et al. 2009, *Astrophys. J.*, 704, 324
69. S. Jogee, S; H. Miller, K. Penner, R. E. Skelton, C. J. Conselice, et al. 2009, *Astrophys. J.*, 697, 1971
70. P. F. Hopkins, J. D. Younger, C. C. Hayward, D. Narayanan, & L. Hernquist 2009, ArXiv:0911.1131
71. T. J. Cox, P. Jonsson, R. S. Somerville, J. R. Primack, & A. Dekel 2008, *Monthly Notices of the RAS*, 384, 386
72. W. de Blok, T. van der Hulst & S. McGaugh, 1996, *Bull. Am. Astr. Soc.*, 28, 1387
73. O. J. Eggen, D. Lynden-Bell, & A. R. Sandage, 1962, *Astrophys. J.*, 136, 748
74. J. Kormendy & R. Kennicutt, *Astronomy and Astrophysics Review* 2004, 42, 603
75. E. Athanassoula, 2009, ArXiv:0910.5180
76. A. Toomre, J. Toomre, 1972, *Astrophys. J.*, 178, 623

Redshift-space distortions in deep redshift surveys as a probe of the invisible Universe

Luigi Guzzo* and Olivier Le Fèvre†

*INAF-Osservatorio Astronomico di Brera, Merate/Milano, Italy
†Laboratoire d'Astrophysique de Marseille, CNRS, Université de Provence, Marseille, France

Abstract. Massive redshift surveys of galaxies beyond the local Universe (i.e. $z > 0.3$) provide an exhaustive probe of the observed acceleration of cosmic expansion. While they have the ability to measure the expansion history $H(z)$ through Baryonic Acoustic Oscillations in the galaxy power spectrum, they can at the same time probe $f(z)$ using the redshift-space distortions introduced in the observed clustering pattern by galaxy peculiar motions. Coupling these two measurements one can in principle distinguish whether cosmic acceleration is due to a new form of 'dark energy' in the cosmic budget, or rather requires a modification of General Relativity. These two radically alternative scenarios are degenerate when considering $H(z)$ alone, as yielded, e.g., by the Hubble diagram of Type Ia supernovae. We review our recent measurements of redshift distortions at $z \sim 1$ based on the VIMOS VLT Deep Survey "Wide" data and discuss the revived interest on this technique in the context of dark energy. Current results are consistent with the simplest cosmological-constant scenario, but error bars are still too large to rule out alternative models. Forecasts based both on extensive simulations and Fisher-matrix computations, show that next-generation deep surveys optimizing the combination of large volumes and good galaxy sampling will be able to use redshift distortions as a key tool to understand the physical origin of cosmic acceleration. Among these, we introduce the newly started VIMOS Public Extragalactic Redshift Survey (VIPERS) at the ESO VLT, which is building at $<z>\sim 0.8$ a sample comparable to current local redshift surveys (10^5 redshifts in a volume 5×10^7 h^{-1} Mpc^3). Finally, we provide an example of the exquisite accuracy that could be reached on measurements of redshift-space distortions (among many others), with a massive 20,000 deg^2 near-infrared spectroscopic survey from space, as foreseen by the EUCLID mission for the ESA 'Cosmic Vision' program.

Keywords: Cosmology, Large Scale Structures, Deep redshift surveys
PACS: 98.80.Es, 98.80.Bp, 98.65.Dx, 98.62.Py,

INTRODUCTION

The current standard cosmological model, concordantly supported by virtually all available observations, tells us that we live in a low-density, expanding Universe with a spatially flat geometry, that appears to have recently entered a phase of accelerated expansion. This latter conclusion emerges naturally when interpreting the observed Hubble diagram of distant Type Ia supernovae within the standard Friedmann-Lemaitre-Robertson-Walker (FLRW) cosmology (Riess et al. 1998; Perlmutter et al. 1999), requiring an extra mass-energy contribution in the Friedmann equation in the form of a fluid with equation of state $w = -1$. This corresponds to adding a *cosmological constant* in the equations of General Relativity (GR), i.e. the term originally introduced by Einstein to obtain a static solution. A constant, vacuum-like equation of state $w = -1$, however, has a few disturbing features, as e.g. that of making the current epoch a special one in which the contributions from matter and cosmological constant are comparable. Among possible remedies, scenarios with evolving 'dark energy' density related to a cosmological scalar field, have been proposed (see e.g. Frieman et al. 2008, for a review). Both the cosmological constant and these more sophisticated variants can be seen as modifications of the source term in the right-hand side of Einstein field equations, i.e. adding extra contributions to the stress-energy tensor. Alternatively, however, one could assume that it is the theory of gravity that needs to be revised and thus modify the left-hand side of the equation. In this case, the apparent acceleration could simply be evidence of our still limited knowledge of the laws of Nature (see e.g. Copeland et al. 2006 for a comprehensive review).

These two radically different explanations cannot be distinguished by measuring only the expansion history $H(z)$ or equivalently an evolution of the equation of state $w(z)$ for the extra component. A way to break this degeneracy is to look at the linear growth of density inhomogeneities. The rate at which matter is assembled by gravity in the expanding Universe can be characterized by the parameter $f = d\ln D/d\ln a$, where D(t) is the time-dependent part of the solution of the linear growth equation (Heath 1977) and $a = (1+z)^{-1}$ is the cosmic scale factor. Models with the

CP1241, *Invisible Universe: Proceedings of the Conference,* edited by J.-M. Alimi and A. Füzfa

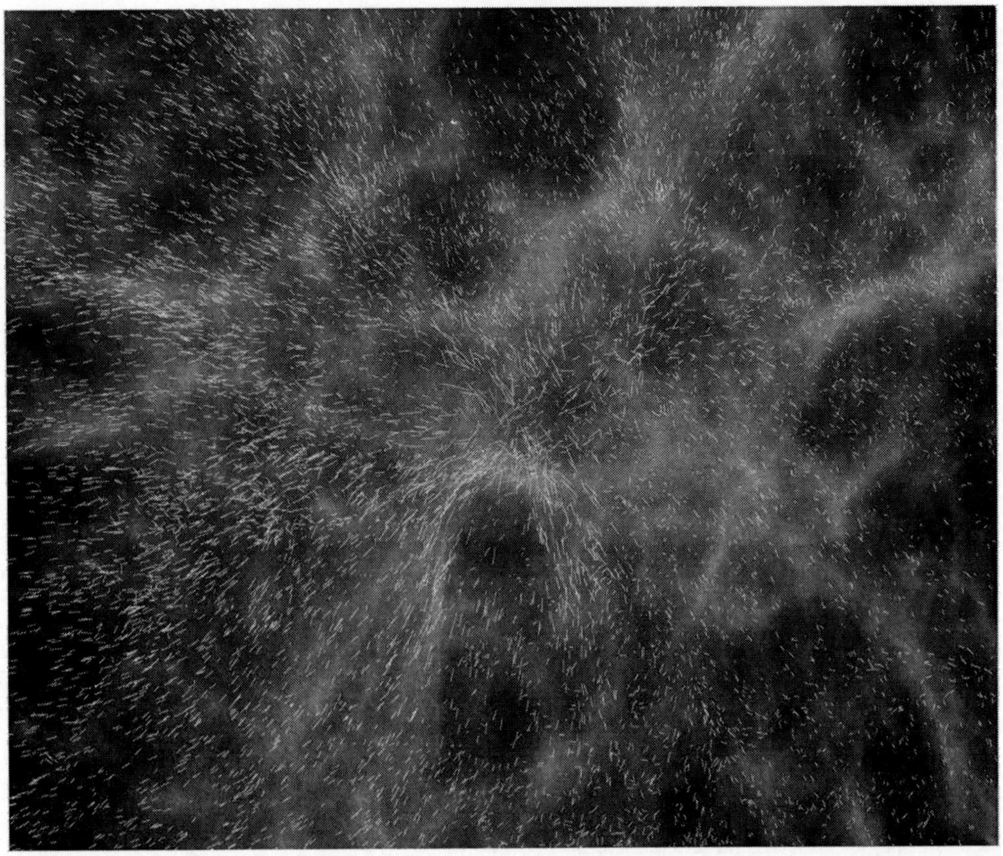

FIGURE 1. The infall velocity field (yellow arrows) around a large-scale overdensity at $z = 1$, as seen in a box of 30 h^{-1} Mpc side extracted from a numerical simulation of structure formation. This gives a visual example of how galaxy peculiar motions trace the growth of structure. (K. Dolag, ESO-PR-0804)

same expansion history $H(z)$, but based on a different gravity theory predict a different growth rate $f(z)$, which is therefore a sensitive probe of the physics behind cosmic acceleration. A convenient approximation for $f(z)$ for a broad range of models is given by the form $f(z) \simeq [\Omega_m(z)]^\gamma$, with γ depending on the gravity theory (Wang & Steinhardt 1998; Linder 2005). GR-based Friedmann models have $\gamma = 0.55$, while for example in the DGP 'braneworld' theory, an extra-dimensional modification of gravity, one has $\gamma = 0.68$ (e.g. Lue et al. 2004).

PROBING DARK ENERGY WITH GALAXY REDSHIFT SURVEYS

Measurements of galaxy clustering from large redshift surveys, quantified through galaxy-galaxy correlations (or their Fourier transform, the power spectrum P(k)), have the interesting property of containing direct information on both $H(z)$ and $f(z)$. Baryonic Acoustic Oscillations (BAO) within the last-scattering surface give rise to a characteristic feature in the galaxy distribution at comoving separations of 150 Mpc (Eisenstein et al. 2005; Blake & Glazebrook 2003; Seo & Eisenstein 2003). This corresponds to the characteristic scale fixed by the comoving sound horizon at recombination and is accurately measured by Cosmic Microwave Bakground anisotropies (Komatsu et al. 2009). Compared to the observed galaxy BAO peak position at different redshifts, this yields a measurement of $H(z)$ in the radial direction, and of the comoving angular diameter distance $D_A(z) = r(z)/(1+z)$ (where $r(z)$ is the comoving distance), in the transverse direction.

The important property of galaxy clustering as measured in redshift space is that of containing, together with the BAO information on $H(z)$, the imprint of the linear growth rate of structure through a measurable anisotropy. Such

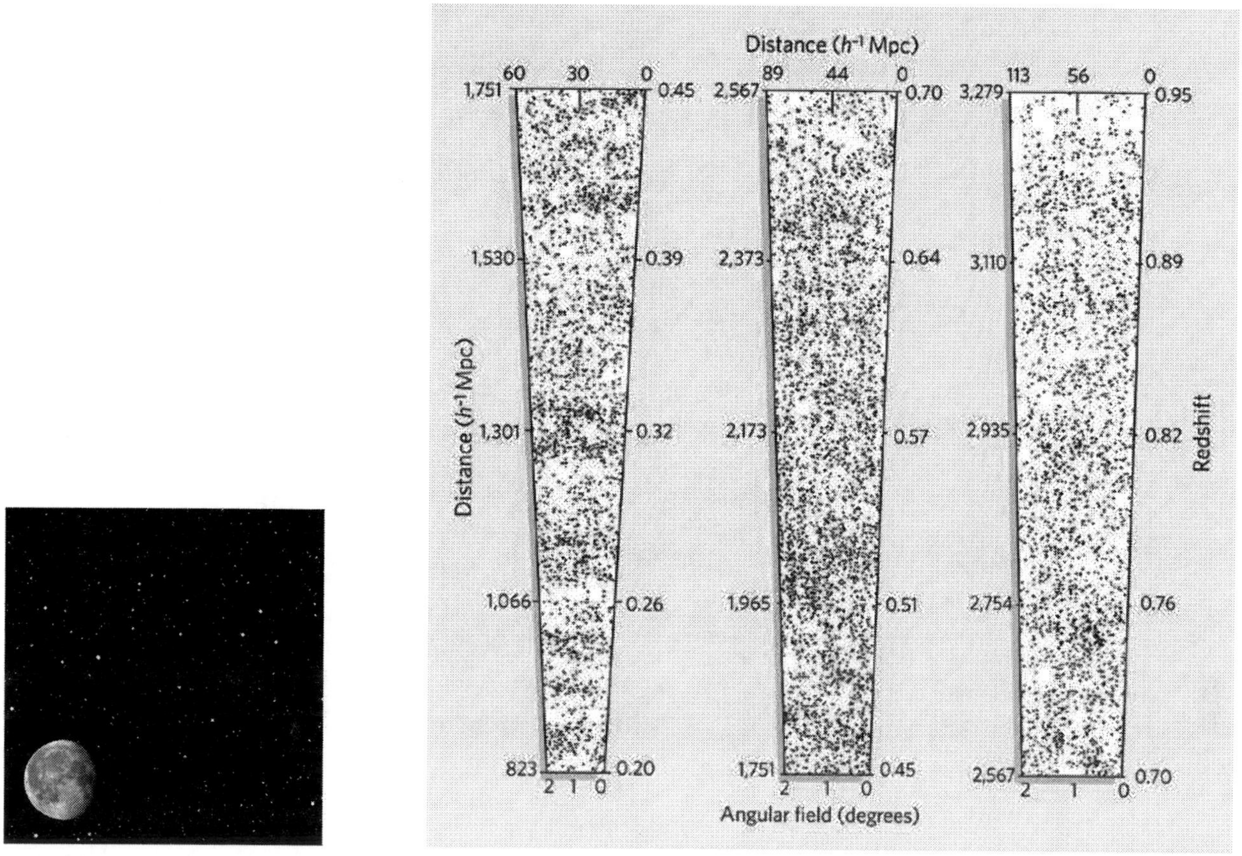

FIGURE 2. *Left:* The VVDS-Wide "F22" field at R.A.$\sim 22^h$, compared to the size of the Moon for reference. *Right:* The corresponding spatial distribution of $\sim$ 10,000 galaxies observed in the same field (Garilli et al. 2008; figure by C. Marinoni and originally presented in Strauss 2008)

redshift-space distortion of the observed clustering is due to the galaxy coherent flows towards overdensities, that trace the cosmic growth rate at different epochs. Galaxy peculiar motions are a direct consequence of the growth of structure (Figure 1). When redshifts are used to measure galaxy distances, the contribution from peculiar velocities produces a distortion of the clustering pattern, which is proportional to $f(z)$. This can be measured by modeling the anisotropy of the redshift-space two-point correlation function $\xi(r_p,\pi)$. This is quantified by a 'compression parameter' β (see Hamilton 1998, for a review), which is directly related to the growth rate as $f(z) = \beta(z)b_L(z)$. The bias factor b_L appearing here, is the ratio of the clustering amplitude of the galaxies used in the measurement, to that of the underlying matter (Kaiser 1984). This can be obtained either directly from the redshift surveys data using higher-order clustering (Verde et al. 2002), or using the information from CMB anisotropy observations. In practice, however, knowledge of b_L is not crucial: what galaxy clustering actually measures is the combination of parameters $f(z) * \sigma_8(z)$, where σ_8 is the *rms* amplitude of *matter* fluctuations when averaged over 8 h^{-1} Mpc spheres (see Song & Percival 2009 for more discussion). In the assumption of linear bias (valid on large scales), one has $b_L \simeq \sigma_8^{gal}/\sigma_8^{mass}$, and thus when measuring β (from the clustering anisotropy) and σ_8^{gal} (from the clustering amplitude), one has in practice an estimate of $\beta(z) \times \sigma_8^{gal}(z) = f(z) \times \sigma_8^{mass}(z)$. This measurement can then be compared to self-consistent model predictions at that specific redshift for $f(z) \times \sigma_8(z)$. (It should be said, however, that to normalize the model one needs to use nevertheless an external measurement, as e.g. $\sigma_8(z_{CMB})$). Using data from the VVDS-Wide survey, Guzzo et al. (2008, see next section) measured β at $z \sim 0.8$ (Figure 4), pointing out the important role redshift distortions can play as a future test of dark energy. This work stimulated renewed interest, prompting a number of investigations of the accuracies reachable with current and future surveys (e.g. Wang 2008; Linder 2008; Nesseris & Perivolaropulos 2008; Acquaviva et al. 2008; Song & Percival 2009; White et al. 2009; Percival & White 2009).

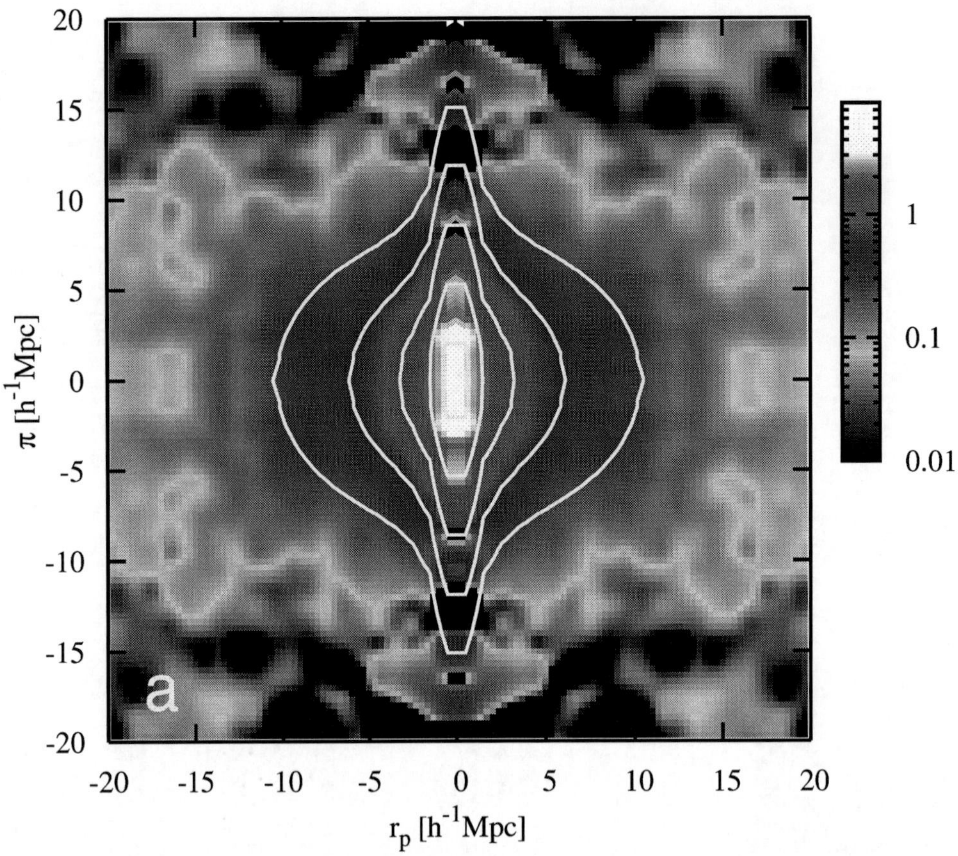

FIGURE 3. $\xi(r_p,\pi)$ at $<z>\simeq 0.77$ from the VVDS-Wide survey, replicated over four quadrants to enhance deviations from circular symmetry. Colors correspond to the level of correlation as a function of the transverse (r_p) and radial (π) separation of galaxy pairs. The effect of galaxy infall due to the growth of large-scale structure is proportional to the flattening of the purple-blue large-scale levels, with the solid contours corresponding to the best-fitting distortion model with $\beta = 0.70$ and $\sigma_{12} = 412$ km s^{-1} (see Guzzo et al. 2008 for details)

THE GROWTH RATE AT $z \sim 1$ FROM THE VIMOS-VLT DEEP SURVEY

The VIMOS-VLT Deep Survey (VVDS) was designed to probe the combined evolution of galaxies and large scale structure to $z \sim 2$ using the VIMOS spectrograph at the ESO VLT. It measured so far ~ 40000 spectra (Le Fèvre et al. 2005; Garilli et al. 2008) over its Deep (0.5 deg^2 to $I_{AB} \leq 24$) and Wide ($\sim$8 deg^2 to $I_{AB} \leq 22.5$) parts (Figure 2).

A sub-sample of 5895 galaxies with $0.6 < z < 1.2$ (volume $V = 6.35 \times 10^6 h^{-3} Mpc^3$) in the 4-deg^2 F22 field of VVDS-Wide was used to obtain a measurement of β at an effective redshift $z = 0.77$ (Guzzo et al. 2008). The observed $\xi(r_p,\pi)$ (Figure 3), was fit with a distortion model including both linear and non-linear distortions. These are described respectively by two parameters, the compression β and the rms pairwise dispersion σ_{12}. The errors on the best-fit estimate were obtained from 100 fully realistic mock realizations of the survey (De Lucia & Blaizot 2006), which have been also used to explore and minimize systematic errors. A value for the linear bias at the effective redshift of the survey was then estimated using the PDF reconstruction method (Marinoni et al. 2005), normalized using the amplitude of mass fluctuations σ_8 from WMAP3 (Spergel et al. 2007).This yielded $b_L = 1.3 \pm 0.1$, which corresponds to a growth rate $f(z = 0.77) = 0.91 \pm 0.36$ (see Guzzo et al. 2008, for details). This value is compared in Fig. 4 to model predictions, together with measurements from the 2dFGRS, SDSS and 2SLAQ surveys.

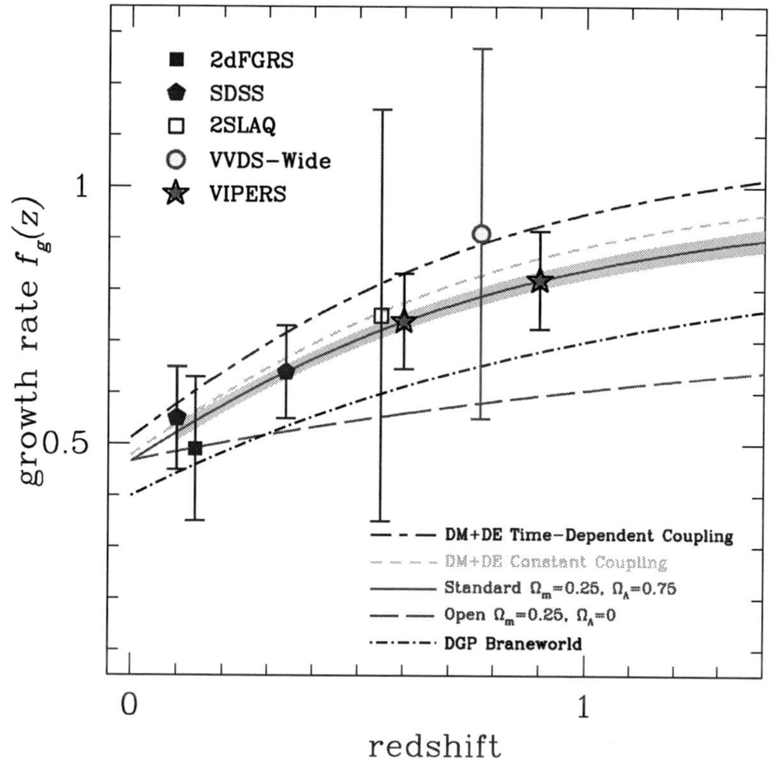

FIGURE 4. Current estimates of the growth rate $f = \beta b_L$ and forecasts for future surveys, compared to predictions from theoretical models. Current observational data are from the 2dFGRS (Hawkins et al. 2003), SDSS (Tegmark et al. 2004; Cabrè & Gaztanaga 2009), 2SLAQ (Ross et al. 2007) and VVDS surveys (Guzzo et al. 2008). The model predictions correspond to: the standard cosmological constant $w = -1$ mode (ΛCDM, solid line); an open $\Omega_\Lambda = 0$ model with the same Ω_m (long-dashed line, for both cases $f(z) \simeq \Omega(z)^{0.55}$); two models in which dark energy is coupled to dark matter (Amendola 2000) (upper dashed curves) ; the DGP braneworld model, an extra-dimensional modification of the gravitation theory (Dvali et al. 2000) for which $f(z) \simeq \Omega(z)^{0.68}$ (dot-dashed curve). The two stars show forecasts for the ongoing VIPERS survey, based on using one single galaxy population to measure β. Given the high sampling rate of this survey, however, using simultaneously two galaxy populations with different bias as suggested by McDonald & Seljak (2009), should allow error bars to be reduced by a factor ~ 5 or even more. The shaded band shows instead the predicted error on $f(z)$ from a typical EUCLID redshift survey of $\sim 10^8$ infrared galaxies with $H < 22$.

EXPECTATIONS FOR THE NEXT GENERATION OF REDSHIFT SURVEYS

Only ~ 6000 redshifts were used to obtain the VVDS result discussed above. There is thus ample room for improvement. Guzzo et al. (2008) calibrated through simulations an empirical scaling formula to forecast the expected errors on β from future surveys, which go as $\sigma_\beta/\beta \propto (< n >^{0.44} V^{0.5})^{-1}$ (i.e. nearly as the inverse square root of the total number of objects). For densities $< 10^{-2}$ h^3 Mpc^{-3} and bias values $b \sim 1$, this formula is in good agreement with the predictions based on the Fisher matrix approach (Percival & White 2009), and shows that it is possible to go below a 10% uncertainty on β at $z \sim 1$ already using current instrumentation. Following these predictions, two years ago it was proposed to perform a new redshift survey with the ESO VLT, aimed at constructing a sample at $z \sim 1$ with statistical properties comparable to local redshift surveys like 2dFGRS and SDSS. The approved program, named VIPERS (VIMOS Public Extragalactic Redshift Survey) started at the end of 2008 . It targets 100,000 galaxies similarly to VVDS-Wide ($I_{AB} < 22.5$), but extending it over ~ 24 deg^2 and with a robust color-color pre-selection that excludes a priori galaxies below $z = 0.5$. Figure 5 shows the galaxy distribution in a mock catalogue of the VIPERS W1 field (8×2 deg^2). VIPERS will produce, at a median redshift $< z > \sim 0.8$, a total sample with volume 5×10^7 h^{-3} Mpc^3, equivalent to the $z \sim 0.1$ 2dFGRS. VIPERS is expected to measure the growth rate in two redshifts bins at $z \sim 0.6$

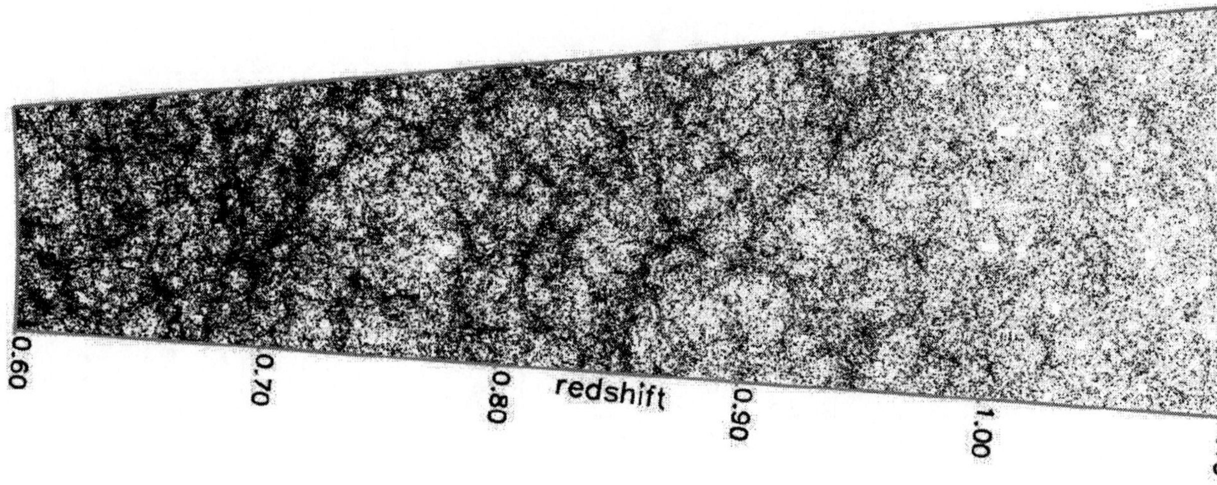

FIGURE 5. Simulated galaxy distribution between $z = 0.6$ and 1.1 in the W1 field of the VIPERS survey, corresponding to an area of $\sim 2 \times 8$ deg^2 within the CFHT Legacy Survey W1 field. (Mock survey obtained from the Millennium simulation, courtesy of J. Blaizot).

and ~ 0.9 with errors on $f\sigma_8$ below 10% (stars in Figure 4), i.e. comparable to local measurements nowadays. This expected value is actually an upper limit; it has been suggested recently (McDonald & Seljak 2009), that the combination in the same volume of measurements from two populations with different bias factors can abate the impact of cosmic variance and reduce errors of a further factor of $5-10$, depending on the relative bias of the two populations. This should not be possible for surveys sparsely covering very large volumes to hunt for BAO (as e.g. WiggleZ and BOSS, see below), but should be feasible for VIPERS, given its average galaxy density of $\sim 2 \times 10^{-3}$ h^3 Mpc^{-3}. In addition to its main cosmological goals, VIPERS will in parallel extend a number of investigations that have just been sketched, at these redshifts, by the VVDS. Several of these are also related to the invisible Universe, specifically to the relation of luminous and dark matter in galaxies. This is the case, for example, of detailed studies of small-scale clustering as a function of galaxy properties at $z \sim 1$, which can be described in terms of Halo Occupation Distribution models (Pollo et al. 2006; Meneux et al. 2008), the accurate shape of the galaxy luminosity and stellar mass functions, the mean density of massive clusters identified through their velocity dispersion, and many others (see the VIPERS web site, http://vipers.inaf.it for a full description of the project goals).

Interesting results from redshift-space distortions are to be expected also from the WiggleZ (http://wigglez.swin.edu.au) and BOSS (http://cosmology.lbl.gov/BOSS/) surveys. These projects were conceived specifically to detect accurately the BAO signal by surveying a very large volume with low-density tracers and use it to measure $H(z)$ at redshifts out to 1.0 and 0.7, respectively. Despite their low mean density of galaxies, the large volume should allow these surveys to obtain average errors on β between 3 and 15% over this range.

For the more distant future (2017-2018), the ESA EUCLID mission, resulting from the merging of the original DUNE (Refregier et al. 2008) and SPACE (Cimatti et al. 2008) concepts proposed to the ESA Cosmic Vision program, should represent the definitive experiment to map the structure of the dark and visible Universe to $z \sim 2$. EUCLID plans to survey 20,000 deg^2 of extragalactic sky both in imaging (optical and infrared) and spectroscopy ($\sim$ 100 million redshifts), to measure simultaneously $H(z)$ and $f(z)$ to percent accuracy in several redshift bins to $z \sim 2$, through a combination of dark-energy probes including BAO, redshift-space distortions and weak gravitational lensing. The expected errors on the growth rate (considering only redshift distortions, which are only part of the huge scientific potential of EUCLID), are indicated by the colored band in Figure 4, that clearly indicates how such a survey could uniquely discriminate modified gravity models from a dark energy or cosmological constant scenario.

ACKNOWLEDGMENTS

We thank the VVDS and VIPERS teams for their support, as well as the EUCLID Near-IR Spectrograph (ENIS) team for many enlightening discussions.

REFERENCES

1. Acquaviva, V., et al., 2008, PhRvD, 78, 043514
2. Amendola, L., 2000, MNRAS, 312, 521
3. Blake, G. & Glazebrook, K., 2003, ApJ, 594, 665
4. Cabre‘, A. & Gaztanaga, E., 2009, MNRAS, 393, 1183
5. Cimatti, A., et al., 2008, Exp. Astron., 23, 39
6. Colless, M., et al., 2001, MNRAS, 328, 1039
7. Copeland, E.J., Sami, M. & Tsujikawa, S., 2006, IJMPD, 15, 1753
8. Davis et al., 2003, SPIE, 4834, 161
9. De Lucia, G. & Blaizot, J., 2006, MNRAS, 375, 2
10. Dvali, G., Gabadadze, G., & Porrati, M., 2000, Phys. Lett. B 485, 208
11. Eisenstein D.J., et al., 2005, ApJ, 633, 560 Frieman, J.A., Turner, M.S. & Huterer, D., ARAA, 46, 385
12. Garilli, B. et al., 2008, A&A, 486, 683
13. Guzzo, L., et al., in 'A Century of Cosmology', Venice, G. Chincarini et al. eds., 2007, NCimB, 122, 1385
14. Guzzo L., Pierleoni, M., Meneux, B. et al., 2008, Nature 451, 541
15. Hamilton A.J.S., 1998, in The Evolving Universe, Kluwer ASSL, vol 231, 185
16. Hawkins, E. et al., 2003, MNRAS, 346, 78
17. Heath, D.J., 1977, MNRAS, 179, 351
18. Ilbert, O., et al., 2006, A&A, 457, 841
19. Kaiser, N., 1984, ApJ, 284, L9
20. Komatsu, E., et al.., 2009, ApJS, 180, 330
21. Le Fèvre O. et al., 2005, A&A, 439, 845
22. Landy, S.D. & Szalay, A.S., 1993, ApJ, 412,
23. Linder, E.V., 2005, Phys.Rev D, 72, 043529
24. Linder, E., 2008, APh, 29, 336
25. Lue, R., Scoccimarro, R. & Starkman, G.D., 2004, Phys. Rev. D, 69, 124015
26. Marinoni, C. et al., 2005, A&A, 442, 801
27. McDonald
28. Meneux, B., et al., 2008, A&A, 478, 299
29. Nesseris, S. & Perivolaropulos, L., 2008, Phys.Rev D, 77, 3504
30. Percival, W.J. & White, M., 2009, MNRAS, 393, 297
31. Perlmutter, S. et al., 1999, ApJ, 517, 565
32. Pollo, A., et al., 2006, A&A, 451, 409
33. Refregier, A., et al., Exp. Astron., 23, 17
34. Riess A.G. et al., 1998, AJ, 116, 1009
35. Ross, N.P. et al., 2007, MNRAS, 381, 573
36. Seo, H. & Eisenstein, D.J., 2003, ApJ, 598, 720
37. Song, Y.-S. & Percival, W.J., 2009, JCAP, 10, 004
38. Spergel D.N. et al., 2007, ApJS, 170, 377
39. Strauss, M.S., et al., 2002, AJ, 124, 1810
40. Strauss, M.S., 2008, Nature, 451, 531
41. Tegmark, M., et al., 2004, ApJ, 606, 702
42. Wang, L. & Steinhardt, P.J., 1998, ApJ, 508, 483
43. Wang, Y., 2008, JCAP, 05, 021
44. White, M., Song, Y.-S. & Percival, W.J., 2009, MNRAS, 397, 1348
45. Verde, L. et al., 2002, MNRAS, 335, 432

Observational Astrophysical Aspect of dark energy and dark matter.

OBSERVATIONAL CONSTRAINTS OF A DIRAC-MILNE UNIVERSE (ABSTRACT)

Aurélien Benoit-Lévy[1], Gabriel Chardin[2]

[1]*Institut d'Astrophysique de Paris, 98 bis bvd Arago, 75014 Paris, France*

[2]*CSNSM, IN2P3-CNRS, Université Paris XI, 91405 Orsay, France*

In this contribution, we present the Dirac-Milne Universe, an alternative cosmological model in which separate domains of matter and antimatter are present in equal quantities, with the further hypothesis that antimatter has a negative active gravitational mass. One consequence of theses hypotheses is the linear evolution of the scale factor throughout the thermal history of the Universe. We investigate this new model and confront it with some of the classical cosmological tests such as primordial nucleosynthesis, Type Ia supernovæ and Cosmic Microwave Background, showing that it is surprisingly concordant.

Keywords: Matter-antimatter cosmology, Big-Bang Nucleosynthesis, Type Ia supernovæ
PACS: 95.36.+x, 98.80.Es, 98.80.Ft

THE EVOLUTION OF MASS PROFILES OF GALAXY CLUSTERS (ABSTRACT)

A. Biviano and B. Poggianti

INAF/Osservatorio Astronomico di Trieste, Italy

We determine the average mass profile of galaxy clusters at two different redshifts and compare its evolution with cosmological model predictions. We use two samples of galaxy clusters spanning similar (evolutionary corrected) mass ranges at different redshifts. The sample of low-redshift ($z \simeq 0.0 - 0.1$) clusters is extracted from the ESO Nearby Abell Cluster Survey (ENACS) catalog. The sample of high-redshift ($z \simeq 0.4 - 0.8$) clusters is mostly made of clusters from the ESO Distant Cluster Survey (EDisCS). We determine the average mass-profiles for these two cluster samples by solving the Jeans equation for hydrostatic equilibrium, using galaxies as tracers. By using two cluster galaxy populations, characterized by the presence and, respectively, absence of emission-lines in their spectra ('ELGs' and 'nELGs' hereafter), we are able to partially break the mass-profile orbital-anisotropy degeneracy.

We find that the mass-profiles of both the nearby and the distant clusters are reasonably well fitted by a Navarro, Frenk & White (NFW) model. The best-fit values of the NFW concentration parameter are as predicted by cosmological numerical simulations; cluster mass-density profiles become more concentrated with time. The evolution of the number-density profile of nELGs proceeds in the opposite sense, becoming less concentrated with time.

In our analysis we also recover the orbital anisotropy of nELGs and ELGs . We find that in low-z clusters nELGs follow almost isotropic orbits and ELGs have more radially-elongated orbits. In high-z clusters both nELGs and ELGs follow radially-elongated orbits. We discuss these results in terms of the predicted secular mass growth of galaxy clusters and the transformation of ELGs into nELGs.

CP1241, *Invisible Universe: Proceedings of the Conference,* edited by J.-M. Alimi and A. Füzfa

Keywords: Galaxies: clusters: general – Galaxies: kinematics and dynamics
PACS: 98.65.Cw, 98.62.Ck, 98.62.Dm, 98.80.Es

THE INTEGRATED SACHS-WOLFE EFFECT: A CONFIRMATION FOR THE CASE OF DARK ENERGY (ABSTRACT)

Tommaso Giannantonio

Argelander-Institut für Astronomie der Universität Bonn, Auf dem Hügel 71, D-53121 Bonn, Germany

We review in these proceedings the status of the integrated Sachs-Wolfe (ISW) effect measurements, our current data and some cosmological applications. After briefly describing how the ISW effect is produced, we will show how we have measured it by cross-correlating in real space the cosmic microwave background (CMB) data from the WMAP satellite with six galaxy catalogues spanning a redshift range $0 < z < 2$. By generating random realisations of the maps with a Monte Carlo technique, we estimated the significance of the combined detection to be $> 4\sigma$ when fitting a single amplitude. If we interpret this detection to be due to the ISW effect, we can conclude that this is one more confirmation that the Universe is presently undergoing a transition to a dark energy phase, independently agreeing with the results from other observations. We then discuss how these data, together with other tests of structure formation such as galaxy counts, weak lensing and peculiar velocities, are a promising tool to break the degeneracies which exist at the background level between different theories of gravity. We work in detail the cases of the Dvali-Gabadadze-Porrati braneworld model and the class of $f(R)$ theories, finding no evidence for a departure from general relativity.

Keywords: Cosmology, CMB, Large Scale Structure
PACS: 98.80.Es, 98.70.Vc, 04.50.Kd

CMB DATA CONSTRAINT ON SELF-ANNIHILATION OF DARK MATTER PARTICLES (ABSTRACT)

Jaiseung Kim and Pavel Naselsky

Niels Bohr Institute, Blegdamsvej 17, DK-2100 Copenhagen, Denmark

The problem of dark matter in the Universe is one of the most important issues of modern cosmology. A lot of effort has been put in constraining different models of dark energy with SNIa data, CMBR, lensing or angular size statistics each of them having their own limitations and systematics. Therefore every complementary cosmological test has its value.

Future generation of interferometric gravitational wave detectors is hoped to provide accurate measurements of the final stages of binary inspirals. The sources probed by such experiments are of extragalactic origin and the observed chirp mass distribution carries information about their redshifts. Moreover the luminosity distance is directly observable is such experiments. This creates the possibility to establish a new kind of cosmological tests, supplementary to more standard ones. The paper highlights the utility of gravity wave experiments for testing dark energy in the Universe.

Keywords: Astrophysics - Cosmology and Extragalactic Astrophysics, High Energy Physics - Phenomenology
PACS: 95.35.+d, 98.70.Vc, 98.80.Cq, 98.80.-k, 98.80.Es

ARE GALAXY CLUSTERS SUGGESTING AN ACCELERATING UNIVERSE? (ABSTRACT)

J. A. S. Lima[1], R. F. L. Holanda[2] and J. V. Cunha[3]

Departamento de Astronomia, Universidade de São Paulo
Rua do Matão, 1226 - 05508-900, São Paulo, SP, Brazil

[1]*limajas@astro.iag.usp.br*

[2]*holanda@astro.iag.usp.br*

[3]*cunhajv@astro.iag.usp.br*

The present cosmic accelerating stage is discussed through a new kinematic method based on the Sunyaev-Zel'dovich effect (SZE) and X-ray surface brightness data from galaxy clusters. By using the SZE/X-ray data from 38 galaxy clusters in the redshift range $0.14 \leq z \leq 0.89$ [Bonament et al., Astrop. J. **647**, 25 (2006)] it is found that the present Universe is accelerating and that the transition from an earlier decelerating to a late time accelerating regime is relatively recent. The ability of the ongoing Planck satellite mission to obtain tighter constraints on the expansion history through SZE/X-ray angular diameters is also discussed. Our results are fully independent on the validity of any metric gravity theory, the possible matter-energy contents filling the Universe, as well as on the SNe Ia Hubble diagram from which the presenting accelerating stage was inferred.

Keywords: acceleration cosmic, Sunyaev-Zeldovich effect, galaxy clusters
PACS: 98.65.-r , 98.80.-k, 98.80.Es

LARGE ANGULAR CORRELATIONS ON THE SKY (ABSTRACT)

Yin-Zhe Ma, George Efstathiou and Duncan Hanson

Institute of Astronomy and Kavli Institute for Cosmology, Madingley Road, Cambridge, CB3 0HA, UK.

The lack of large angular correlations of the temperature fluctuation on the cosmic microwave background (hereafter CMB) sky have been considered by some to be the serious problems for the standard inflationary Λ cold dark matter cosmology (hereafter ΛCDM). In this work, we analyze the angular correlation function of the CMB sky. We present a method of reconstructing the multipole coefficient a_{lm} of temperature fluctuation on the entire sky from the CMB maps that include a sky-cut (e.g. KQ75 mask). From these reconstructions we calculate the angular correlation function of the temperature fluctuations. We apply an optimal estimator of the a_{lm} and show that it is possible to compute the angular correlation function at large angular scales from our realization of the sky even if a sky cut is applied to the data.

Keywords: correlation function,cosmic microwave background, anomaly
PACS: 98.80.Es, 98.70.Vc

STELLAR POPULATIONS AND MORPHOLOGY ON THE RED SEQUENCE AT Z≈1 (ABSTRACT)

Simona Mei[1], Brad P. Holden[2], John P. Blakeslee[3], Holland C. Ford[4], Marijn Franx[5], Garth D. Illingworth[6], Myungkook J. Jee[7], Roderik Overzier[8], Marc Postman[9], Piero Rosati[1]0, Arjen Van der Wel[1]1 and James G. Bartlett[1]2

[1]*University of Paris Denis Diderot, 75205 Paris Cedex 13, France - GEPI, Observatoire de Paris, Section de Meudon, 5 Place J. Janssen, 92195 Meudon Cedex, France, simona.mei@obspm.fr*

[2]*Lick Observatory, University of California, Santa Cruz, CA 95064*

[3]*Herzberg Institute of Astrophysics, National Research Council of Canada, Victoria, BC V9E2E7, Canada*

[4]*Department of Physics and Astronomy, Johns Hopkins University, Baltimore, MD 21218*

[5]*Leiden Observatory, Postbus 9513, 2300 RA Leiden, Netherlands*

[6]*Lick Observatory, University of California, Santa Cruz, CA 95064*

[7]*University of California, Davis, CA 95616*

[8]*Max-Planck-Institut fur Astrophysik, Karl-Schwarzschild-Strasse 1, D-85748 Garching bei Munchen, Germany*

[9]*Space Telescope Science Institute, 3700 San Martin Drive, Baltimore, MD 21218*

[10]*European Southern Observatory, Karl-Schwarzschild-Str. 2, D-85748 Garching, Germany*

[11]*Max-Planck Institute for Astronomy, Königstuhl 17, D-69117, Heidelberg, Germany*

[12]*APC, 10 rue Alice Domon et Léonie Duquet, 75205 Paris Cedex 13, France (UMR 7164, CEA, CNRS, Obs. de Paris, Univ. Paris Diderot)*

We present results from a detailed study of cluster red sequence at $z \approx 1$ from the ACS Intermediate Redshift Cluster Survey (Mei et al. 2009). Our analysis shows that the red sequence is well defined at $z \approx 1$ and elliptical and lenticular galaxies lie on similar color-magnitude relations. We analyze the parameters of the early–type color-magnitude relations – scatter, slope and zero–point – as a function of redshift, galaxy properties and cluster mass. Our results suggest that bright elliptical galaxies in cluster cores are on average older than S0 galaxies and peripheral elliptical galaxies (by about 0.5 Gyr, using a simple, single burst solar metallicity Bruzual & Charlot (2003) stellar population model). The red sequence does not show significant evolution out to redshift $z \approx 1.3$ nor significant dependence on cluster mass. The fraction of morphological early–type galaxies on the red sequence is 80% to 90% of the total red sequence population for most of our clusters. In the highest redshift, low mass cluster of our sample, early–type/late–type fractions on the red sequence are similar ($\approx 50\%$), with most of the late–type population composed of galaxies classified as S0/a. This trend is not correlated with the cluster's X–ray luminosity, nor with its velocity dispersion, and could be a real evolution with redshift.

Keywords: Galaxies; Galaxy clusters
PACS: 90

CUSPED DENSITY PROFILES OF GRAVITATIONAL LENS OBJECT (ABSTRACT)

P. T. Mutka

Department of Physical Sciences, Theoretical Physics Division, University of Oulu, P.B. 3000, 90014 Oulun Yliopisto, Finland, EU

We have developed an analytic formulation for axially symmetric GNFW lens model with parametrized cusp slope (α). The lensing theory has several implications, for example strong lensing is very difficult without cusped mass profile. Required cusp strength for strong lensing depends on the lens object mass and concentration. Exceedingly high concentrations are required for profiles, that have $\alpha > -1$ in order to produce multiple lensed images.

We study mass profiles of lens objects with double image lenses, since they are resilient against deviations from axial symmetry, perturbations from microlensing, and halo substructure. The statistics of the observed image flux ratios is connected to the general properties of the of the lens mass density profiles. Our analysis is based on a limiting value for the shallowest cusp slope α_{CSL} able to produce the observed flux ratio with any lens geometry and lens–source alignment. The cusp slope limit (CSL) does not depend on cosmology, total lens mass, concentration or redshifts of the the lens and the lensed object. In case of axial symmetry the limiting value is depending only on the magnification ratio (observed flux ratio of the images). This removes uncertainties in the lens and source distributions from the statistical analysis.

Distribution of these threshold values reveals existence of halo population(s) with similar profiles in the sample; most of the halos have cusp slope $\alpha = -1.95 \pm 0.02$. We have also found an imprint of a second population with a cusp slope value $\alpha = -1.49 \pm 0.09$. There is about 99 per cent estimated probability, that the observed feature in the distribution is produced by the second population of lenses, with their own characteristic density profile. We analyze error sources in our analysis with mock catalogues, and discuss about alternative explanations for the second population signature.

Keywords: Cosmology; dark energy; time variation of physical constants
PACS: 98.80.-k, 95.36.+x, 98.80.Es

NEW CONSTRAINT ON THE VARYING FINE STRUCTURE CONSTANT (ABSTRACT)

M. Nakashima[1,2], K. Ichikawa[3], R. Nagata[2] and J. Yokoyama[2,4]

[1] *Department of Physics, Graduate School of Science, The University of Tokyo, Tokyo 113-0033, Japan*

[2] *Research Center for the Early Universe (RESCEU), Graduate School of Science, The University of Tokyo, Tokyo 113-0033, Japan*

[3] *Department of Micro Engineering, Kyoto University, Kyoto 606-8501, Japan*

[4] *Institute for the Physics and Mathematics of the Universe(IPMU), The University of Tokyo, Kashiwa, Chiba, 277-8568, Japan*

Some physical imprints of quintessence scalar fields on dark matter (DM) clustering are illustrated, and a comparison with the concordance model ΛCDM is highlighted. First, we estimate the cosmological parameters for two quintessence models, based on scalar fields rolling down the Ratra-Peebles or Sugra potential, by a statistical analysis of the Hubble diagram of type Ia supernovae. Then, the effect of these realistic dark energy models on large-scale DM clustering is established through N-body simulations. Various effects like large-scale distribution of DM, cluster mass function and halos internal velocities are illustrated. It is found that realistic dark energy models lead to quite different DM clustering, due to a combination of the variation of the equation of state and differences in the cosmological parameters, even at z=0. This conclusion contradicts other works in the recent litterature and the importance of considering more realistic models in studying the impact of quintessence on structure formation is highlighted.

Keywords: Cosmology; dark energy; time variation of physical constants
PACS: 98.80.-k, 95.36.+x, 98.80.Es

GENETIC ALGORITHMS AND SUPERNOVAE TYPE IA ANALYSIS, CONSTRAINTS ON DARK ENERGY(ABSTRACT)

Charalampos Bogdanos[1] and Savvas Nesseris[2]

[1]*LPT, Université de Paris-Sud-11, Bât. 210, 91405 Orsay CEDEX, France*

[2]*The Niels Bohr International Academy, Blegdamsvej 17, DK-2100, Copenhagen ŕ, Denmark*

We introduce genetic algorithms as a means to analyze supernovae type Ia data and extract model-independent constraints on the evolution of the Dark Energy equation of state $w(z) \equiv \frac{P_{DE}}{\rho_{DE}}$. Specifically, we will give a brief introduction ?DE to the genetic algorithms along with some simple examples to illustrate their advantages and finally we will apply them to the supernovae type Ia data. We find that genetic algorithms can lead to results in line with already established parametric and non-parametric reconstruction methods and could be used as a complementary way of treating SNIa data. As a non-parametric method, genetic algorithms provide a model-independent way to analyze data and can minimize bias due to premature choice of a dark energy model.

Keywords: dark energy, supernovae, cosmology, genetic algorithms
PACS: 98.80.Es,98.65.Dx,98.62.Sb

THE NEARBY SUPERNOVA FACTORY DATASET - IMPROVING SNE IA AS DARK ENERGY PROBES (ABSTRACT)

R. Pereira1, G. Aldering[2], P. Antilogus[3], C. Aragon[2], S. Bailey[3], C. Baltay[4], S. Bongard[3], C. Buton[1], M. Childress[2], N. Chotard[1], Y. Copin[1], E. Gangler[1], S. Loken[2], P. Nugent[2], R. Pain[3], E. Pecontal[5], S. Perlmutter[2,6], D. Rabinowitz[4], G. Rigaudier[5], K. Runge[2], R. Scalzo[4], G. Smadja[1], H. K. Fakhouri[2,6], C. Tao[7], R. C. Thomas[2] and C. Wu[3]

[1]*Université de Lyon, F-69622 France; Université de Lyon 1; CNRS/IN2P3, Institut de Physique Nucléaire de Lyon*

[2]*Physics Division, Lawrence Berkeley National Laboratory, 1 Cyclotron Road, Berkeley, CA 94720*

[3]*Laboratoire de Physique Nucléaire et des Hautes Energies, CNRS/IN2P3; Universités Paris VI et Paris VII, 4 place Jussieu, Tour 43, 75252 Paris Cedex 05*

[4]*Department of Physics, Yale University, New Haven, CT 06250-8121*

[5]*Université de Lyon, F-69622 France; Université de Lyon 1; CRAL, Observatoire de Lyon, F-69230 Saint Genis Laval; CNRS, UMR 5574; ENS de Lyon*

[6]*Department of Physics, University of California Berkeley, Berkeley, CA 94720*

[7]*CPPM, 163 av. Luminy, 13288 Marseille Cedex 09*

During the 2004–2008 period the Nearby Supernova Factory discovered ∼1000 supernovæ, and obtained detailed spectrophotometric time series of ∼190 SNe Ia with the SuperNova Integral Field Spectrograph. These supernovæ are located in the nearby Hubble flow ($0.03 < z < 0.08$), allowing relative distance measurements with minimal uncertainties from peculiar velocities and bulk flows. Each SNe was spectrophotometrically observed over a period of ∼50 days, with a median number of 15 observations per target. The first spectrum is typically obtained at -4 days, and in some instances as early as -14 days before maximum light. Each spectrum (3200Å – 10000Å) is photometrically calibrated, with simultaneous observations of field stars used to correct for atmospheric extinction variability.

The first results, impacting the use of SNe Ia as dark energy probes, are presented. We discuss how the SNfactory dataset can be used to find correlations between spectroscopic and photometric properties of SNe Ia, and demonstrate a new spectral indicator able to standardize SNe Ia with a better accuracy than light curve width and colour. We also show how this dataset can be used to build light curves without K correction-induced uncertainties in the photometric system best matched for combining with high-z data, thus addressing a major source of systematic error in current analyses.

Keywords: supernovae, survey, cosmology, spectrophotometry
PACS:98.80.Es, 95.36.+x, 97.60.Bw, 95.75.Fg

CONSTRAINING THE DARK ENERGY EQUATION OF STATE USING ALTERNATIVE HIGH-Z COSMIC TRACERS (ABSTRACT)

M. Plionis[1,2], R. Terlevich[2], S. Basilakos[3], F. Bresolin[4], E. Terlevich[2], J. Melnick[4] and R. Chavez[2]

[1] *Institute of Astronomy and Astrophysics, National Observatory of Athens, Greece*

[2] *INAOE, Puebla, Mexico*

[3] *Institute for Astronomy of the University of Hawaii, Honolulu, USA*

[4] *ESO, Santiago, Chile*

We propose to use alternative cosmic tracers to measure the dark energy equation of state and the matter content of the Universe [$w(z)$ & Ω_m]. Our proposed method consists of two components: (a) tracing the Hubble relation using HII galaxies which can be detected up to very large redshifts, $z \sim 4$, as an alternative to supernovae type Ia, and (b) measuring the clustering pattern of X-ray selected AGN at a median redshift of ~ 1. Each component of the method can in itself provide interesting constraints on the cosmological parameters, especially under our anticipation that we will reduce the corresponding random and systematic errors significantly. However, by joining their likelihood functions we will be able to put stringent cosmological constraints and break the known degeneracies between the *dark energy* equation of state (whether it is constant or variable) and the matter content of the universe and provide a powerful and alternative route to measure the contribution to the global dynamics and the equation of state of *dark energy*. A preliminary joint analysis of X-ray selected AGN (based on the largest to-date XMM survey; the 2XMM) and the currently largest SNIa sample (Hicken et al.), using as priors a flat universe and the WMAP5 normalization of the power-spectrum, provides: $\Omega_{\rm m} = 0.27 \pm 0.02$ and $w = -0.96 \pm 0.07$.

Keywords: Cosmology; Dark Matter; Dark Energy

ON THE ROLE OF Λ IN GRAVITATIONAL LENSING (ABSTRACT)

Mauro Sereno

Institut für Theoretische Physik, Universität Zürich, Winterthurerstrasse 190, CH-8057 Zürich, Switzerland.

The cosmological constant Λ affects gravitational lensing phenomena. Related effect can be studied in the framework of the Schwarzschild-de Sitter space-time in the weak deflection limit. The main contribution of Λ can be accounted for by a proper use of angular diameter distances. For near lenses and static observers, unresolved images are slightly demagnified, the radius of the Einstein ring decreases and the time delay increases. The effect is however negligible. Two novel contributions, which show up in a cosmological context, are also discussed. First, a term $\delta\hat{\alpha}_\Lambda = 2mb\Lambda/3$ has to be added to the bending angle, where m is the lens mass and b the impact parameter. Second, Λ brings about a difference in the redshifts of multiple images. Both effects are quite small for real astrophysical systems, $\delta\hat{\alpha}_\Lambda \lesssim 0.1\mu$arcsec and $\Delta z_{\rm s} \lesssim 10^{-7}$.

Keywords: Classical black holes; Gravitational Lensing
PACS: 95.30.Sf, 04.70.Bw, 98.62.Sb

SEARCHING FOR HIDDEN UNEXPECTED FEATURES IN THE SNIA DATA (ABSTRACT)

A. Shafieloo[1] and L. Perivolaropoulos[2]

[1]*Department of Physics, University of Oxford, 1 Keble Road, Oxford, OX1 3NP, UK*

[2]*Department of Physics, University of Ioannina, Greece*

It is known that χ^2 statistic and likelihood analysis may not be sensitive to the all features of the data. Despite of the fact that by using χ^2 statistic we can measure the overall goodness of fit for a model confronted to a data set, some specific features of the data can stay undetectable. For instance, it has been pointed out that there is an unexpected brightness of the SnIa data at $z > 1$ in the Union compilation. We quantify this statement by constructing a new statistic, called Binned Normalized Difference (BND) statistic, which is applicable directly on the Type Ia Supernova (SnIa) distance moduli. This statistic is designed to pick up systematic brightness trends of SnIa data points with respect to a best fit cosmological model at high redshifts. According to this statistic there are 2.2%, 5.3% and 12.6% consistency between the Gold06, Union08 and Constitution09 data and spatially flat ΛCDM model when the real data is compared with many realizations of the simulated monte carlo datasets. The corresponding realization probability in the context of a $(w_0, w_1) = (-1.4, 2)$ model is more than 30% for all mentioned datasets indicating a much better consistency for this model with respect to the BND statistic. The unexpected high z brightness of SnIa can be interpreted either as a trend towards more deceleration at high z than expected in the context of ΛCDM or as a statistical fluctuation or finally as a systematic effect perhaps due to a mild SnIa evolution at high z.

Keywords: Dark Energy, Cosmological Constant, Statistical Analysis
PACS: 98.80.Es, 98.62.Sb, 98.65.Dx

TENTATIVE EVIDENCE FOR SLOWING DOWN OF COSMIC ACCELERATION FROM RECENT SMALL REDSHIFT SUPERNOVAE AND BAO DATA (ABSTRACT)

A. Shafieloo[1], V. Sahni[2] and A. A. Starobinsky[3,4]

[1]*Department of Physics, University of Oxford, 1 Keble Road, Oxford, OX1 3NP, UK*

[2]*Inter University Centre for Astronomy and Astrophysics, Post Bag 4, Ganeshkhind, Pune, 411007, India*

[3]*Landau Institute for Theoretical Physics, Moscow 119334, Russia*

[4]*RESCEU, Graduate School of Science, The University of Tokyo, Tokyo 113-0033, Japan*

An investigation of dark energy (DE) using the Constitution SnIa sample (which includes recent CfA data at low redshifts) reveals a slight inconsistency (more than 1σ) with the standard spatially flat ΛCDM model, if the assumption of a constant equation of state (w) for DE is dropped. This effect, which is most clearly seen using the recently introduced Om diagnostic, corresponds to an increase of Om(z) and w(z) at redshifts z < 0.3. In geometrical terms, this suggests that cosmic acceleration may have already peaked and that we are currently witnessing its slowing down. Interestingly, such DE behaviour also provides a better fit to baryon acoustic oscillation (BAO) data. Including the cosmic microwave background (CMB) "shift parameter" as another independent observable, we show that the well known CPL parametrization strains to fit the data simultaneously at low and high redshifts. This could either be because of the presence of systematics, or because the CPL ansatz is not versatile enough to catch the rapid variation in DE suggested by the data. The effect we observe could correspond to DE decaying into dark matter (or something else). A toy model which mimics this process agrees well with the combined SnIa+BAO+CMB data.

Keywords: Dark Energy, Cosmological Constant, Cosmological Parameter Estimation
PACS: 98.80.Es, 98.62.Sb

COSMOLOGICAL TESTS OF GENERAL RELATIVITY WITH TOMOGRAPHIC SURVEYS (ABSTRACT)

Alessandra Silvestri (in collaboration with: GongBo Zhao, Levon Pogosian and Joel Zylberberg)

Kavli Institute for Astrophysics and Space Research, MIT, Cambridge, MA 02139, USA

Future cosmological surveys, combining galaxy counts and weak lensing measurements, will map the evolution of matter perturbations and gravitational potentials from the matter dominated epoch until today. In addition to tightening the constraints on allowed expansion histories, the combination of these measurements will test the relationships between matter overdensities, local curvature, and the Newtonian potential. These relationships can be modified in alternative theories of gravity and by exotic forms of dark energy. I will present the results by Zhao et al. 2009 of an analysis on the potential of upcoming and future tomographic surveys, such as DES and LSST, with the aid of CMB and supernovae data, to detect departures from the growth of perturbations expected within General Relativity with a cosmological constant.

Keywords: Cosmology, Gravity, Large Scale Structure
PACS: 98.80.-k, 98.65.Cw, 95.36.+x

THE CROSS-CORRELATION OF THE LYMAN-α FOREST AND WEAK LENSING OF CMB: A NEW COSMOLOGICAL TOOL (ABSTRACT)

Alberto Vallinotto

Center for Particle Astrophysics, Fermi National Accelerator Laboratory,
P.O. Box 500, Kirk Rd. & Pine St., Batavia, IL 60510-0500 USA, (email:avalli@fnal.gov)

I present estimates for the cross-correlation of the Lyman-α flux (and flux variance) with CMB convergence maps. These cross-correlations should be detectable by upcoming experiments that are already taking data (Planck and SDSS-III). I also show that the two correlators are sensitive to different wavelengths of the density field and potentially represent a new useful cosmological tool to probe intermediate-to-small scales and to successfully constrain the Lyman-α flux-matter relation.

Keywords: Cosmology, large scale structure of the universe, Lyman-α forest, CMB lensing, neutrinos.
PACS: 98.62.Ra, 98.70.Vc, 95.30.Sf

EXPERIMENTAL PARTICLE PHYSICS AND DARK MATTER CANDIDATES

Dark Matter Candidates: A Status Report

Lars Bergström

The Oskar Klein Centre for Cosmoparticle Physics
Department of Physics, Stockholm University
AlbaNova, SE-106 91 Stockholm, Sweden

Abstract. A brief review is given of various dark matter candidates. In particular, some of the most studied candidates like axions, inert Higgs doublet, sterile neutrinos, supersymmetric particles and Kaluza-Klein particles are discussed. In particular, indirect detection methods are reviewed, with gamma-ray detection being a particularly important method. The situation also for indirect detection through antimatter cosmic rays has recently become quite interesting with new results having dark matter as one of the possible explanations. Problems of this explanation and possible solutions are discussed, and the importance of new measurements is emphasized. If the explanation is indeed dark matter, some unusual and unexpected properties would be needed. One should always keep the possible "conventional" astrophysical explanations, like electron and positron radiation from pulsars in mind. To get a conclusive case for dark matter, detection in other channels will likely be needed, such as direct detection or detection of angular and spectral signatures in gamma-rays.

Keywords: Dark matter, indirect detection, models
PACS: 14.80.Ly, 14.80.Rt, 95.35.+d, 95.85.Pw

INTRODUCTION

The dark matter problem was pointed out by Zwicky more than 75 years ago [1], but it is only with the advent of precision cosmology that the problem has been identified as one of the most pressing one in astroparticle physics and cosmology (for alternatives, with their own problems especially at the scale of galaxy custers or larger, see [2]). There exist several extensive reviews of particle dark matter [3, 4, 5, 6, 7] as well as a recent book [8], in particular covering the prime candidate which has become something of a template for dark matter, namely the lightest supersymmetric particle. Here I will very briefly review the main suspects of dark matter, and then focus on recent developments, concerning supersymmetric dark matter, but will also mention some new interesting - though speculative, types of dark matter models that may perhaps explain the suprising new measurements of a large flux of positrons in the cosmic rays [9, 10]. The enhanced cross sections needed in these models, in particular the so-called Sommerfeld enhancement, will also be briefly discussed.

THE DARK MATTER PROBLEM

The dark matter problem has as mentioned been a part of astrophysics for a long time – since Zwicky's observation of a large velocity dispersion of the members of the Coma galaxy cluster [1]. Similarly, the problem of galactic rotation curves - the stars rotate "too fast" to be bound by Newtonian gravity if all matter is visible - can be traced back to Bacbcock's measurements of the Andromeda galaxy 1939 [11]. It took, however, several decades before it was recognized as a real problem, and in its modern form it goes back to the late 1970's and early 1980's when the so-called cold dark matter paradigm appeared [12] (in this context, cold means matter moving with non-relativistic velocities when structure formed in the universe). Today, a wealth of impressive data from studies of the microwave background radiation, supernova distance measurements, and large-scale galaxy surveys have together solidified the Standard Model of cosmology, where structure formed through gravitational amplification of small density perturbations with the help of cold dark matter. Without the existence of dark matter the density contrast seen in the universe today could not have formed, given the small amplitude of density fluctuations inferred from anisotropies of the cosmic microwave background.

Present-day cosmology of course also has another, mysterious component: a cosmological constant Λ or a similar agent (such as time-varying *quintessence* exerting negative pressure such that the expansion of the universe is today accelerating). For the purpose of this article, however, this *dark energy* plays little role other than to fix the background

CP1241, *Invisible Universe: Proceedings of the Conference,* edited by J.-M. Alimi and A. Füzfa

TABLE 1. Properties of various Dark Matter Candidates

Type	Particle Spin	Approximate Mass Scale
Axion	0	μeV-meV
Inert Higgs Doublet	0	50 GeV
Sterile Neutrino	1/2	keV
Neutralino	1/2	10 GeV - 10 TeV
Kaluza-Klein UED	1	TeV

metric and thus influencing late-time structure formation. In fact, most large-scale N-body simulations are now carried out in this cosmological Standard Model, the ΛCDM model. Modern models of cosmology contain a brief period of enormously accelerated expansion, inflation, which gives a nearly scale-invariant spectrum of primordial fluctuations, which together with the fact that the universe observationally appears to be very flat (i.e., the total energy density is equal to the critical density) are cornerstones of the Standard Model of cosmology.

Models for Dark Matter

Almost all current models of dark matter use the standard concept of quantum field theory to describe the properties of elementary particle candidates (for exceptions, see for instance [13, 14]). This means that they can be characterized by the mass and spin of the dark matter particle. The mass of proposed candidates spans a very large range, as illustrated in Table 1.

The density of cold dark matter (CDM) is now given to an accuracy of a few percent. With h being the Hubble constant today in units of 100 $\text{kms}^{-1}\text{Mpc}^{-1}$, the density derived with the help of the 5-year WMAP data [15] is

$$\Omega_{\text{CDM}}h^2 = 0.1131 \pm 0.0034, \tag{1}$$

with the estimate $h = 0.705 \pm 0.0134$.

Using the simplest type of models of thermally produced dark matter (reasonably far away from thresholds and branch cuts) this corresponds to an average of the annihilation rate at the time of chemical decoupling of [3]

$$\langle \sigma_A v \rangle = 2.8 \cdot 10^{-26}\ \text{cm}^3\text{s}^{-1}. \tag{2}$$

The fact that this corresponds to what one computes using a weak interaction cross section for particles of mass around the electroweak scale around a few hundred GeV is sometimes coined the "WIMP miracle" (WIMP standing for Weakly Interacting Massive Particle), but it may of course be a coincidence. However, most of the detailed models proposed for the dark matter are in fact containing WIMPs as dark matter particles.

The rate in Eq. (2) is a convenient quantity to keep in mind, but it has to be remarked that this is the value needed at the time of freeze-out, when the temperature was typically of the order of $(0.05 - 0.1)M_X$ (with M_X the mass of the dark matter particle) and the velocity $v/c \sim 0.2 - 0.3$. There are now publicly available computer codes [16, 17] that solve the Boltzmann equation numerically, taking various effects into account, such as co-annihilations which may change the effective average annihilation cross section appreciably if there are other states than the one giving the dark matter particle which are nearly degenerate in mass. There are also computer packages available (e.g., [18]) that can perform likelihood analyses of the probability distribution of combinations of parameters, in particular for supersymmetric dark matter models. As the parameter space, even for extremely simplified supersymmetric models is extremely large (typically, 7-dimensional or larger) more and more powerful methods have been developed to scan parameter space.

As we will see later, the simple Eq. (2) may be modified by orders of magnitude in the halo today, for example by the Sommerfeld enhancement – if there are zero velocity bound states in the annihilating system (cold dark matter particles should today move with typical Galactic velocities of $v/c \sim 10^{-3}$). One should also be aware of the large astrophysical uncertainties present when estimating the observable annihilation rate today, as it may be influenced by the presence of substructure in the dark matter distribution, such as indicated in large simulations of structure formation [19, 20].

The computation of the boost factor in realistic astrophysical and particle physics scenarios is a formidable task, which has so far only been partially addressed. It may be anticipated that this will be one of the main problem areas of future indirect detection studies of dark matter. For direct detection, there is no corresponding enhancement of the scattering rate. However, the detailed small-scale structure of the local region of the dark matter halo may play a role [21]. Also, it has been speculated that the galactic disk may be accompanied by a co-rotating dark matter disk [22], something that could make gravitational capture in the planets and the Sun easier. This could be of importance both for direct detecttion and indirect detection of neutrinos (see next Section).

Indirect Detection of Dark Matter

Dark matter may in principle be discovered in three different ways:

- At accelerators, like the soon to be deployed LHC accelerator at CERN, or at on-going searches with the Fermilab Tevatron collider.
- Directly, as particles making up the dark matter scatter in terrestrial detectors.
- Indirectly, but observing annihilation products in the cosmic rays. Particularly important are positrons and antiprotons, as ordinary comsic ray sources should give rather small fractions of antimatter, whereas in most cases (and in particular for supersymmetric particles) the initial state is CP-symmetric, as also the final state then should be. However, there is still a to large extent unknown background of these charged particles, coming from violent processes in the local universe. Characteristic features of annihilation may be smeared out due to propagation effects such as diffusion due to magnetic fields, and for positrons also by bremsstrahlung and energy losses due to inverse Compton scattering on CMB photons or IR and optical light coming from starlight.
  This problem is not shared by gamma-rays, which travel essentially without absorption through the galaxy. Thus a characteristic signal both in angular variation (which should reflect line-of-sight integral of the square of the dark matter density distribution) and energy should be searched for. The latter would reflect the mass of the annihilating particle - in particular, no gamma-rays should be produced at an energy larger than the dark matter particle mass, remembering that the annihilating particles move non-relativistically in the dark matter halo.
  Neutrinos also travel in stright lines from their sources, but rates from the dark matter halos are expected to be low. However, for favourable values of parameters a detectable signal from dark matter particles graviationally in the Earth, and in particular the Sun, may be searched for, e.g. in the IceCube detector at the South Pole, with its dedicated insert detector DeepCore [23, 25].

Let us now focus on indirect detection of gamma rays in the Galactic halo. The annihilation rate towards a direction making the angle Ψ with respect to the galactic centre is conveniently given by an expression which is factorized in one part that depends on particle theory, and another part which is related to the distribution of dark matter [24]

$$\Phi_\gamma(\psi) = 0.94 \cdot 10^{-9} \left(\frac{N_\gamma \langle \sigma v \rangle}{10^{-29}\,\mathrm{cm^3 s^{-1}}} \right) \left(\frac{100\,\mathrm{GeV}}{M_\chi} \right)^2 J(\psi)\ \mathrm{m^{-2}\,s^{-1}\,sr^{-1}} \tag{3}$$

where the dimensionless astrophysical function $J(\Psi)$ is given by

$$J(\psi) = \frac{1}{8.5\,\mathrm{kpc}} \cdot \left(\frac{1}{0.3\,\mathrm{GeV/cm^3}} \right)^2 \int_{line\ of\ sight} \rho^2(l)\, dl(\psi)\,, \tag{4}$$

with $\rho(l)$ being the local dark matter density following the line of sight $l(\Psi)$.

The particle physics factor $N_\gamma \langle \sigma v \rangle$, which is the angle and velocity averaged annihilation rate times the number of photons created per annihilation, can usually be rather accurately computed for a given dark matter candidate. In particular, for cross sections containing an s-wave piece, usually σv does not depend on velocity to an excellent approximation, for the typical small Galactic velocities $v/c \sim 10^{-3}$ in the halo. However, the value of σv may in some cases, in particular for the Sommerfeld enhanced models, depend on velocity or rather, after the angular average, on the velocity dispersion v_{disp}, in the simplest case like $1/v_{\mathrm{disp}}$. Also, $\rho^2(l)$ may vary rapidly along the line of sight if there exists substructure in the halo dark matter distribution. Therefore, in general the integration does not factorize as in Eq. (3) and has to be performed over the full phase space of the dark matter distribution, which for a given halo – like that of the Milky Way – unfortunately is poorly known.

The procedure in this situation has been to introduce a "boost factor", which unfortunately does not seem to have a unique definition. In particular, the boost of the detection rates for a given model will depend on the particle of

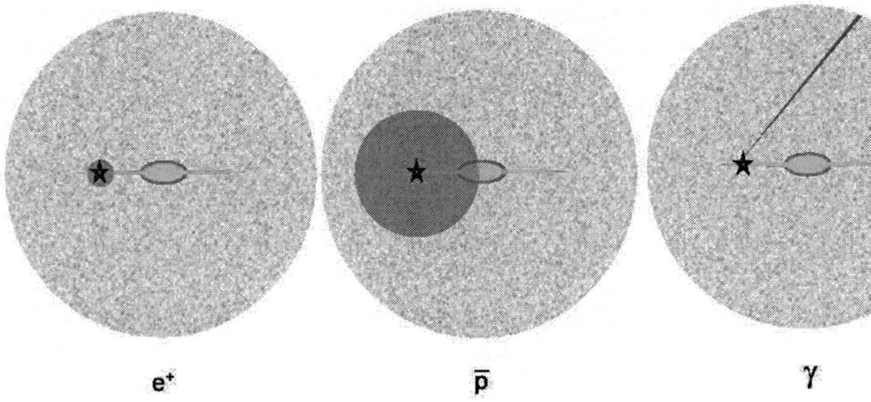

FIGURE 1. Illustration of the volumes in the solar neigbourhood entering the calculation of the average boost factor in the dark matter halo. Theses are schematic pictures for 100 GeV dark matter particles annihilating into, from left to right, positrons, antiprotons, and gamma-rays. The difference in size for antiprotons and positrons depends on the different energy loss properties, as positrons at these energies radiate through synchrotron and inverse Compton emission much faster than do antiprotons. For gamma-rays, large boosts can occur in certain directions, for example if the line of sight passes through a clumpy of dark matter. Also, the direction close to the galactic center is generally expected to give the highest gamma-ray yields (but the backgrounds may also be more severe).

detection, and the energy. One possible definition of the boost factor would be

$$B_{\rm tot} = B_\rho \times B_{\sigma v} = \left(\frac{\langle \rho^2(r)\rangle_{\Delta V}}{\langle \rho_0^2(r)\rangle_{\Delta V}}\right) \times \left(\frac{\langle \sigma v\rangle_{v\simeq v_{\rm disp}}}{\langle \sigma v\rangle_{v\simeq v_F}}\right)_{\Delta V}, \tag{5}$$

with $v_{\rm disp}$ the velocity dispersion in the object under study and v_F the typical velocity at freeze-out.

The boost factor contains the typical averaging volume ΔV, and this is different for different particle species. For electrons and positrons, due to their rapid energy loss due to synchrotron radiation and inverse Compton scattering, the averaging volume is quite small, being less than a kpc for 100 GeV particles. The corresponding volume for antiprotons is much larger, and, finally, for gamma-rays, the volume is just given by the cone given by the angular acceptance of the instrument (which generally depends on energy). See Fig. 1.

In fact, a factorized expression like that in Eq. (5) may not even be possible, since the cross section may depend strongly on the dispersion velocity (such as in Sommerfeld-enhanced models), and small substructures in the halo would have a lower velocity dispersion, and therefore higher boost, than the ambient halo. One would thus need to do a full N-body calculation to estimate the average boost factor, including its angular dependence.

Since the dark matter density in most models (as is verified by N-body simulations) should peaked near the galactic center, lines of sight that cross or go near the galactic center would a priori be advantageous when searching for a gamma-rays signal of dark matter. However, as substructure which may boost the signal is most likely destroyed by tidal interactions near the dense center [19], and as the galactic center may have a number of other unresolved gamma-ray sources (like the HESS source [26]), probably of more mundane astrophysical origin, the optimal search strategy may be more subtle [27]. The Fermi gamma-ray satellite has taken data for more than a year, and one may expect limits (or claimed signals) which should be an order of magnitude more sensitive than those of the predecesor EGRET

[28], and spanning energies up to several hundred GeV. As the gamma-ray sky would be rather complex due to sources like pulsars, active galactic nuclei and others, the extraction of a signal will be a rather delicate work. Probably both the angular distribution and the energy spectrum would have to be simultaneously fitted. Also, since a large part of a possible signal would come from the Milky Way halo, but also contain a more isotropic extragalactic component [29], the analysis of galactic and extragalactic gamma-ray distributions should be performed simultaneously, perhaps most optimally using characteristic anisotropy signals [30] in addition to the spectral information (as low redshifts should dominate the extragalactic signal [29], the effects of redshifting that signal should be small). To extract the expected spectral distribution of gamma-rays, it is important to go beyond leading order when computing annihilation rates, since very large effects may be generated by inner bremsstrahlung corrections [31].

The complexity of the analysis should not be underestimated, as an example one should note with caution the preliminary apparent excess of gamma-rays seen by Fermi (but much smaller than that inferred by EGRET data [32]), as modeling of other sources by the Fermi Collaboration is still at an early stage. However, if the excess would persist, and if it follows what one typically expects for a dark matter halo (having a cuspy density $\propto 1/r$ close to the center), a very careful analysis could be called for.

Axions

Although at times not very much in focus of dark matter phenomenologists and experimentalists, the axion remains one of the earliest suggestions of a viable particle candidate for dark matter, and in fact one of the most attractive. This is not least due to the fact that its existence was motivated by solving the strong *CP* problem in particle physics, and its possible role for dark matter comes as an extra bonus. A disadvantage in the cosmological context is, however, that the axion needed to solve the *CP* problem only solves the dark matter problem for a small range of masses – thus some fine-tuning mechanism seems to be needed. For a recent review of the field, see [33].

The original idea of Peccei and Quinn was to make the CP violating phase dynamical [34] by introducing a global symmetry, $U(1)_{PQ}$, which is spontaneously broken. The Goldstone boson of this broken global symmetry is the axion, which however gets a non-zero mass from the QCD anomaly, which can be interpreted as a mixing of the axion field with the π and η mesons [35, 36].

The earliest attemps, using only the standard model particles but with an enlarged Higgs sector (and which did not address at all the dark matter problem), were soon ruled out experimentally and the "invisible axion" was invented [37, 38] with a very high mass scale of symmetry breaking and with very massive fermions carrying PQ charge. This means that only a feeble strong or electromagnetic interaction leaks out to the visible sector through triangle loop diagrams.

The phenomenology of the axion is determined, up to numerical factors, by one number only - the scale f_a of symmetry breaking. In particular, the mass is given by

$$m_a = 0.62 \text{ eV} \left(\frac{10^7 \text{ GeV}}{f_a} \right). \tag{6}$$

A naive expectation, from, e.g., Grand Unified Theories, is that f_a is related to the unification scale $\sim 10^{16}$ GeV, which would make the expected mass fall in the sub-μeV range. It turns out, however, that such a light axion would in general overclose the Universe and thus is not viable [33]. There are some hypothetical mechnisms in string theory [39], however, that could make the mass scale smaller. It is also possible that the Peccei-Quinn symmetry breaks before inflation, in which case no strong fine-tuning is required to achieve a large f_a. The density of axions would then depend on a cosmic random number (a very small misalignment angle) and anthropic selection could be unavoidable (see, e.g., [40] and references therein). A possible signature of this mechanism would be primordial isocurvature fluctuations in the cosmic microwave background [40].

The experimentally important coupling to two photons is due to the effective Lagrangian term

$$\mathscr{L}_{a\gamma\gamma} = \left(\frac{\alpha_{\text{em}}}{2\pi f_a} \right) \kappa \left(\vec{E} \cdot \vec{B} \right) a, \tag{7}$$

where $\vec{E}$ is the electric field, $\vec{B}$ is the magnetic field and κ is a model-dependent parameter of order unity.

The axion has been gradually more and more constrained by laboratory searches, stellar cooling and the supernova dynamics to indeed be very light, $m_a < 0.01$ eV [41]. It then couples so weakly to other matter [42] that it never was in thermal equilibrium in the early universe and it would behave today as cold dark matter. There is an acceptable range between around 10^{-5} and 10^{-2} eV where axions pass all observational constraints and would not overclose the universe.

There is a considerable uncertainty in the relation between mass and relic density, depending on the several possible sources of axion production such as vacuum misalignment, emission from cosmic strings etc. For a recent discussion of the relic density of axions in various scenarios, see [43].

The coupling in Eq. (7) implies that resonant conversion between a galactic axion and an electric photon mode may take place in the presence of a strong magnetic field - not even the "invisible axion" may be undetectable [44], since the number density of these light particles in the Galaxy has to be enormous if axions are to make up the dark matter.

There are now a couple of experiments (for a recent review, see [45]) which have had the experimental sensitivity to probe, and so far rule out, only a tiny part of the interesting region. The expected potential of the significantly upgraded Livermore experiment ADMX [45], will allow a deep probe into the interesting mass window where axions are indeed a main fraction of dark matter.

There have recently been laboratory searches [46] for light axions emitted from processes in the Sun. Although not in a mass and coupling constant range directly relevant for dark matter, the exclusion region covered is quite impressive [46]. There could also be other interesting mechanisms, like axion-photon conversion, that could possibly influence cosmological measurements in interesting ways [47].

For the time being the axion remains undetected, but if it exists in the appropriate mass range it is still one of the prime candidates for the dark matter.

Inert Higgs

We now turn to massive particles, WIMPs, and start with one of the most minimal extensions of the Standard Model (for an earlier, even simpler one, see [48]). It was noted already in 1978 that a model with two Higgs doublets containing a discrete symmetry could contain a state, the lightest neutral scalar or pseudoscalar boson, which is stable [49]. Almost three decades later, the model reappeared [50] as a way to obtain improved naturalness with a Higgs that could be rather massive, larger than 300 GeV.

The possibility of one of the lighter neutral states in the enlarged Higgs sector to be the dark matter was also pointed out, and soon the basic properties of this "inert" Higgs candidate for dark matter were investigated [51]. It turns out that this model contains a dark matter candidate, a particle that does not couple directly to Standard Model fermions and is stable due to the discrete symmetry of the model (hence its relative inertness). Rates for indirect detection (i.e. the observation of products of pair annihilation in the halo [51], or in the Earth or Sun [52]) would then appear to be suppressed, unless its mass would be larger than the W mass. However, if it is just below the W mass, the virtual creation af a W pair which then converts to $\gamma\gamma$ or $Z\gamma$, would give rather spectacular rates for these observationally interesting line processes [53]. These would populate an energy region which is particularly favourable for detection in the Large Area Gamma-ray Telescope of the Fermi satellite [54]. The first results on dark matter searches from Fermi should appear soon, for estimates of its potential for dark matter detection made before launch, see [55].

The inert Higgs doublet model of dark matter is compatible with existing accelerator bounds [50, 56] and is an interesting, very minimal model with interesting phenomenology.

Neutrinos

The neutrino was a favoured particle dark matter candidate in the period starting in the end of the 1970's, with the first calculations of the relic density for massive neutrinos [57, 58, 59, 60]. Of the many candidates for non-baryonic dark matter proposed, neutrinos are often said to have the undisputed virtue of being known to exist. The direct mass limits from accelerators are not very useful for cosmological neutrinos, given the small mass differences measured in neutrino oscillation experiments. The only direct limit of relevance is the one on the electron neutrino, 2 eV [61], which taken together with mass differences inferred from neutrino observations can still allow 6 eV for the sum of neutrino masses. However, there are observational limits from cosmology on the mass range allowed for this sum, which are much more restrictive. From an analysis of the WMAP 5-year data [15] a bound derived on the allowed amount of a hot component translates to 0.63 eV for the sum of neutrino masses [62]. If one is willing to also trust modelling of the Ly-α forest, a significantly better limit can be obtained [63]. There has recently been progress in the rather difficult problem of treating the structure formation problem including light neutrinos in an accurate way [64].

Since the contribution to the dark matter density is

$$\Omega_\nu h^2 = \frac{\sum_i m_{\nu_i}}{94 \text{ eV}}, \tag{8}$$

we see that, given the cosmological bound of around 0.6 eV for the sum of neutrino masses, light neutrinos (which behave as hot, not cold, dark matter) is at most roughly one tenth of the cold dark matter.

There is a fundamental objection to having a massive but light neutrino (or in fact, any fermion or boson that was once in thermal equlibrium) as the dominant constituent of dark matter on all scales where it is observationally needed. This has to do the restrictions on density evolution given by Liouville's theorem. Quantitatively, Tremaine and Gunn found [65] that to explain the dark matter of a dwarf galaxy of velocity dispersion σ (usually of order 100 km/s) and core radius r_c (typically 1 kpc), the neutrino mass has to fulfill

$$m_\nu \geq 120 \text{ eV} \left(\frac{100 \text{ km/s}}{\sigma}\right)^{\frac{1}{4}} \left(\frac{1 \text{ kpc}}{r_c}\right). \tag{9}$$

Recently, this bound has been improved somewhat [66], and gives a lower limit of roughy 400 eV for the sterile neutrino mass.

Sterile neutrinos do not interact through standard weak interactions [67], but communicate with the rest of the neutrino sector through fermion mixing (for recent reviews of such models, see [68, 69]). They are limited by a variety of observational data [70], but it seems that, e.g, a region below 10 keV for mixing angles smaller than $\sin^2\theta \sim 10^{-10}$ is allowed. Again one has to impose some unknown tuning mechanism to match the WMAP data on relic density. A sterile neutrino of this mass would be warm dark matter, i.e., intermediate between hot and cold dark matter, and this would have some beneficial effects on some possible problems with the CDM scenario such as the absence of a predicted cusp in the central regions of some galaxies, or the lack of substructure in the form of dwarf galaxies bound to the Milky Way. Actually, the latter problem does not seem as serious now, as new data from the Sloan Digital Sky Survey and the Keck telescope have revealed a number of new, faint satellite galaxies [71, 72], and the mass dependence of the number distribution seems to agree well with CDM simulations [73].

In principle, one could also have had a cold dark matter standard model neutrino of mass around 3 GeV, but this window was closed long ago by the LEP experiment at CERN. The pioneering papers which worked out the dark matter phenomenology of such massive neutrinos (e.g., [57, 58, 59, 60]) were important, however, since they showed that a weakly interacting, massive particle ("WIMP") could serve as cold dark matter with the required relic density.

To conclude this section about neutrinos, it seems that it is very plausible that they make up some of the dark matter in the universe (given the experimental results on neutrino oscillations), but most of the dark matter is probably of some other form. Particle physics offers several other promising candidates for this.

Supersymmetry

Supersymmetry has since its invention [74] fascinated a generation of theoretical physicists, and motivated many experimentalists like those now involved in CERN's LHC project, likely to produce data on multi-TeV proton-proton collisions next year.

Supersymmetry is an ingredient in many superstring theories which attempt to unite all the fundamental forces of nature, including gravity. In most versions of the low-energy theory there is, to avoid, for example, excessive baryon number violating processes, a conserved multiplicative quantum number, R-parity:

$$R = (-1)^{3(B-L)+2S}, \tag{10}$$

where B is the baryon number, L the lepton number and S the spin of the particle. This implies that $R = +1$ for ordinary particles and $R = -1$ for supersymmetric particles. This means that supersymmetric particles can only be created or annihilated in pairs in reactions of ordinary particles. It also means that a single supersymmetric particle can only decay into final states containing an odd number of supersymmetric particles. In particular, this makes the lightest supersymmetric particle stable, since there is no kinematically allowed state with negative R-parity which it can decay to. In fact, this is similar to the discrete parity mentioned for the inert Higgs model. It seems that most (but not all) models for dark matter have to rely on a similar discrete symmetry (which in the simplest case is just a Z_2 symmetry), so maybe one should nowadays, when the possibility of explaining dark matter is one of the main motivations when constructing new particle physics models, generically introduce a multiplicative discrete "D-symmetry" (with D standing for Dark) with

$$D = 1 \quad \text{Standard Model Sector} \tag{11}$$

$$D = -1 \quad \text{New Particle Sector} \tag{12}$$

Since this is a multiplicative quantum number, it means that particles in the $D = -1$ sector can only be pair-annihilated or -produced, and the lightest particle with $D = -1$ is stable. If it is electrically neutral, it is then a dark matter candidate.

Thus, pair-produced neutralinos χ in the early universe which left thermal equilibrium as the universe kept expanding should have a non-zero relic abundance today. If the scale of supersymmetry breaking is related to that of electroweak breaking, χ will be a WIMP and Ω_χ will be of the right order of magnitude to explain the non-baryonic cold dark matter. It would indeed appear as an economic solution if two of the most outstanding problems in fundamental science, that of dark matter and that of the unification of the basic forces, would have a common element of solution - supersymmetry.

The idea that supersymmetric particles could be good dark matter candidates became attractive when it was realised that breaking of supersymmetry could be related to the electroweak scale, and that, e.g., the supersymmetric partner of the photon (the photino) would couple to fermions with electroweak strength [75]. Then most of the phenomenology would be similar to the (failed) attemps to have multi-GeV neutrinos as dark matter. After some early work along these lines [76, 77, 78, 79, 80], the first more complete discussion of the various possible supersymmetric candidates was provided in [81], where in particular the lightest neutralino was identified as perhaps the most promising one.

A disadvantage of a full supersymmetric model (even making the particle content minimal, the Minimal Supersymmetric Standard Model, MSSM) is that the number of free parameters is excessively large - of the order of 100. Therefore, most treatments have focused on constrained models, such as minimal supergravity (mSUGRA) models [82], where one has the opportunity to explain electroweak symmetry breaking by radiative corrections caused by running from a unification scale down to the electroweak scale (for a detailed analysis of dark matter in mSUGRA models, see [83]).

Supersymmetric particles

Let us now focus on the lightest supersymmetric particle, which if R-parity is conserved, should be stable. In some early work, a decaying photino [76] or a gravitino [77] were considered, but for various reasons [81] the most natural supersymmetric dark matter candidate was decided to be the lightest neutralino χ. In fact, especially the decaying gravitino option has recently been revived with considerable interest [84]. In view of the need for very large boost factors to explain the new PAMELA [9] and ATIC [10] data, decaying gravitino scenarios may provide an alternative [85].

Returning to the neutralino χ, it is a mixture of the supersymmetric partners of the photon, the Z and the two neutral CP-even Higgs bosons present in the minimal extension of the supersymmetric standard model (see, e.g., [86]). It has gauge couplings and a mass which for a large range of parameters in the supersymmetric sector imply a relic density in the required range to explain the observed $\Omega_M h^2 \sim 0.1$. As we will see, its couplings to ordinary matter also means that its existence as dark matter in our galaxy's halo may be experimentally tested. For an extensive review of the literature on supersymmetric dark matter up to mid-1995, see Ref. [3]. Some improvements were discussed in [4], and the most recent, rather full discussion can be found in [5, 8].

The phenomenology has not changed very much since these reviews. The neutralino remains a very promising candidate, with possibilities for discovery in direct detection [87] and in various channels of indirect detection [88].

Here we just point out two recent developments. It was noticed in [24] that the indirect process of annihilation to $\gamma\gamma$ and $Z\gamma$ [24], although often with too small branching ratios to be observable, has a remarkable behaviour as the annihilating particles are either electroweak doublets (pure higgsinos) or triplets (pure gauginos). Namely, the cross section tends to a constant value proportional to $1/m_W^2$ instead as $1/m_\chi^2$ as could be expected on dimensional grounds. This means that the unitarity limit [89]

$$\sigma_{\text{unitarity}} < \frac{4\pi}{v m_\chi^2} \tag{13}$$

will eventually be violated at very high masses. This led Hisano, Matsumoto and Nojiri [90] to investigate the behaviour of the amplitude near that limit. They discovered that including perturbatively higher order corrections, they would get a value slightly higher than that found in [24], but more importantly, unitarity was restored. A crucial step forward was then taken in [91] by non-perturbatively summing up in the ladder approximation to all orders the attractive t-channel exchange diagrams. The result is a zero-energy bound state for some particular dark matter masses and typical galactic velocities. The appearance of the bound state makes the cross section increase two to three orders of magnitude, compared to that when velicities were corresponding to the freeze-out temperature $T \sim m_\chi/20$ (corresponding to $v \sim 0.3 - 0.4$)

This phenomenon, thus discovered in [91], and verified in [92, 93, 94, 95] (see also [96, 97, 98]) gives the possibility of very strong indirect (in particular, γ-ray) signals for particular masses (usually in the TeV region). It is analogous to what happens for positronium near bound state thresholds, as originally discussed by Sommerfeld [99]. It is today the well-known "Sommerfeld enhancement" of the annihilation rate, and may be a generic phenomenon. Of course, supersymmetric TeV particles interacting through standard model gauge bosons may have difficulty to give the required relic density, unless one tolerates some fine-tuning, as is explicitly done in "split SUSY" models [100]. The Sommerfeld enhancement is today extensively discussed in connection with the surprising new results on the high positron flux at high energies, see later. If Sommerfeld enhancement is active for positrons, one would also expect large, perhaps detectable, signals in radio waves and gamma-rays [101, 102, 103]. Rather important bounds on the enhancement follow from the early structure formation and effects on the diffuse gamma-ray background or the cosmic microwave background, especially if there is no saturation of the effect at very small velocities [104].

Another example of a recent development of supersymmetric dark matter phenomenology (although its history goes back to the late 1980's [105]) is the change of helicity structure caused by QED radiative corrections to low-velocity annihilation.

This has recently proposed as a method to detect an otherwise undetectable leptonic dark matter candidate [106]. It has also been applied to MSSM and mSUGRA models, and found to be very important [107], causing sometimes large boosts to the highest energy end of the γ-ray spectrum. In particular, it has been shown to increase the potential for γ-ray detection from dwarf satellite galaxies [108].

A Majorana fermion (as many dark matter candidates are) suffers a helicity suppression for S-wave annihilation [109], such that the amplitude contains a factor of fermion mass m_f, meaning that, e.g., the e^+e^- final state is highly suppressed. However, by emitting a photon from an internal (t-channel) charged leg, which only costs a factor of α_{em}/π, the helicity suppression may be avoided. The effect will be that these radiative corrections, instead of as usual being a percent of the lowest order process, may instead give enhancement *factors* of several thousand to million times the suppressed lowest order, low-velocity, rate [105]. The resulting spectra will have a characteristic very sharp drop at the endpoint $E_\gamma = m_\chi$ of both the γ-ray and positron spectrum, see Fig. 2.

Kaluza-Klein Particles in Universal Extra Dimensions

The lightest KK particle (LKP) is an interesting, viable particle dark matter candidate arising from extra dimensional extensions of the standard model (SM) of particle physics. It appears in models of universal extra dimensions (UED) [110, 111] (for the first proposal of having TeV sized extra dimensions see [112]), where all SM fields propagate in the higher dimensional bulk, and is stable due to conserved KK parity, a remnant of KK mode number conservation. This again an example of a D-symmetry analogous to the R-parity of supersymmetry, meaning the the LKP is a WIMP. Contrary to the case of supersymmetry however, the unknown parameter space is quite small and will be scanned entirely by, e.g., LHC.

Consider the simplest, five dimensional model with one UED compactified on an S^1/Z_2 orbifold of radius R. All Standard Model fields are then accompanied by a tower of KK states; at tree-level, the nth KK mode mass is given by

$$m^{(n)} = \sqrt{(n/R)^2 + m_{\rm EW}^2}\,, \tag{14}$$

where $m_{\rm EW}$ is the corresponding zero mode mass. The LKP can be shown in the minimal treatment of radiative corrections to be the first KK mode $B^{(1)}$ of the weak hypercharge gauge boson. (For a non-minimal version, see [113].) The spectrum and the $B^{(1)}$ relic density were first computed in [111, 114]. Depending on the exact form of the mass spectrum and the resulting coannihilation channels, the limit from WMAP [15] of $\Omega_{\rm CDM}h^2 = 0.1131 \pm 0.0034$ corresponds to a mass of the dark matter candidate $B^{(1)}$ between roughly 0.5 to 1 TeV. Collider measurements of

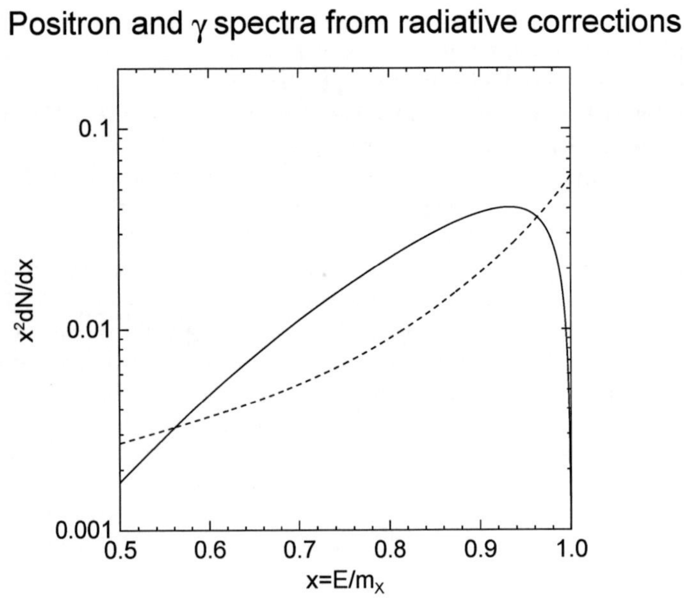

FIGURE 2. The γ-ray (solid) and positron (dashed) spectrum obtained by internal bremsstrahlung of the highly suppressed lowest order process $\chi\chi \to e^+e^-$ [105, 107]. Here $x = E_\gamma/m_\chi$ or E_{e^+}/m_χ, and the t-channel fermion, from which the radiation originates, is assumed to be 10 % heavier than the neutralino.

electroweak observables give a current constraint of $R^{-1} > 0.3$ TeV, whereas LHC should probe compactification radii up to 1.5 TeV (for a review of the detailed phenomenology of KK dark matter, see [115]).

The most interesting aspect of KK dark matter is that it provides an example of a spin-1 dark matter candidate. This means that the helicity structure of the matrix elements for annihilation will change, in particular the explicit factor of fermion mass that appears in the s-wave matrix element for slow Majorana particles annihilating in the halo is not present. This means that new interesting direct annihilation modes will appear, such as $\nu\bar{\nu}$, or e^+e^-, which are usually severely suppressed for neutralino annihilation, for example. The LKP was early recognized [116, 117] as a potentially important source of positrons, and was used to investigate the HEAT excess [118], which now has been superseded by the much more convincing PAMELA positron excess [9]. In fact, in the recent ATIC paper claiming evidence for a possible peak around 600 GeV perhaps related to dark matter [10], a KK model was shown as an example of a model that would give a good fit to the shape of the electron plus positron spectrum (although there the normalization was fitted to an arbitrary, high value). This model was revisited by Hooper and Zurek [119], who concluded that a boost factor of the order of several hundred would be needed, which may be difficult to explain using current models of the dark matter halo (especially as no Sommerfeld enhancement is expected for these models). They point out, however, that such a high boost factor need not necessarily conflict with other data. On the other hand, as we will see in the next Section, some doubts have been cast on the strength of the ATIC peak.

Models with Enhanced Annihilation Rate

We have already mentioned the dramatic change of focus that has taken place in the dark matter community, triggered by the new positron data. In the late summer of 2008, the first of these exciting new data were reported from the PAMELA satellite, designed to measure the content of antimatter in the cosmic rays. The results were first communicated at conferences, and later a paper was put on the arXiv and was later published in Nature [9]. An unexpectedly high ratio of positrons over electrons was measured, in particular in the region between 10 and 100 GeV, where previously only weak indications of an excess had been seen [120]. This new precision measurement of the cosmic ray positron flux, which definitely disagrees with a standard background [121] has opened up a whole new field of speculations about the possible cause of this positron excess. As mentioned, a similar, difficult to explain excess, a "bump", in the electron plus positron spectrum was reported by the balloon experiment ATIC [10]. Simultaneously, other data from PAMELA indicate that the antiproton flux is in agreement with standard expectations [122].

There are a variety of astrophysical models proposed for the needed extra primary component of positrons, mainly based on having nearby pulsars as a source [123]. Although pulsars with the required properties like distance, age, and energy output are known to exist, it turns out not to be trivial to fit both ATIC and PAMELA with these models (see, for example, [124, 125]). For this and other reasons, the dark matter interpretation, which already had been applied to the much more uncertain HEAT data [126] has been one of the leading hypotheses.

It was clear from the outset that to fit the PAMELA and ATIC positron data with a dark matter model a high mass is needed (reflecting the bump at around 600 GeV of ATIC). However, since the local average dark matter density is well-known to be around 0.3 - 0.4 GeV/cm^3, the number density decreases as $1/M_X$ and therefore the annihilation rate as $1/M_X^2$ with M_X the mass of the annihilating particle. This means that with $\langle\sigma v\rangle = 3\cdot 10^{-26}$ cm^3/s, which is the standard value of the annihilation rate in the halo for thermally produced WIMPs (see Eq. (2)), the rate of positrons, even for a contrived model which annihilates to e^+e^- with unit branching ratio is much too small to explain the measured result.

To a good approximation, the local electron plus positron flux for such a model is given by, assuming an energy loss of $10^{-16}E^2$ GeVs^{-1} (with E in GeV) from inverse Compton and synchrotron radiation,

$$E^3\frac{d\phi}{dE} = 6\cdot 10^{-4}E\left(\frac{1\ \text{TeV}}{M_X}\right)^2\theta(M_X - E)B_{\text{tot}}\ \text{m}^{-2}\text{s}^{-1}\text{sr}^{-1}\text{GeV}^2, \tag{15}$$

which means that the boost factor $B_{\text{tot}} \sim 200$ (see Eq.(5) for the definition of B_{tot}) for a 600 GeV particle, that may otherwise explain the ATIC bump. Similar boost factors seem to be generic, also for supersymmetric models giving e^+e^- through internal bremsstrahlung [127].

Returning to Eq. (5), we see that such a boost factor can be given by a large inhomogeneity which has to be very local, since positrons and electrons of several hundred GeV do not diffuse very far before losing essentially all their energy, see Fig. 3. Although not excluded [128] (see, however, [129]), this would seem to be extremely unlikely in most structure formation scenarios. Therefore, most models rely on the second factor in Eq. (5), i.e., the Sommerfeld enhancement factor. This means arguably also a non-negligible amount of fine-tuning of the mass spectrum, in particular also for the degeneracy between the lightest and next-to-lightest particle in the new sector. For a detailed discussion of the required model-builing, see [94]. Similar fine-tuning is needed for the decaying dark matter scenario, where the decay rate has to be precisely tuned to give the measured flux. Since the antiproton ratio seems to be normal according to the PAMELA measurements [122], the final states should be mainly leptons (with perhaps intermediate light new particles decaying into leptons). For an interesting such model, which may in fact contain an almost standard axion, see [130].

It seems that at present it is possible to construct models of the Sommerfeld enhanced type [131, 101, 132] which do marginally not contradict present data [102, 103, 133]. It seems, however, that a non-standard halo (not as cuspy as NFW), and carefully chosen annihilation channels have to be used. Decaying dark matter is also a possibility [134], with again some tuning of the mass and liftime necessary.

The recent Fermi satellite data [135], for the summed electron and positron spectrum up to one TeV shows some interesting features. The very high excess around 600 GeV seen by ATIC is not verified (although ATIC have presented new data [10] which support their previous findings). One the other hand, there still seems to be a milder "bump" in the spectrum (see Fig. 3), which may either be produced by dark matter annihilation into muons (the shape does not fit annihilation directly into electrons and positrons) [136] or, of course, may be caused by local astrophysical sources like pulsars [137]. It seems, however, that the non-observation of gamma-rays from galaxy clusters again puts contraints on the allowed amount of substructure [138].

In [136] it was found that an excellent fit could be obtained by considering a model of mass 1.6 TeV, annihilating into muon pairs. The troublesome fact is, however, that again a very high boost factor is needed. The energy loss of positrons or electrons created in the muon decays is largely unknown, but usually it is given by $\tau_0 = 10^{16}$ s/GeV^2. As the diffusion length is then quite small for TeV particles, only the local dark matter density ρ_0 is important (i.e., the production rate is largely independent of the shape of the halo). It is found that an enhancement factor of

$$E_F = \left(\frac{\rho_0}{0.3\ \text{GeV cm}^{-3}}\right)^2\left(\frac{\tau_0}{10^{16}\text{s GeV}^{-2}}\right)B_F, \tag{16}$$

is needed, where B_F is the boost factor from substructure and/or Sommerfeld enhancement. In [136] it was found that for the canonical values of ρ_0 and τ_0, an enhancement factor of $E_F = B_F = 1100$ is needed to fit that data with the excellent result shown in Fig. 3.

It will be interesting to see whether this puzzling situation will be resolved as more data are presented. Particularly important will be the launch in 2010 of the AMS experiment [139] to the International Space Station. AMS will have a

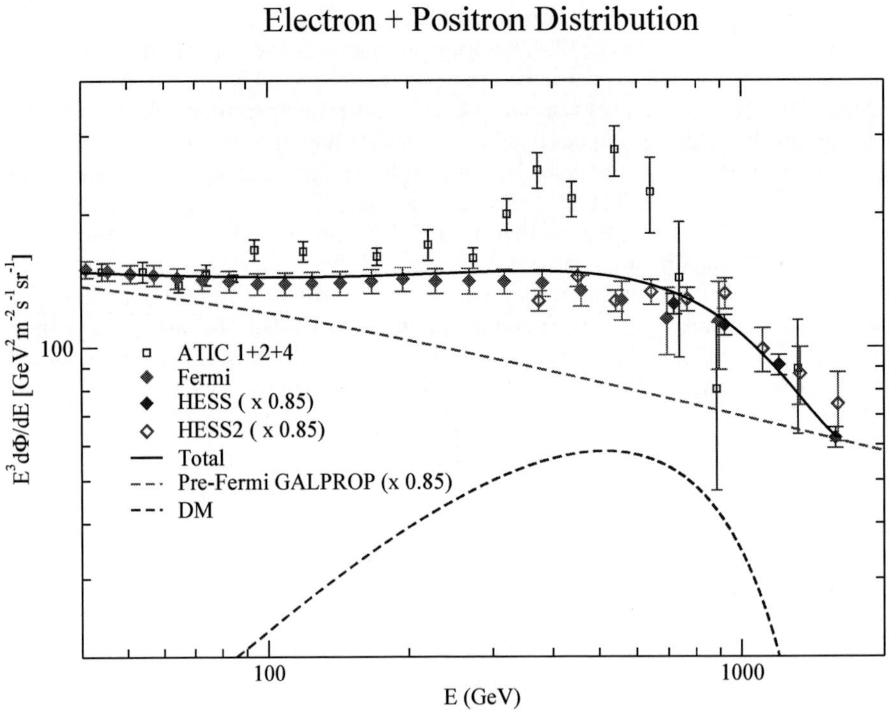

FIGURE 3. Compilation of the measured fluxes of electron plus positrons from Fermi [135], ATIC [10] and HESS [140], compared with a standard background curve [141], and a theoretical curve which has a massive dark matter candidate annihilating into muon pairs with large boost factor. See [136] for details.

superconductiong magnet, allowing particle identification up to 1 TeV. In particular, by combining the measurements of antiproton and positron fluxes, many of the presently proposed explanations for the observed antimatter excesses can be tested.

ACKNOWLEDGMENTS

This work was supported by the Swedish Research Council (VR). I wish to thank colleagues at the Oskar Klein Centre in Stockholm for stimulating discussions, and Jean-Michel Alimi for perfect Conference arrangements.

REFERENCES

1. F. Zwicky, Helv. Phys. Acta **6** (1933) 110.
2. M. Milgrom, these Proceedings.
3. G. Jungman, M. Kamionkowski and K. Griest, Phys. Rep. **267** (1996) 195.
4. L. Bergström, Rept. Prog. Phys. **63**, 793 (2000) [arXiv:hep-ph/0002126]; New J. Phys. **11**, 105006 (2009) [arXiv:0903.4849 [hep-ph]].
5. G. Bertone, D. Hooper and J. Silk, Phys. Rept. **405**, 279 (2005) [arXiv:hep-ph/0404175].
6. M. Kamionkowski, arXiv:0706.2986 [astro-ph].
7. L. Bergström, New J. Phys. **11**, 105006 (2009) [arXiv:0903.4849 [hep-ph]].
8. G. Bertone (*ed.*), *Particle Dark Matter*, Cambridge University Press, 2009.
9. O. Adriani *et al.* [PAMELA Collaboration], Nature **458**, 607 (2009) [arXiv:0810.4995 [astro-ph]].
10. J. Chang *et al.*, Nature **456** (2008) 362.
11. H.W. Babcock, Lick Observatory bulletin **498** (1939), 41.
12. P. J. E. Peebles, Astrophys. J. **263** (1982) L1.
13. T. Kikuchi and N. Okada, Phys. Lett. B **665**, 186 (2008) [arXiv:0711.1506 [hep-ph]].
14. A. Kusenko and M. E. Shaposhnikov, Phys. Lett. B **418**, 46 (1998) [arXiv:hep-ph/9709492].

15. E. Komatsu *et al.* [WMAP Collaboration], Astrophys. J. Suppl. **180**, 330 (2009) [arXiv:0803.0547 [astro-ph]].
16. P. Gondolo, J. Edsjö, P. Ullio, L. Bergström, M. Schelke and E. A. Baltz, JCAP **0407**, 008 (2004) [arXiv:astro-ph/0406204].
17. G. Belanger, F. Boudjema, A. Pukhov and A. Semenov, Comput. Phys. Commun. **174**, 577 (2006) [arXiv:hep-ph/0405253].
18. R. R. de Austri, R. Trotta and L. Roszkowski, JHEP **0605**, 002 (2006) [arXiv:hep-ph/0602028]; R. Trotta, F. Feroz, M. P. Hobson, L. Roszkowski and R. Ruiz de Austri, JHEP **0812**, 024 (2008) [arXiv:0809.3792 [hep-ph]]; Y. Akrami, P. Scott, J. Edsjö, J. Conrad and L. Bergström, arXiv:0910.3950 [hep-ph].
19. V. Springel *et al.*, Nature **456N7218** (2008) 73.
20. J. Diemand, M. Kuhlen and P. Madau, Astrophys. J. **657**, 262 (2007) [arXiv:astro-ph/0611370].
21. M. Kamionkowski and S. M. Koushiappas, Phys. Rev. D **77**, 103509 (2008) [arXiv:0801.3269 [astro-ph]].
22. T. Bruch, J. Read, L. Baudis and G. Lake, Astrophys. J. **696**, 920 (2009) [arXiv:0804.2896 [astro-ph]].
23. F. Halzen and D. Hooper, New J. Phys. **11** 105019.
24. L. Bergström, P. Ullio and J. H. Buckley, Astropart. Phys. **9**, 137 (1998) [arXiv:astro-ph/9712318].
25. M. R. Buckley, K. Freese, D. Hooper, D. Spolyar and H. Murayama, arXiv:0907.2385 [astro-ph.HE].
26. F. Aharonian *et al.* [The HESS Collaboration], Astron. Astrophys. **425**, L13 (2004) [arXiv:astro-ph/0408145].
27. E. A. Baltz *et al.*, JCAP **0807**, 013 (2008) [arXiv:0806.2911 [astro-ph]].
28. S. D. Hunter *et al.*, Astrophys. J. **481** (1997) 205.
29. L. Bergström, J. Edsjö and P. Ullio, Phys. Rev. Lett. **87**, 251301 (2001) [arXiv:astro-ph/0105048]; P. Ullio, L. Bergström, J. Edsjö and C. G. Lacey, Phys. Rev. D **66**, 123502 (2002) [arXiv:astro-ph/0207125].
30. S. Ando and E. Komatsu, Phys. Rev. D **73**, 023521 (2006) [arXiv:astro-ph/0512217]; A. Cuoco, J. Brandbyge, S. Hannestad, T. Haugboelle and G. Miele, Phys. Rev. D **77**, 123518 (2008) [arXiv:0710.4136 [astro-ph]];
31. T. Bringmann, L. Bergström and J. Edsjö, JHEP **0801**, 049 (2008) [arXiv:0710.3169 [hep-ph]].
32. W. de Boer, C. Sander, V. Zhukov, A. V. Gladyshev and D. I. Kazakov, Astron. Astrophys. **444**, 51 (2005) [arXiv:astro-ph/0508617].
33. M. P. Hertzberg, M. Tegmark and F. Wilczek, Phys. Rev. D **78**, 083507 (2008) [arXiv:0807.1726 [astro-ph]].
34. R. Peccei and H.R. Quinn, Phys. Rev. Lett. **38** (1977) 1440.
35. S. Weinberg, Phys. Rev. Lett. **40** (1978) 223.
36. F. Wilczek, Phys. Rev. Lett. **40** (1978) 279.
37. M. Dine, W. Fischler and M. Srednicki, Phys. Lett. **B104** (1981) 199; A.R. Zhitnitsky, Sov. J. Nucl. Phys. **31** (1980) 260.
38. J.E. Kim, Phys. Rev. Lett. **43** (1979) 103; M.A. Shifman, A.I. Vainshtein and V.I. Zakharov, Nucl. Phys. **B166** (1980) 493.
39. P. Svrcek and E. Witten, JHEP **0606**, 051 (2006) [arXiv:hep-th/0605206].
40. J. Hamann, S. Hannestad, G. G. Raffelt and Y. Y. Y. Wong, arXiv:0904.0647 [hep-ph].
41. H. Murayama, G. Raffelt, C. Hagmann, K. van Bibber, and L.J. Rosenberg, Eur. Phys. J. **C3** (1998) 264.
42. J.E. Kim, Phys. Rev. Lett. **43** (1979) 103; M.A. Shifman, A.I. Vainshtein and V.I. Zakharov, Nucl. Phys. **B166** (1980) 493; M. Dine, W. Fischler and M. Srednicki, Phys. Lett. **104B** (1981) 199.
43. L. Visinelli and P. Gondolo, arXiv:0903.4377 [astro-ph.CO].
44. P. Sikivie, Phys. Rev. Lett. **48** (1982) 1156.
45. S. J. Asztalos, L. J. Rosenberg, K. van Bibber, P. Sikivie and K. Zioutas, Ann. Rev. Nucl. Part. Sci. **56** (2006) 293.
46. K. Zioutas *et al.* [CAST Collaboration], Phys. Rev. Lett. **94**, 121301 (2005) [arXiv:hep-ex/0411033].
47. C. Csaki, N. Kaloper and J. Terning, Phys. Rev. Lett. **88**, 161302 (2002) [arXiv:hep-ph/0111311]; L. Östman and E. Mörtsell, JCAP **0502**, 005 (2005) [arXiv:astro-ph/0410501]; A. Mirizzi, G. G. Raffelt and P. D. Serpico, Lect. Notes Phys. **741**, 115 (2008) [arXiv:astro-ph/0607415].
48. H. Davoudiasl, R. Kitano, T. Li and H. Murayama, Phys. Lett. B **609**, 117 (2005) [arXiv:hep-ph/0405097].
49. N. G. Deshpande and E. Ma, Phys. Rev. D **18**, 2574 (1978); E. Ma, Phys. Rev. D **73** (2006) 077301 [hep-ph/0601225].
50. R. Barbieri, L. J. Hall and V. S. Rychkov, Phys. Rev. D **74** (2006) 015007 [hep-ph/0603188].
51. L. Lopez Honorez, E. Nezri, J. L. Oliver and M. H. G. Tytgat, JCAP **0702**, 028 (2007) [arXiv:hep-ph/0612275].
52. S. Andreas, M. H. G. Tytgat and Q. Swillens, JCAP **0904**, 004 (2009) [arXiv:0901.1750 [hep-ph]].
53. M. Gustafsson, E. Lundström, L. Bergström and J. Edsjö, Phys. Rev. Lett. **99**, 041301 (2007) [arXiv:astro-ph/0703512].
54. The Fermi-LAT collaboration, see e.g. W. B. Atwood *et al.* [LAT Collaboration], arXiv:0902.1089 [astro-ph.IM].
55. E. A. Baltz *et al.*, JCAP **0807**, 013 (2008) [arXiv:0806.2911 [astro-ph]].
56. E. Lundström, M. Gustafsson and J. Edsjö, Phys. Rev. D **79**, 035013 (2009) [arXiv:0810.3924 [hep-ph]].
57. P. Hut, Phys. Lett. **69B** (1977) 85.
58. B.W. Lee and S. Weinberg, Phys. Rev. Lett. **39** (1977) 165.
59. M.I. Vysotsky, A.D. Dolgov and Ya. B. Zel'dovich, JETP Lett. **26** (1977) 188.
60. J. E. Gunn, B.W. Lee, I. Lerche, D.N. Schramm and G. Steigman, Astrophys. J. **223** (1978) 1015.
61. C. Amsler et al., Phys. Lett. **B667**, 1 (2008).
62. S. Hannestad, A. Mirizzi, G. G. Raffelt and Y. Y. Y. Wong, JCAP **0804**, 019 (2008) [arXiv:0803.1585 [astro-ph]].
63. U. Seljak *et al.* [SDSS Collaboration], Phys. Rev. D **71**, 103515 (2005) [arXiv:astro-ph/0407372].
64. J. Brandbyge and S. Hannestad, arXiv:0812.3149 [astro-ph].
65. S. Tremaine and J.G. Gunn, Phys. Rev. Lett. **42** (1979) 407.
66. D. Gorbunov, A. Khmelnitsky and V. Rubakov, JCAP **0810**, 041 (2008) [arXiv:0808.3910 [hep-ph]]; A. Boyarsky, O. Ruchayskiy and D. Iakubovskyi, JCAP **0903**, 005 (2009) [arXiv:0808.3902 [hep-ph]].
67. S. Dodelson and L. M. Widrow, Phys. Rev. Lett. **72**, 17 (1994) [arXiv:hep-ph/9303287]; X. D. Shi and G. M. Fuller, Phys. Rev. Lett. **82**, 2832 (1999) [arXiv:astro-ph/9810076]; T. Asaka, M. Shaposhnikov and A. Kusenko, Phys. Lett. B **638**, 401

(2006) [arXiv:hep-ph/0602150].
68. K. N. Abazajian, arXiv:0903.2040 [astro-ph.CO].
69. A. Boyarsky, O. Ruchayskiy and M. Shaposhnikov, Ann. Rev. Nucl. Part. Sci. **59**, 191 (2009) [arXiv:0901.0011 [hep-ph]].
70. S. H. Hansen, J. Lesgourgues, S. Pastor and J. Silk, Mon. Not. Roy. Astron. Soc. **333**, 544 (2002) [arXiv:astro-ph/0106108]; K. Abazajian and S. M. Koushiappas, Phys. Rev. D **74**, 023527 (2006) [arXiv:astro-ph/0605271]; D. Boyanovsky, H. J. de Vega and N. Sanchez, Phys. Rev. D **77**, 043518 (2008) [arXiv:0710.5180 [astro-ph]]; A. Boyarsky, J. Lesgourgues, O. Ruchayskiy and M. Viel, Phys. Rev. Lett. **102**, 201304 (2009) [arXiv:0812.3256 [hep-ph]].
71. J. D. Simon and M. Geha, Astrophys. J. **670**, 313 (2007) [arXiv:0706.0516 [astro-ph]].
72. L. E. Strigari, S. M. Koushiappas, J. S. Bullock, M. Kaplinghat, J. D. Simon, M. Geha and B. Willman, arXiv:0709.1510 [astro-ph].
73. A. V. Maccio', X. Kang, F. Fontanot, R. S. Somerville, S. E. Koposov and P. Monaco, arXiv:0903.4681 [astro-ph.CO].
74. J. Wess and B. Zumino, Nucl. Phys. B **70**, 39 (1974).
75. P. Fayet, Phys. Lett. **86B** (1979) 272.
76. N. Cabibbo, G. Farrar and L. Maiani, Phys. Lett. **105B** (1981) 155.
77. H. Pagels and J.R. Primack, Phys. Rev. Lett. **48** (1982) 223.
78. S. Weinberg, Phys. Rev. Lett. **50** (1983) 387.
79. H. Goldberg, Phys. Rev. Lett. **50** (1983) 1419.
80. L.M. Krauss, Nucl. Phys. **B227** (1983) 556.
81. J. Ellis, J.S. Hagelin, D.V. Nanopoulos, K. Olive, M. Srednicki, Nucl. Phys. **B238** (1984) 453.
82. A. H. Chamseddine, R. L. Arnowitt and P. Nath, Phys. Rev. Lett. **49**, 970 (1982); R. Barbieri, S. Ferrara and C. A. Savoy, Phys. Lett. B **119**, 343 (1982); L. J. Hall, J. D. Lykken and S. Weinberg, Phys. Rev. D **27** (1983) 2359.
83. J. Edsjö, M. Schelke, P. Ullio and P. Gondolo, JCAP **0304**, 001 (2003) [arXiv:hep-ph/0301106]; H. Baer, E. K. Park and X. Tata, New J. Phys. **11**, 105024 (2009) [arXiv:0903.0555 [hep-ph]].
84. F. Takayama and M. Yamaguchi, Phys. Lett. B **485**, 388 (2000) [arXiv:hep-ph/0005214];
W. Buchmuller, M. Endo and T. Shindou, JHEP **0811**, 079 (2008) [arXiv:0809.4667 [hep-ph]].
85. K. Hamaguchi, F. Takahashi and T. T. Yanagida, arXiv:0901.2168 [hep-ph];
M. Endo and T. Shindou, arXiv:0903.1813 [hep-ph];
S. L. Chen, R. N. Mohapatra, S. Nussinov and Y. Zhang, arXiv:0903.2562 [hep-ph]; K. Ishiwata, S. Matsumoto and T. Moroi, arXiv:0903.3125 [hep-ph];
A. Ibarra and D. Tran, JCAP **0902**, 021 (2009) [arXiv:0811.1555 [hep-ph]].
86. H.E. Haber and G.L. Kane, Phys. Rep. **117** (1985) 75.
87. J. R. Ellis, K. A. Olive, Y. Santoso and V. C. Spanos, Phys. Rev. D **71**, 095007 (2005) [arXiv:hep-ph/0502001].
88. J. L. Feng, K. T. Matchev and F. Wilczek, Phys. Rev. D **63**, 045024 (2001) [arXiv:astro-ph/0008115].
89. M. Jacob and G. C. Wick, Annals Phys. **7** (1959) 404 [Annals Phys. **281** (2000) 774].
90. J. Hisano, S. Matsumoto and M. M. Nojiri, Phys. Rev. D **67**, 075014 (2003) [arXiv:hep-ph/0212022].
91. J. Hisano, S. Matsumoto and M. M. Nojiri, Phys. Rev. Lett. **92**, 031303 (2004) [arXiv:hep-ph/0307216].
92. M. Cirelli, N. Fornengo and A. Strumia, Nucl. Phys. B **753**, 178 (2006) [arXiv:hep-ph/0512090].
93. M. Cirelli, A. Strumia and M. Tamburini, Nucl. Phys. B **787**, 152 (2007) [arXiv:0706.4071 [hep-ph]].
94. N. Arkani-Hamed, D. P. Finkbeiner, T. R. Slatyer and N. Weiner, Phys. Rev. D **79**, 015014 (2009) [arXiv:0810.0713 [hep-ph]].
95. M. Lattanzi and J. I. Silk, arXiv:0812.0360 [astro-ph].
96. J. March-Russell, S. M. West, D. Cumberbatch and D. Hooper, JHEP **0807**, 058 (2008) [arXiv:0801.3440 [hep-ph]].
97. J. D. March-Russell and S. M. West, arXiv:0812.0559 [astro-ph].
98. T. Hambye, F. S. Ling, L. L. Honorez and J. Rocher, arXiv:0903.4010 [hep-ph].
99. A. Sommerfeld, Annalen der Physik **403**, 257 (1931).
100. G. F. Giudice and A. Romanino, Nucl. Phys. B **699**, 65 (2004) [Erratum-ibid. B **706**, 65 (2005)] [arXiv:hep-ph/0406088].
101. I. Cholis, G. Dobler, D. P. Finkbeiner, L. Goodenough and N. Weiner, arXiv:0811.3641 [astro-ph].
102. G. Bertone, M. Cirelli, A. Strumia and M. Taoso, arXiv:0811.3744 [astro-ph].
103. L. Bergström, G. Bertone, T. Bringmann, J. Edsjö and M. Taoso, arXiv:0812.3895 [astro-ph].
104. M. Kamionkowski and S. Profumo, Phys. Rev. Lett. **101**, 261301 (2008) [arXiv:0810.3233 [astro-ph]].
105. L. Bergström, Phys. Lett. B **225**, 372 (1989).
106. E. A. Baltz and L. Bergström, Phys. Rev. D **67**, 043516 (2003) [arXiv:hep-ph/0211325].
107. T. Bringmann, L. Bergström and J. Edsjö, JHEP **0801**, 049 (2008) [arXiv:0710.3169 [hep-ph]].
108. T. Bringmann, M. Doro and M. Fornasa, JCAP **0901**, 016 (2009) [arXiv:0809.2269 [astro-ph]].
109. H. Goldberg, Phys. Rev. Lett. **50**, 1419 (1983).
110. T. Appelquist, H. C. Cheng, and B. A. Dobrescu, Phys. Rev. D **64**, 035002 (2001), [hep-ph/0012100].
111. H. C. Cheng, K. T. Matchev and M. Schmaltz, Phys. Rev. D **66**, 036005 (2002), [hep-ph/0204342].
112. I. Antoniadis, Phys. Lett. B **246**, 377 (1990).
113. T. Flacke, A. Menon and D. J. Phalen, arXiv:0811.1598 [hep-ph].
114. G. Servant and T. M. P. Tait, Nucl. Phys. B **650**, 391 (2003), [hep-ph/0206071].
115. D. Hooper and S. Profumo, Phys. Rept. **453**, 29 (2007) [arXiv:hep-ph/0701197].
116. H. C. Cheng, J. L. Feng and K. T. Matchev, Phys. Rev. Lett. **89** (2002) 211301 [hep-ph/0207125].
117. D. Hooper and G. D. Kribs, Phys. Rev. D **70**, 115004 (2004) [arXiv:hep-ph/0406026]; D. Hooper and J. Silk, Phys. Rev. D **71**, 083503 (2005) [arXiv:hep-ph/0409104].

118. S.W. Barwick *et al.*, Phys. Rev. Lett. **75**, 390-393 (1995).
119. D. Hooper and K. Zurek, arXiv:0902.0593 [hep-ph].
120. D. Müller and K. K. Tang, Astrophys. J. **312**, 183-194 (1987); R.L. Golden *et al.*,Astrophys. J. **457**, L103-L106 (1996); H. Gast, J. Olzem, and S. Schael, Proc. XLIst Rencontres de Moriond, Electroweak Interactions and Unified Theories, 421-428 (2006); S.W. Barwick *et al.*, Astrophys. J. **482**, L191-194 (1997); J.J. Beatty, *et al.*, Phys. Rev. Lett. **93**, 241102-241105 (2004).
121. I. V. Moskalenko and A. W. Strong, Astrophys. J. **493**, 694 (1998) [arXiv:astro-ph/9710124].
122. O. Adriani *et al.*, Phys. Rev. Lett. **102**, 051101 (2009) [arXiv:0810.4994 [astro-ph]].
123. F.A. Aharonian, A.M. Atoyan and H.J. Völk, Astron. Astrophys. **294** L41 (1995); L. Zhang, and K.S. Cheng, Astron. Astrophys. **368**, 1063-1070 (2001); I. Büsching, O.C. de Jager, M.S. Potgieter and C. Venter, Astrophys. J. **78**, L39-L42 (2008).
124. S. Profumo, arXiv:0812.4457 [astro-ph].
125. D. Malyshev, I. Cholis and J. Gelfand, Phys. Rev. D **80**, 063005 (2009) [arXiv:0903.1310 [astro-ph.HE]].
126. E. A. Baltz and J. Edsjö, Phys. Rev. D **59**, 023511 (1999) [arXiv:astro-ph/9808243]; E. A. Baltz, J. Edsjö, K. Freese and P. Gondolo, Phys. Rev. D **65**, 063511 (2002) [arXiv:astro-ph/0109318]; G. L. Kane, L. T. Wang and T. T. Wang, Phys. Lett. B **536**, 263 (2002) [arXiv:hep-ph/0202156]; H. C. Cheng, J. L. Feng and K. T. Matchev, Phys. Rev. Lett. **89** (2002) 211301 [hep-ph/0207125]; P. Brun, G. Bertone, J. Lavalle, P. Salati and R. Taillet, Phys. Rev. D **76**, 083506 (2007) [arXiv:0704.2543 [astro-ph]].
127. L. Bergström, T. Bringmann and J. Edsjö, Phys. Rev. D **78**, 103520 (2008) [arXiv:0808.3725 [astro-ph]].
128. D. Hooper, A. Stebbins and K. M. Zurek, arXiv:0812.3202 [hep-ph].
129. T. Bringmann, J. Lavalle and P. Salati, arXiv:0902.3665 [astro-ph.CO].
130. Y. Nomura and J. Thaler, Phys. Rev. D **79**, 075008 (2009) [arXiv:0810.5397 [hep-ph]].
131. M. Cirelli, M. Kadastik, M. Raidal and A. Strumia, Nucl. Phys. B **813**, 1 (2009) [arXiv:0809.2409 [hep-ph]].
132. E. Borriello, A. Cuoco and G. Miele, arXiv:0903.1852 [astro-ph.GA].
133. J. Hisano, M. Kawasaki, K. Kohri, T. Moroi and K. Nakayama, Phys. Rev. D **79**, 083522 (2009) [arXiv:0901.3582 [hep-ph]].
134. E. Nardi, F. Sannino and A. Strumia, JCAP **0901**, 043 (2009) [arXiv:0811.4153 [hep-ph]]; A. Arvanitaki, S. Dimopoulos, S. Dubovsky, P. W. Graham, R. Harnik and S. Rajendran, Phys. Rev. D **80**, 055011 (2009) [arXiv:0904.2789 [hep-ph]].
135. A. A. Abdo *et al.* [The Fermi LAT Collaboration], Phys. Rev. Lett. **102**, 181101 (2009) [arXiv:0905.0025 [astro-ph.HE]].
136. L. Bergström, J. Edsjö and G. Zaharijas, Phys. Rev. Lett. **103**, 031103 (2009) [arXiv:0905.0333 [astro-ph.HE]]; L. Bergström, talk given at Rencontres de Blois, 2009, to appear in the Proceedings.
137. D. Grasso *et al.* [FERMI-LAT Collaboration], Astropart. Phys. **32**, 140 (2009) [arXiv:0905.0636 [astro-ph.HE]].
138. A. Pinzke, C. Pfrommer and L. Bergström, Phys. Rev. Lett. **103**, 181302 (2009) [arXiv:0905.1948 [astro-ph.HE]].
139. S. P. Ahlen *et al.*, Nucl. Instrum. Meth. A **350**, 351 (1994).
140. F. Aharonian *et al.* [H.E.S.S. Collaboration], Phys. Rev. Lett. **101**, 261104 (2008) [arXiv:0811.3894 [astro-ph]]; H. E. S. Aharonian, arXiv:0905.0105 [astro-ph.HE].
141. A. W. Strong, I. V. Moskalenko and O. Reimer, Astrophys. J. **613**, 962 (2004) [arXiv:astro-ph/0406254].
142. J. M. Siegal-Gaskins, JCAP **0810**, 040 (2008) [arXiv:0807.1328 [astro-ph]].
143. A. Arvanitaki, S. Dimopoulos, S. Dubovsky, P. W. Graham, R. Harnik and S. Rajendran, Phys. Rev. D **80**, 055011 (2009) [arXiv:0904.2789 [hep-ph]]; A. Ibarra and D. Tran, JCAP **0902**, 021 (2009) [arXiv:0811.1555 [hep-ph]].
144. P. Meade, M. Papucci, A. Strumia and T. Volansky, arXiv:0905.0480 [hep-ph].

Dark Matter (mostly Direct) Detection

G. Chardin

CSNSM, Université Paris-Sud, Orsay, France and CNRS/IN2P3

Abstract. After briefly summarizing the motivations for the Λ-CDM model, I will argue rather provocatively that the many candidates already reported by indirect detection experiments are a strong indication that it is nearly impossible to distinguish a WIMP annihilation signal from astrophysical background using the indirect detection method. This structural uncertainty argues strongly in favor of direct detection of WIMP interactions in the controlled environment of the laboratory and I will describe the main strategies developed to detect WIMPs using the direct detection technique in underground sites, protected from the cosmic-ray induced background. I will show that most of the present experimental efforts is now focused on three identification techniques of nuclear recoils, i.e. charge-phonon, light-phonon and charge-scintillation simultaneous detection. Finally, since direct detection has also its own candidate, I will comment on the DAMA annual modulation candidate, and on the lessons that can be learned from the long controversy over this result.

Keywords: Dark Matter, WIMP detection, cryogenic detectors, scintillation detectors, semiconductor detectors
PACS: 07.57.Kp; 07.85.Nc; 29.40.Mc; 61.82.Fk; 72.20.Jv; 95.35.+d

MOTIVATIONS

Over the last ten years, a number of different and apparently remarkably convergent experiments have evidenced a very surprising composition of the universe, with ordinary nuclear matter only representing 4% of the universe energy content, while the remaining 96% appear to be constituted of Dark Matter and Dark Energy. A non dissipative matter component, Dark Matter, represents approximately 23% of the universe content, and nearly 85% of its matter content. Dark Energy, on the other hand, appears to represent approximately 73% of the Universe energy content under the form of a mysterious and rather featureless vacuum component. The resulting universe is named the Concordance or Λ-CDM model.

Both the particle, astroparticle physics and astrophysics communities define among their top priorities the determination of the universe composition. Although a number of alternative solutions to the Dark Matter and Dark Energy problems have been described in this conference, I will here assume that the Λ-CDM (Cold Dark Matter) description in terms of its three main components is correct. Truly enough, a number of CDM candidates can be considered, and most notably the stable remnants of supersymmetric particles if a multiplicative conserved quantum number such as R-parity imposes the stability of the lightest supersymmetric particle (LSP). The success of WIMP as a Dark Matter candidate is related to the so-called "WIMP miracle", the fascinating coincidence that, for a relatively large mass range, the interaction cross-sections producing a remnant dark matter density $\Omega_m \sim 0.23$, as observed, are characteristic of electroweak interactions. In this sense, particle physics provides a natural explanation for the Cold Dark Matter problem —but no convincing explanation for the Λ component. On the other hand, we must admit that there are more than 100 free parameters in supersymmetry, and that the constraints on the mass of supersymmetric particles are mostly phenomenological, with the LEP accelerator data imposing rather severe constraints on the lower bound of SUSY particle masses of ≈ 35 GeV under the Minimal Supersymmetric Standard Model (MSSM) hypothesis.

In the following, apart from a critical discussion of the indirect detection method and of the DAMA annual modulation candidate, I will therefore review the experimental efforts and progress towards WIMP detection.

DIRECT OR INDIRECT DETECTION ?

I will argue rather provocatively that it appears impossible for indirect detection to ascertain the existence of a WIMP signal. Paradoxically, this is due to the large number of candidates already claimed by several groups using the indirect detection technique. In effect, the observation of Dark Matter signals have been claimed by a number of authors using data from the INTEGRAL [1], CACTUS [2], HEAT [3], CANGAROO [4], PAMELA [5, 6], ATIC [7], EGRET [8]

CP1241, *Invisible Universe: Proceedings of the Conference*, edited by J.-M. Alimi and A. Füzfa and FERMI [9] experiments. Event with this rather impressive list, I am certainly missing several indirect detection WIMP candidates and a few hundred references, and I apologize for my lack of completeness.

The mass range of these tentative WIMP candidates is wide, from MeV energies for the INTEGRAL signal, to several tens of GeV for the EGRET WIMP candidate, and even up to a few TeV for the CANGAROO candidate. Given the variety and number of these claims, it is difficult to believe that all of them are really associated to WIMP annihilation signals. It probably means on the other hand that uncertainties, notably in the estimate of the gamma-ray background of conventional astrophysical sources, lead to a corresponding uncertainty in the flux that can be attributed, or not, to a WIMP candidate. In particular, we know that diffuse gamma-ray emission is produced by interactions of cosmic rays on interstellar gas, but the *prediction* of its flux is uncertain, if at all possible, since we are not able to probe other than indirectly astrophysical objects as complex as the Galactic Center.

Another important source of uncertainty, and of possible adjustment between observations and models of WIMP annihilation, comes from the boost factor, related to the fact that the annihilation rate of Majorana particles depends on the square of the density and not on the density itself. Since the density profile of the WIMP accumulation at the center of a galaxy is largely unknown and a subject of some controversy, it seems again doubtful that we can predict a WIMP annihilation signal without an important degree of uncertainty.

There is one method, on the other hand, where indirect detection could bring a smoking gun of a WIMP annihilation signal: energetic neutrinos pointing from the core of the Sun would indeed provide a very convincing signature for a Dark Matter candidate. Detectors such as Superkamiokande [10] have already established some constraints by the non observation of such events. Exploring a large fraction of the SUSY parameter space, on the other hand, would require truly enormous detectors, beyond the next generation of proton decay detectors at the megaton scale. Still, this method provides a convincing signature, and is complementary to the direct detection technique by its better sensitivity to axial (spin-dependent) rather than scalar couplings.

DIRECT DETECTION

The principle of direct detection is very simple, if very difficult to implement in practice due to the very small expected number of WIMP interactions with ordinary matter for a large fraction of SUSY models. Supposing WIMPs are virialized in the potential well their host galaxies, their typical velocity is of the order of 10^{-3} c in a galaxy such as our Milky Way. For WIMPs with mass between a few tens of GeV to a few hundreds of GeV, as considered by most SUSY models compatible with accelerator constraints, the recoil energy associated with a collision between a WIMP and an electron leads to energies in the electron-volt range, much below the energy threshold of massive detectors.

On the other hand, it is simple to check that, since the mismatch between the mass of WIMP and of a nucleus (typically a few GeV to a few tens of GeV) is much reduced compared to the WIMP-electron case, WIMP interactions involving a nuclear recoil will lead to recoil energies in the few keV to a few tens of keV range, which can be detected by several techniques. For this reason, with the possible exception of the DAMA/LIBRA experiment [13, 14], using the annual modulation signature of events detected in sodium iodide detectors, nearly all present WIMP direct detection experiments are now attempting to discriminate nuclear recoils among the overwhelming background of gamma-rays and beta-rays interactions.

In the following, we will describe the three main techniques —charge-phonon, light-phonon, charge-scintillation— presently used by leading WIMP direct detection experiments in order to reach the level of sensitivity required to probe a significant fraction of SUSY models.

Charge-phonon detectors

Two experiments, CDMS [11] and EDELWEISS [12], use the simultaneous detection of the charge and phonon signals associated with a particle interaction as a powerful method to discriminate the tiny signal of WIMP-induced nuclear recoils from the radioactive background. Apart from a few exotic semiconductors, this technique is therefore limited to silicon and germanium nuclear targets. These two materials benefit from the huge purification effort realized in semiconductor industry. Since it is easier to realize large germanium detectors than silicon detectors with a very low level of uncompensated charged impurities, and since the expected number of WIMP interactions per kilogram of nuclear target is about an order of magnitude larger for germanium than for silicon —when scalar interactions are present at a decent level— most of the efforts of CDMS and EDELWEISS have focused on germanium detectors.

FIGURE 1. Photograph of a Z-sensitive ionisation-phonon (ZIP) detector of the CDMS experiment. The photolithographically-fabricated thin film on the surface is the phonon sensor, trapping non-equilibrium phonons in Al fins. The aluminum fins are coupled to tungsten sensors in the superconductor-normal transition. Silicon and germanium ZIPs, weighing 100 g and 250 g respectively, are presently operated in the Soudan (Minnesota) underground laboratory

The main disadvantages of the cryogenic germanium technique are its complexity and its cost: a tonne-scale Dark Matter experiment using germanium will probably have a cost somewhere between 50 Meuro and 100 Meuro, significantly larger than the cost advertized for a tonne-scale argon or xenon direct detection experiment. But the quality of their identification of nuclear recoils and their impressive energy resolution —sub-keV at the energies of interest— still make it a detector of choice for the identification of WIMP interactions. Indeed, for particle interactions occurring in the volume of the detector, i.e. at a distance larger than a few tens of microns from the surface of the detector, the background rejection factor of electron recoils compared to nuclear recoils appears nearly infinite.

For surface interactions, on the other hand, electrons and holes created during a particle interaction may diffuse and be trapped by the surface electrode before they are able to drift across the detector. Since the charge signal is very nearly proportional to the drift length in the detector, a trapped charge carrier will not contribute to the charge signal. Recall that the charge-phonon discrimination technique is based on the fact that a nuclear recoil produces approximately three times less ionization than an electron recoil of the same energy. Using a simple diffusion model, we can estimate (and we measure experimentally) [15] that, for any particle interaction occurring very close to the surface, more than 50% of the charge will be trapped. Radioactivity interactions occurring close to the surface may therefore produce events extremely difficult to distinguish from nuclear recoil interactions. For this reason, most of the development effort of the CDMS and EDELWEISS experiments has focused on solving the challenge posed by surface interactions.

Despite their apparent similarities, the CDMS and EDELWEISS experiments use rather different detectors, CDMS collecting the phonon signal using athermal phonons, while EDELWEISS measures the phonon signal using nearly thermal phonons, and have also rather different charge collection schemes. I describe the specific characters of these two experiments in the following.

The CDMS experiment

At the moment of writing, the most sensitive WIMP direct detection experiment is CDMS (Cryogenic Dark Matter Search) (if we limit ourselves to SUSY models compatible with accelerator constraints). Like its competitors, the CDMS experiment is protected from the background induced by cosmic-ray interactions in an underground laboratory, in this case an iron mine in Minnesota (USA).

The CDMS detectors present several innovative features. The phonon sensor used in the CDMS detectors uses the steep variation in impedance of a tungsten line in the normal to superconductor transition (TES, or Transition Edge Sensor) [16, 17]. Since the heat capacity of such an element at the transition temperature is of the order of a metal at the same temperature, it is impossible to cover the surface of the detector with such TES sensors without creating a catastrophic heat capacity. The CDMS experiment has therefore used the phonon collection scheme developed by Booth [18], where thin aluminum films are used to collect phonons with an energy sufficient to break the superconductivity of aluminum Cooper pairs, i.e. with an energy larger than about 2×10^{-4} eV. Thousands of such coupled tungsten-aluminum elements constitute a CDMS phonon sensor, read by SQUID-based electronics, and covering a large fraction of the detector surface (see Fig. 1). The sharing of the phonon signal among four quadrant sensors provides a measurement of the transverse X-Y position of the interaction with a precision of the order of one to a few millimeters. The charge readout, on the other hand, is using conventional FET-based electronics, with a sensitivity to interactions down to a few keV for both charge and phonon channels.

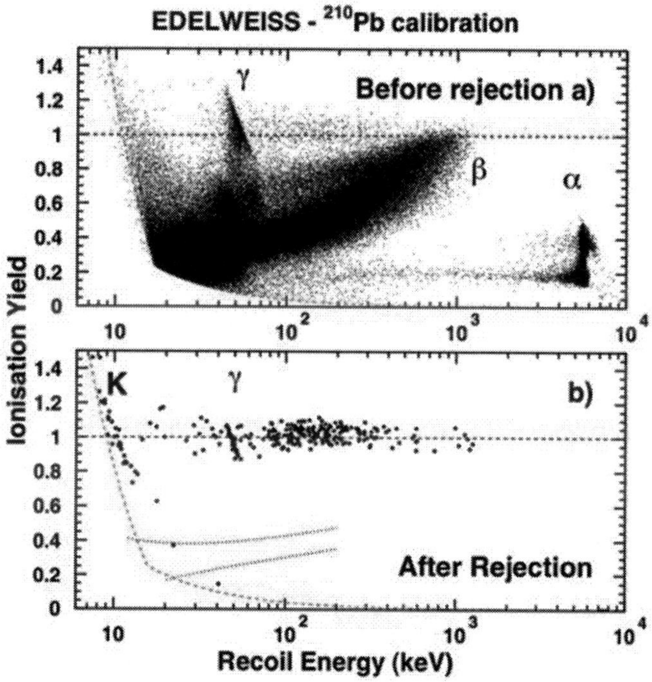

FIGURE 2. Scatter plots of the ionisation yield versus the recoil energy recorded in a 200 g InterDigit detector of the EDELWEISS experiment exposed to surface events generated by a ^{210}Pb sources, (a) before and (b) after applying the cuts to reject near-surface events. The ionisation yield is normalised to unity for electron recoil events with full charge collection. The total number of events in (a) is 185,000, out of which 50,000 are due to the 5.3 MeV alphas (labelled (α)) and 11,000 are due to the 46 keV gammas from ^{210}Pb (labelled (γ)). After the cuts, most of the remaining events originate from the gamma background of the experiment and the 10 keV (K) line from cosmogenic ^{68}Ge and ^{65}Zn. 90% of the nuclear recoils are expected between the two dotted lines in panel (b) (after Ref. [25]).

After initial attempts that only partially alleviated the problem of incomplete charge collection by using amorphous underlayers of silicon or germanium sputtered on the electrode surface before the metal deposition [15, 21, 22], a powerful discrimination method allowing to identify a large fraction of the surface events was found by CDMS using the relative timing information between the very fast charge and the slower phonon signals, measured with a precision of a few microseconds. Although the surface event identification is not perfect, the level of rejection can be adjusted in a blind analysis using very large statistics gamma-ray and beta-ray calibrations so that the number of leakage events misidentified as nuclear recoils is of the order of one event for the low-background data taking under analysis, leading to a nearly optimal rejection.

This background rejection scheme has been used successfully by the CDMS experiment, leading to the best present sensitivity to WIMP interactions, at least for the region of parameter space compatible with accelerator constraints. The main problem of this technique is that its complexity resulted in a consolidated cost exceeding 10^6 dollars per kilogram

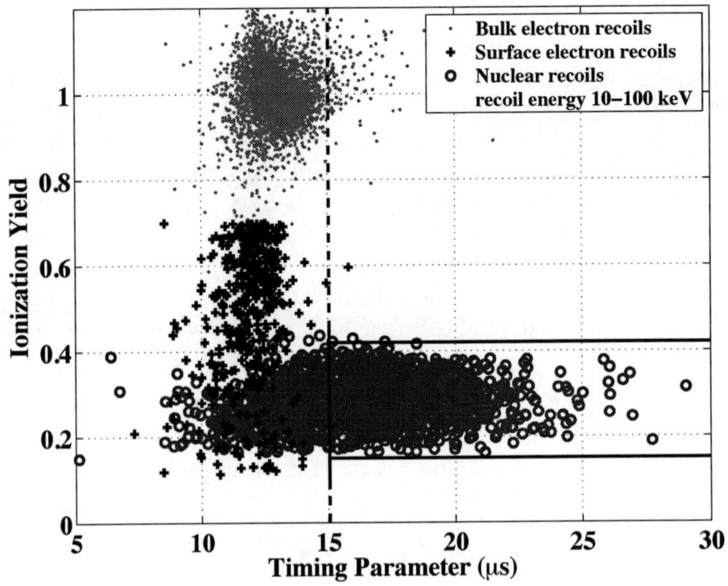

FIGURE 3. Discrimination method using the relative timing between the ionization and the phonon signals recorded in the CDMS ZIP detectors. Ionization yield versus timing parameter for calibration data for bulk-electron recoils (red dots ; upper part of the figure) ; surface electron recoils (crosses (+)) ; neutron-induced nuclear recoils from 252Cf source (o) (Color online.) (after Ref. [11])

for the present generation of detectors. In order to solve this crucial problem for the realization of a tonne-scale experiment, the CDMS experiment has recently greatly improved the reliability of its fabrication process and reduced the number of adjustment steps of the transition temperature of its tungsten TES sensors, biased at a temperature selected of the order of $\approx$ 70 mK, while the bulk transition temperature of tungsten is of the order of 10 mK.

A cost assessment of a large scale experiment based on the CDMS technology, together with a demonstration that surface interactions can be rejected without a strong loss in efficiency, are the two challenges faced at present by CDMS. This cost assessment will have to take into account the InterDigit detectors developed by the EDELWEISS experiment (see below), which can be produced at a lower cost than the present CDMS detectors, since it is quite probable that at most one large-scale germanium experiment will be funded in the forseeable future.

The EDELWEISS experiment

While the first generation of EDELWEISS detectors had initially no specific rejection scheme for surface interactions and was rapidly limited by this background [12], apart from the amorphous underlayers reducing this problem, a powerful identification method has been developed by EDELWEISS using interdigitized electrodes, along the idea of Frisch's grid [23] first tested on a cryogenic detector by Brink [24].

The impressive identification and rejection of surface interactions by the EDELWEISS InterDigit detector, of the order of 10^5 (see Fig. 2), probably constitutes a major step towards the definition of a tonne-scale Dark Matter detector, with zero background [25]. Developments realized along the same line in CDMS to combine the ZIP structure with interleaved electrodes have recently led to extremely encouraging results [26]. Recently, EDELWEISS further improved the design of its interdigitized detector [27], resulting in the FID (Full InterDigit detector), of mass 800 gram and with a fiducial volume exceeding 80% of the detector (Fig. 4). With its ease of fabrication, its relatively low cost and the quality of its background identification and rejection, this detector appears as a solution of choice for the next generation of Dark Matter experiments. First results of the EDELWEISS-II experiment using these detectors are expected in the near future.

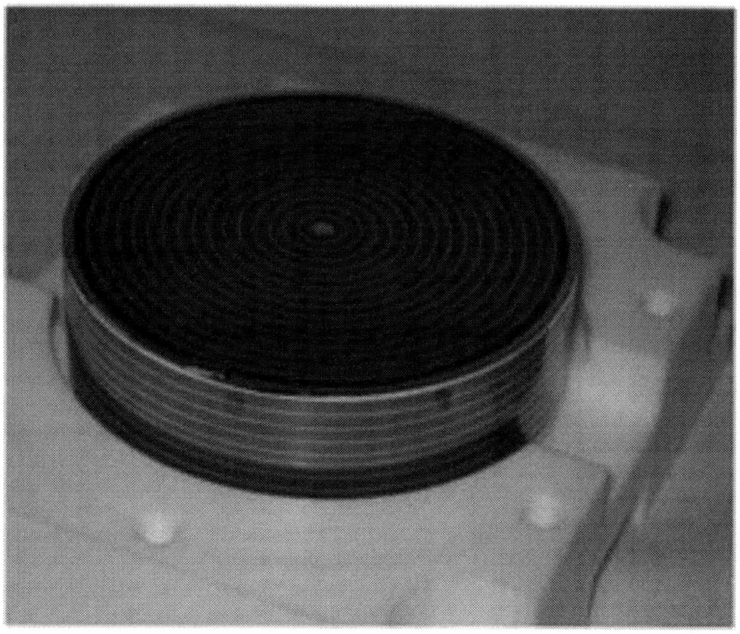

FIGURE 4. Photograph of a Full InterDigitized (FID) detector of the EDELWEISS experiment. The set of interdigitized electrodes covers the whole surface of the detector, leading to a fiducial volume exceeding 80% of the detector. This FID detector presents a level of background rejection performances similar to the initial InterDigit detector (see also Fig. 2 above) (after Ref. [27])

Scintillation-phonon detectors

The simultaneous detection of scintillation and phonons is another powerful technique allowing to discriminate events due to radioactivity from WIMP-induced nuclear recoils. This technique has been developed by the CRESST [29] and ROSEBUD [19] collaborations, joining their efforts in the EURECA project [30]. At millikelvin temperatures, it is not possible to operate photomultipliers to detect scintillation, since the dissipation would be far too high (typically, the total dissipation of the whole set of detectors must remain below a few microwatt for the base temperature to remain below 10 mK, as requested by the CRESST detectors). Instead, a very thin calorimeter such as a silicon or germanium wafer coupled to a thermistor is used to measure by a calorimetric measurement the amount of light emerging from the massive target. Fig. 5 shows an actual detector of the CRESST experiment involving a massive target of 300 gram of $CaWO_4$, with a light detector as a thin silicon wafer (on the left of the photograph), placed above the crystal in running conditions.

The separation between electron and nuclear recoils is excellent above 20 keV of recoil energy and the scintillation-phonon technique does not appear to suffer from surface effects. It might then seem that this technique, with a variety of nuclear targets at its disposal, would be the detector of choice for WIMP detection. What are then the limitations of this technique ?

The main problem of the scintillation-phonon discrimination scheme is related to the relatively small light yield produced in a particle interaction: in effect, only between 2 and 3 percent of the initial energy of an electron recoil is converted in scintillation for the best known scintillators at low temperatures. This fraction is further reduced by the quenching factor, since nuclear recoils are usually highly inefficient to produce scintillation, compared to electrons of the same energy. For example, in a calcium tungstate $CaWO_4$ crystal, taking as reference the light emission of an electron recoil, the scintillation yield for a recoil of the same energy is about 5 times smaller for an alpha particle, about 10 times smaller for an oxygen recoil, and 40 times smaller for a tungsten recoil. This has for consequence that the scintillation produced by a tungsten recoil in $CaWO_4$, one of the best scintillators at low temperatures, only represent 0.05 % of the recoil energy of a tungsten nucleus. This means that unless the light detector has an energy threshold of a few electron-volts, most of the tungsten recoils will have a scintillation signal compatible with noise. This is perfectly fine if it is possible to guarantee that there are no ordinary radioactivity or dissipation events with no associated production of light. If this is not the case, any such event will be indistinguishable from a WIMP interaction.

FIGURE 5. Photograph of a light-phonon detector of the CRESST experiment. The phonon signal is determined by the temperature variation of a transparent 300 gram $CaWO_4$ crystal, with its light detector a 30 mm diameter $\times$ 0.4 mm silicon wafer, both detectors using TES tungsten thermometers. The CRESST experiment is operated in the Gran Sasso underground laboratory (Italy) (after Ref. [28])

It should be also noted that even in the total absence of dissipative events —i.e. with production of phonons (heat) but no light— there exists a natural source of background that produces events extremely difficult to distinguish from WIMP-induced nuclear recoils. These events are due to heavy nuclear recoils associated with alpha radioactivity, for example a polonium nucleus emitted back to back an undetected alpha particle escaping the detector. To identify such events, the CRESST detector holders have been covered with a scintillating material emitting light when traversed by an alpha particle. In this sense, light-phonon detectors also suffer from a crucial surface event problem.

Fig. 6 shows the events recorded during the commissioning run of the CRESST-II experiment. Two events are observed in the window of interest (delimited by the full red line and the two vertical dashed lines) where a large fraction of tungsten nuclear recoils are expected. Although the limitation described in the previous paragraph may make it difficult to reach sensitivities much below the present CRESST sensitivity unless a way is found to distinguish the vast majority of heavy nuclear recoils from spurious phonon events, the scintillation phonon technique must be pursued during at least a few years since it is the only technique benefiting from a large variety of nuclear targets.

When a first clear WIMP nuclear recoil signal will be identified, or if supersymmetry is discovered at LHC, the pressure will be high to determine the mass and interaction cross-section characteristics of the stable WIMP remnants, and the scintillation-phonon technique will then be pushed to its limits to contribute to the determination of the WIMP parameters.

Liquid/gas noble detectors

Probably the most serious contender to discriminating cryogenic detectors for WIMP detection is the charge-scintillation technique, using noble gas/liquid targets. Xenon, due to its high nuclear mass and high scintillation efficiency, is the presently preferred target, but argon, with its very low cost, can be also considered as a strong outsider. Most of these detectors are operated in a double phase TPC (Time Projection Chamber) mode, although there are some attempts to use the simpler experimental approach of a single-phase detector.

At present, the best sensitivities have been obtained by double phase (gas/liquid) TPCs. The principle of operation is common to argon and xenon gas-liquid TPCs. The initial scintillation produced in an interaction in the liquid target is detected by photomultipliers, usually placed inside the liquid in order to maximize the light collection efficiency and to decrease the energy threshold. The charge associated with the interaction is then drifted inside the liquid using an axial electric field in a cylindrical geometry. The purity of the liquid, in order to recover most of the charge signal and obtain an homogeneous detector response, is of course a crucial factor. While charge drift over distances of several meters has been achieved in liquid argon, notably in the ICARUS experiment [31], the challenge appears more difficult for xenon,

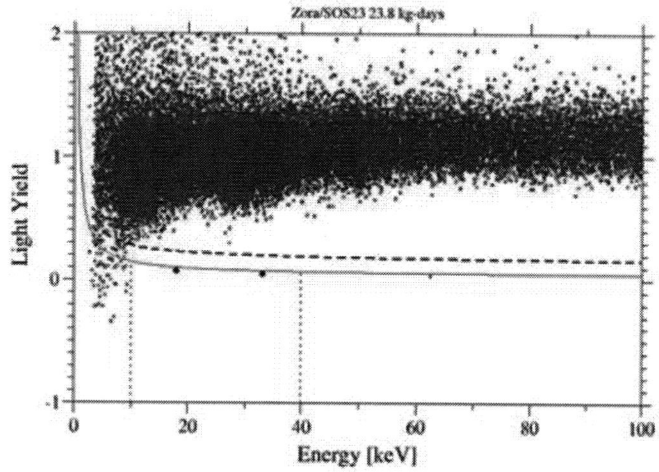

FIGURE 6. Projection in the (E_{Recoil}, Q) plane of the events recorded in a 320 g $CaWO_4$ light-phonon prototype detector of the CRESST experiment during the low-background data taking of the commissioning run. Two events are observed at the border of the signal nuclear recoil zone, between 10 and 40 keV recoil energy (after Ref. [29])

where the temperature —and the difficulty to reduce the level of impurities— is higher and the charge drift length has until now been limited to a few tens of centimeters. When arriving at the surface, the charge signal is extracted from the liquid surface and multiplied in the gas phase using a strong electric field. The secondary scintillation emitted in the process is used as a measure of the ionization produced in the interaction. The identification of nuclear recoils and their separation from natural radioactivity rests then on the primary scintillation, charge and position parameters reconstructed for the interaction.

Xenon TPCs

Liquid xenon, with its high atomic mass number, appears as a target of choice for scalar (spin-independent) couplings. In addition, xenon benefits also from isotopes with non-zero spin, and can also be used to probe models with axial couplings. The XENON-10 [32] and ZEPLIN-III [33] experiments have already obtained very significant results, within typically a factor two of the sensitivity reached by the CDMS experiment for models compatible with accelerator constraints (see Fig. 10 for a comparison of experimental sensitivities).

Figure 7 shows the events recorded by the XENON-10 experiment [32], combining the discrimination based on the charge/scintillation ratio, as measured by the ratio of the primary to secondary scintillation, and complemented by a fiducial cut on the interactions selected. It can be seen on this figure that background rejection, although excellent ($\approx 99.8\%$), is probably not total. It is then necessary to combine the charge/scintillation information with the position information and to define a fiducial volume where a few nuclear recoil candidates are observed. Note in particular that Figure 7 presents events *below* the nuclear recoil zone. A similar analysis has been achieved by the ZEPLIN III experiment.

For both these experiments, candidate events are observed and the question of their origin is raised. Should they be considered as real WIMP candidates, or are they expected from the limitations of the present stage of the experiments ? To answer this question, ZEPLIN-III plans a new stage with photomultipliers with much reduced uranium contamination and associated neutron background, while XENON has started in 2009 its 100-kg phase, with effective physics data taking expected to start somewhere during winter 2009-2010. The expected sensitivity of the XENON-100 experiment is 2×10^{-9} picobarn, assuming no candidate events are observed after one year of operation. The success of this 100-kg stage is therefore essential to define the best strategy for the next generation of tonne-scale Dark Matter experiments, with the XENON-1ton and the LUX projects in line.

Problems faced by this new ambitious generation of direct detection experiments include the uranium contamination of materials surrounding the liquid xenon target, and particularly photomultipliers, and the internal radioactivity due

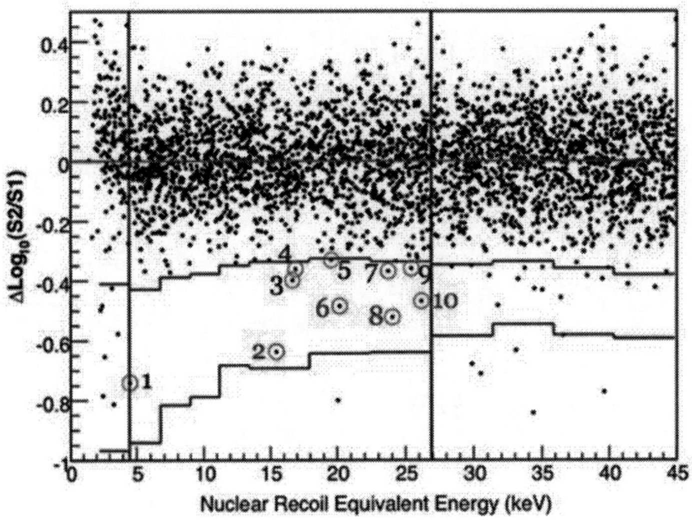

FIGURE 7. Results from 58.6 live-days of WIMP-search in the 5.4 kg liquid Xe target of the XENON-10 experiment. The WIMP search window is defined by the two vertical lines (4.5 to 26.9 keV nuclear recoil equivalent energy) and the blue lines (about 50% nuclear recoil acceptance) (after Ref. [32]). Note the logarithmic scale of the vertical axis, where the ratio of secondary to primary scintillation is plotted.

to the small fraction of krypton still present in xenon, which can be removed by distillation. With their relatively low cost per unit target mass, at least when compared to cryogenic detectors, liquid xenon experiments will probably be a priority for the first tonne-scale direct detection experiment if the results of the 100-kg stage follow the expectations and are not limited by background.

An important caveat concerns the quenching factor measurements for xenon recoils, which are necessary to define the absolute energy scale for recoil energies, and still suffer at present from large uncertainties at low energy (a few keV) of interest for Dark Matter searches. The most precise measurements have been realized in the context of the LUX [43] and the XENON [42] experiments, but these two measurements differ significantly from one another. The previous indications of very high values of the quenching factor (≈ 0.65) reported by the DAMA group [44] now appears completely off scale by a factor > 3 from the previous two measurements. The initial report of a quenching factor increasing at low energies by the ZEPLIN collaboration [45] appears similarly contradicted by the previous two measurements and by the ZEPLIN-III analysis.

Argon TPCs

Double phase Time Projection Chambers (TPCs) using argon as a target follow an experimental scheme very similar to that of xenon TPCs. The WArP experiment [34] has already obtained some results, while the WArP-140 and ArDM experiments are in preparation, which should involve fiducial target mass in the 100-kg and the tonne range respectively.

Being present in the atmosphere at the level of 1%, argon benefits from a very low cost per kilogram, even when the quantity is considered in tens or even hundreds of tons. Extensive developments by the ICARUS collaboration have shown that charge drift lengths of several meters, and TPCs of several hundred tons can be realized and operated over long periods of time.

Comparing the discrimination strategies of argon and xenon TPCs, an important difference comes from the scintillation time constants for electron and nuclear recoil interactions, associated with the relative time difference lifetimes of singlet and triplet states. With a much better separation in argon (≈ 7 ns for the singlet state versus $\approx 1.6\ \mu$s for the triplet state) compared to xenon, this leads to an excellent discrimination by the temporal analysis of the primary scintillation. On the other hand, argon has a cosmogenically produced isotope, ^{39}Ar, which cannot be separated easily, apart from a rather expensive isotopic separation, from the main and stable ^{40}Ar isotope. Although present in a

FIGURE 8. Artist view of the 100 liter prototype of the WArP experiment, using argon TPC immersed in a veto of the same liquid (after Ref. [34]).

relatively small proportion, the presence of ^{39}Ar, combined with its lifetime of 269 years, has limited until now the sensitivity of the WArP experiment.

A second stage of the WArP experiment, using a fiducial argon mass in the 100-kg range, is presently starting with WArP-140, but the main challenge is probably not so much the increase in mass, but the reduction in the internal argon radioactivity. The WArP collaboration has found an interesting alternative to isotopic separation, that should allow an important progress in sensitivity. Argon from oil wells is now used, where cosmogenic production of ^{39}Ar is reduced by more than two orders of magnitude compared to atmospheric argon. At present, the separation of this depleted argon is realized at a slow rate of approximately 0.5 kg per day by the Princeton group, while WArP-140 will require several hundred kilograms since its active veto uses the same argon container as its fiducial target. Still, once this problem is solved, argon TPCs may become a very dangerous competitor. The success of the WArP-140 stage, and the demonstration that it is not background limited, will be essential to validate the experimental strategy for the argon target.

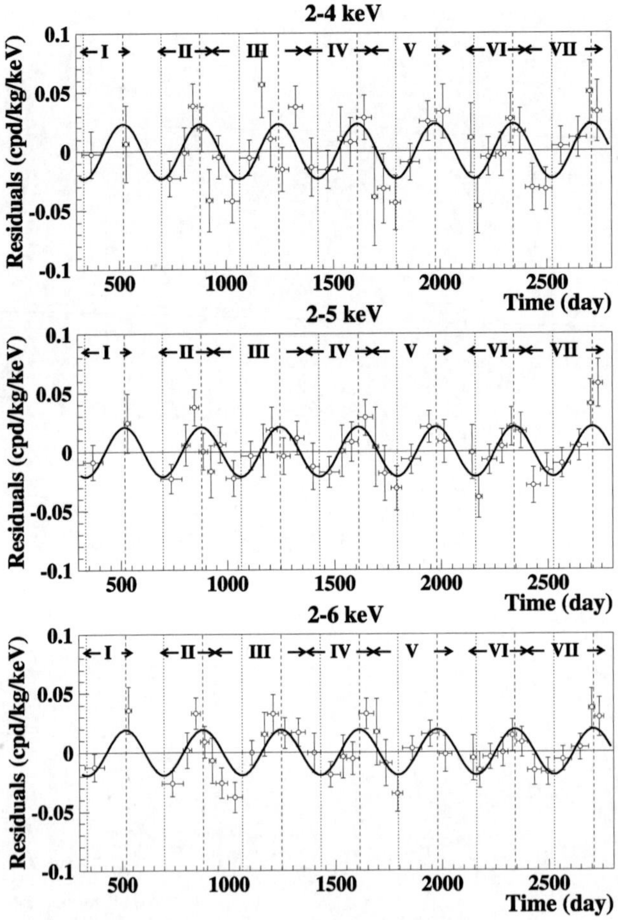

FIGURE 9. Annual modulation amplitude reported by the DAMA experiment using a $\sim 10^5$ kg $\times$ day data sample, recorded over a period of seven years. The statistical significance of the detected modulation is ~ 6.2 standard deviations (from Bernabei et al. Ref. [14])

Single phase noble gas/liquid detectors

In addition to the double phase TPCs, simpler single-phase detectors, such as XMASS [36], using liquid xenon as a target, and DEAP/CLEAN [37], using liquid argon or liquid neon, have been proposed as WIMP detectors with potentially high sensitivity. Since the charge generated in the interaction is not measured by these detectors, the necessary background rejection must rely on the timing information using both the difference in scintillation time constants between electron and nuclear recoil interactions, and the selection of a fiducial volume using the timing information from the photomultipliers covering the surface of the detector. An outstanding purity of the liquid target is a prerequisite for such experiments. Initial attempts by XMASS using a 100-kg detector have not led to competitive sensitivities. Future stages, such as XMASS-800 [36], will determine the competitiveness of such a approach.

The DAMA candidate

After having argued that indirect detection is unable to demonstrate convincingly the existence of a WIMP signal by the very multiplicity of its WIMP candidates, it is necessary to admit that the direct detection technique has also a —presently unique— WIMP candidate, forcefully defended by the DAMA/LIBRA experiment, which has used the annual modulation technique to discriminate a tentative WIMP signal from the radioactive background.

In 2000, the DAMA group published an analysis involving a 160 kg $\times$ year data sample recorded over three full annual cycles [13]. In 2003, the group published the analysis of three additional annual cycles [14], followed by the subsequent analysis of an additional data sample collected by the LIBRA detector, with a total mass of ≈ 250 kg of sodium iodide crystals. Taken at face value, the DAMA observation now presents a $> 8\sigma$ statistical significance, with both phase and amplitude consistent over a period of more than seven years with a WIMP signature (Fig. 9). Assuming standard halo parameters [38] and interpreted in terms of a WIMP candidate, the annual modulation would correspond to a WIMP mass $\sim (52 \pm 10)$ GeV and WIMP–nucleon cross–section $\sim (7 \pm 1)10^{-6}$ picobarn.

Under the assumption of spin-independent interactions, it is extremely difficult, if not impossible, to reconcile the DAMA result, interpreted as a WIMP signal, with the experimental results of the CDMS [11], XENON-10 [32] and KIMS [39] experiments. The DAMA group has contested this contradiction, invoking the uncertainty in the WIMP halo parameters. But if we suppose that the potential WIMP signal is due to interactions with iodine nuclei, the DAMA result is then in direct contradiction with the result from the KIMS experiment [39], using also the iodine nuclear target in CsI crystals. The results of an improved version of the KIMS experiment in the Jeombong underground laboratory (Korea), with still lower background, are awaited, but from the published sensitivity of KIMS, it seems excluded that the DAMA result could be attributed to iodine nuclear recoils.

Using different nuclear targets (germanium and xenon), the sensitivity reached by the CDMS and XENON experiments cannot be considered to be in direct contradiction with the DAMA/LIBRA result, as exotic interaction models (e.g. , interactions on sodium nuclei instead of iodine, low energy WIMPs trapped in the Solar system, channelling, etc.) can be invoked to reconcile the apparently conflicting experimental results. But it should be noted that the sensitivity reached by the CDMS and XENON experiments are of the order of 10^{-2} event/kg/day, two orders of magnitude smaller than the event rate associated with the DAMA/LIBRA signal if it is due to sodium recoils.

It appears therefore extremely probable that the DAMA/LIBRA annual modulation signal is not due to a WIMP signal but to a spurious annual modulation effect, although its cause is not identified. Another problem related to the DAMA annual modulation candidate is that, despite rather severe inconsistencies in the original presentation of the DAMA results [40, 41], a blind analysis or open data policy has never been followed by the DAMA collaboration. I therefore fully agree with the recommendations made by John Bahcall in his summary talk of the 2006 TAUP conference in Seattle:

"DAMA sees a modulation at 6.3 σ" (today $> 8\sigma$, i.e. statistical certainty).

- "Potentially, this is extremely important.
- Existing experiments cannot check this result directly.
- Therefore,
  - Appoint blue-ribbon committee with subpoena power
  - If no mistakes found, repeat experiment but better."

This advice has not been followed by the Italian scientific authorities, despite the recommendations in this direction by the Gran Sasso scientific committee. To reduce the possibility of such long controversies in the future, it is proposed, and accepted by the main direct detection experiments, to follow an open data policy, recommended by the ILIAS and ASPERA astroparticle european coordination. This means that data should be made public after typically one year of private analysis, as in the astrophysics community.

CONCLUSIONS

The agreement between the cosmological observations and the Λ-CDM model is impressive, but the necessary fine-tuning required by the astonishing composition of this Λ-CDM Universe should remind us that there is very probably something fundamentally wrong in this description.

If we use nevertheless the working hypothesis that WIMPs do exist and represent the largest part of Dark Matter, the leading direct detection experiments are now reaching sensitivities that will test the bulk of benchmark models within two years (with sensitivities to WIMP-nucleon cross-sections of 10^{-8} picobarn and below).

At present, there is a strong competition between cryogenic detectors and noble liquid/gas targets. Concerning cryogenic detectors, germanium detectors have at present the best sensitivities, and are still leading the race over the preferred range of parameters for supersymmetric (SUSY) models. On the noble gas/liquid side, liquid xenon, due to its high nuclear mass and low energy threshold, is presently the preferred target, with argon as an outsider, while neon appears more as a control target, where the number of WIMP interactions is expected to be very low.

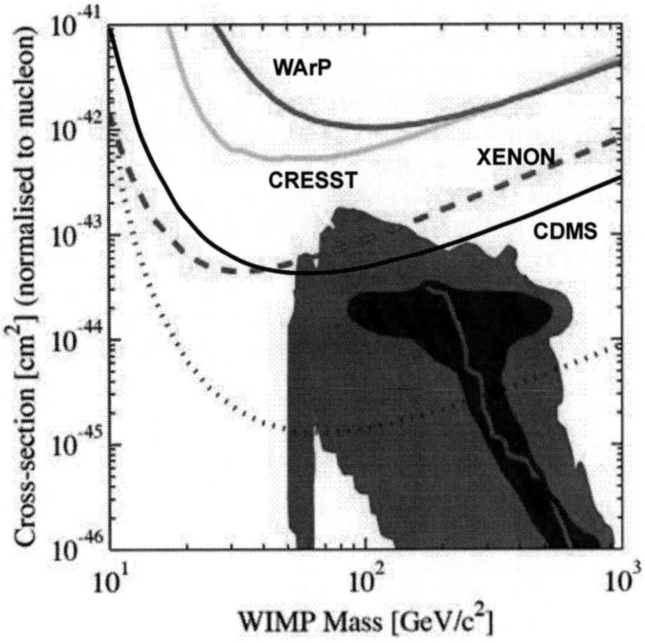

FIGURE 10. Experimental sensitivities of some of the main WIMP direct detection experiments (adapted from DMTools [35]). The CDMS experiment (full black line) presents the best sensitivity for models compatible with accelerator constraints, at the level of $\approx 5\times10^{-44}$ cm^2 or 5×10^{-8} picobarn for a WIMP mass of 100 GeV, followed by the XENON-10 and ZEPLIN-III experiments at about the same level (for clarity only the XENON result is presented). The CRESST, WArP and EDELWEISS experiments have rather similar sensitivities at the level of 5×10^{-7} picobarn. The expected sensitivity of the present generation of direct detection experiments is represented by the dotted line at the level of $\approx 1-2\times10^{-45}$ cm^2 at maximum sensitivity. Regions allowed by various SUSY models are represented by the green and blue regions, with preferred WIMP-nucleon cross-sections in the region of $\sim 10^{-44}$ cm^2 or below.

The second generation of noble gas/liquid experiments are presently starting taking data with active targets in the 100-kg range, with XENON-100 and WArP-140 as the most advanced contenders. With stages in the 10-50 kg range, cryogenic WIMP detection experiments are therefore trying to validate their strategies and to justify their higher cost per kilogram of active target. CDMS and EDELWEISS have made important progress towards the elimination of surface events, which appeared until now as the limiting background of an otherwise extremely powerful technique to identify nuclear recoils induced by WIMPs among the huge number of radioactive background interactions. CRESST is pursuing the light-phonon approach, and expects to reach a level of sensitivity of $\approx 10^{-8}$ picobarn in 2010. But the progress of this technique is threatened by the high suppression ("quenching") factor of light emission for heavy nuclei recoils, leading to a large fraction of recoils with zero detected light.

In this respect, it is important to note that testing a large fraction of SUSY models will almost necessarily require an active target mass in the tonne range. For cryogenic detectors, which are not readily expandable to very high mass, this represents a significant challenge, with an apparently easier upscalability for gas/liquid noble targets towards the tonne scale.

But a significant detection and identification of WIMP interactions will almost necessarily require a fair number of well identified nuclear recoils on at least two different nuclear targets. For this reason, germanium and xenon target nuclei appear at present as the leading contenders, with argon and light-phonon detectors as outsiders.

Finally, once the mass and interaction parameters for a WIMP candidate are measured, we will want to demonstrate the galactic origin and anisotropy of the WIMP signal using the daily (and not annual) modulation using a low density directional detector, as proposed by in the MIMAC or the CYGNUS projects [46].

ACKNOWLEDGMENTS

It is a pleasure to acknowledge stimulating discussions with E. Aprile, B. Cabrera, C. Montanari, F. Proebst, B. Sadoulet, D. Santos and the members of the EDELWEISS collaboration. Needless to say, these people are not responsible for the errors and omissions contained in this paper.

REFERENCES

1. C. Boehm, D. Hooper, J. Silk, M. Casse and J. Paul, *Phys. Rev. Lett.* **92**, 101301 (2004)
2. S. Profumo and M. Kamionkowski, *J. Cosmol. Astropart. Phys.* (JCAP) **0603** (2006) 003
3. E. A. Baltz, J. Edsjo, K. Freese and P. Gondolo, *Phys. Rev.* D **65**, 063511 (2002)
4. K. Tsuchiya, R. Enomoto, L.T. Ksenofontov, M. Mori, T. Naito, CANGAROO-II collaboration, *Astrophys. J.* **606** (2004) L115-L118
5. O. Adriani, *et al.*, *Phys. Rev. Lett.* **102** (2009) 051101 and arXiv:0810.4994
6. O. Adriani, *et al.*, *Nature* **458** (2009) 607 and arXiv:0810.4995
7. J. Chang, *et al.*, *Nature* **456** (2008) 362
8. W. de Boer, C. Sander, V. Zhukov, A.V. Gladyshev, D.I. Kazakov, *Phys. Rev. Lett.* **95** (2005) 209001 ; arXiv:astro-ph/0602325
9. L. Goodenough and D. Hooper, arXiv:0910.2998
10. A. Habig, for the Super-Kamiokande Collaboration, hep-ex/0106024.
11. Z. Ahmed *et al.*, *Phys. Rev. Lett.* **102**, 011301 (2009).
12. V. Sanglard, *et al.*, *Phys. Rev.* D **71**, 122002 (2005).
13. R. Bernabei *et al.*, *Phys. Lett.* **B480**, 23 (2000)
14. R. Bernabei *et al.*, *Riv. Nuovo Cim.* **26** (2003) 1 ; astro-ph/0307403
15. T. Shutt, *et al.*, *Nucl. Instr. Meth.* A **444**, 340 (2000).
16. B. Cabrera, *et al.*, *Appl. Phys. Lett.* **73**, 735 (1998).
17. K.D. Irwin, *et al.*, *Appl. Phys. Lett.* **66**, 1998 (1995).
18. N. Booth, *Appl. Phys. Lett.* **50**, 293 (1987).
19. S. Cebrian, *et al.*, *Astropart. Phys.* **15**, 79 (2001); astro-ph/0004292.
20. J. Llacer, E.E. Haller, and R.C. Cordi, *IEEE Trans. Nucl. Sci.*, **24**, 53 (1977).
21. P.N. Luke, C.P. Cork, N.W. Madden, C.S. Rossington, and M.F. Wesela, *IEEE Trans. Nucl. Sci.* **39**, 590 (1992).
22. P.N. Luke, C.S Rossington, M.F. Wesela, *IEEE Trans. Nucl. Sci.* **41**, 1074 (1994).
23. O. Frisch, *British Atomic Energy Report*, BR-49 (1944).
24. P.L. Brink, *et al.*, *Nucl. Instr. and Meth.* A **559**, 4148 (2006).
25. A. Broniatowski, *et al.*, *Phys. Lett.* B 681 (2009), 305 ; arXiv:0905.0753.
26. M. Pyle, *et al.*, in *Proceedings 13th International Workshop on Low Temperature Detectors*, (Stanford, CA, USA, 2009), American Institute of Physics Conference Proceedings, in press.
27. S. Marnieros, *et al.*, in *Proceedings 13th International Workshop on Low Temperature Detectors*, (Stanford, CA, USA, 2009), American Institute of Physics Conference Proceedings, in press.
28. R. F. Lang, *et al.*, arXiv:0903.4687.
29. G. Angloher, *et al.*, *Astropart. Phys.* **31**, 270 (2009).
30. H. Kraus, *et al.*, *J. Phys. Conf. Ser.* **39**, 139 (2006).
31. S. Amerio, *et al.*, *Nucl. Instr. Methods* A **527** (2004) 329.
32. J. Angle, *et al.*, *Phys. Rev. Lett.* **100**, 21303 (2008).
33. V. N. Lebedenko, *et al.*, *Phys. Rev.* D **80** 052010 (2009).
34. R. Brunetti, *et al.*, *New Astronomy Reviews* **49**, 265 (2005).
35. R. Gaitskell and V. Mandic, Interactive WIMP Limits Plotter, http://dmtools.berkeley.edu/limitplots/
36. Y.D. Kim, the XMASS collaboration, *Phys. Atom. Nucl.*, **69** (2006) 1970.
37. A. Hime, in *Proceedings of the 9th Conference on the Intersections of Particle and Nuclear Physics*, AIP Conf. Proc. **870** (2006) 205.
38. J.D. Lewin and P.F. Smith, *Astropart. Phys.* 6, 87 (1996)
39. H.S. Lee *et al.*, *Phys. Rev. Lett.* **99** (2007) 091301.
40. G. Gerbier, J. Mallet, L. Mosca, C. Tao, arXiv:astro-ph/9710181.
41. G. Gerbier, J. Mallet, L. Mosca, C. Tao, arXiv:astro-ph/9902194.
42. P. Sorensen, et al., *Nucl. Instr. Meth.* A **21** (2004) 667.
43. A. Manzur, A. Curioni, L. Kastens, D.N. McKinsey, K. Ni and T. Wongjirad, arXiv:0909.1063.
44. R. Bernabei, et al., *Phys. Lett.* B **436** (1998) 379.
45. V. Chepel et al., *Astropart. Phys.* **26** (2006) 58.
46. F. Mayet, *et al.*, *J. Phys.* Conf.Ser.**179** (2009) 012011.

SUPERSYMMETRY, DARK MATTER AND NEW PARTICLES

Pierre FAYET

Laboratoire de Physique Théorique de l'ENS (UMR 8549 CNRS)
24 rue Lhomond, 75231 Paris Cedex 05, France

Abstract. Supersymmetric theories introduce superpartners with an odd R-parity which makes the lightest neutralino stable, leading to a missing energy signature and a natural dark matter candidate. They require new charged and neutral Higgs bosons and may also include, beyond the minimal model (MSSM), an extra singlet coupled through a $\lambda H_1 H_2 S$ superpotential (N/nMSSM), with possibly an extra $U(1)$ gauge symmetry (USSM). New light neutral particles very weakly coupled to ordinary ones, such as a light U boson or light dark matter particles, may also exist.

Keywords: supersymmetry, dark matter

INTRODUCTION

Is there a "superworld" of new particles ? Could half of the particles at least have escaped our observations ? Do new states of matter of exist ? After the prediction of antimatter by Dirac, supersymmetric extensions of the standard model led us to anticipate the existence, next to the electron and positron, of two spin-0 *selectrons* $\tilde{e}^-$ with antiparticles $\tilde{e}^+$ [1]. The gluons, photon and other gauge bosons should also be associated with new superpartners, leading to *gluinos, neutralinos* and *charginos*. These new states are characterized by a quantum number called R-parity, related to baryon number B and lepton number L. The lightest superpartner, presumably a neutralino, expected to be stable from R-parity conservation, is one of the best candidates for the non-baryonic dark matter of the Universe.

Next to these particles required by supersymmetric theories there could also exist *new light particles, neutral and very weakly coupled*, such as a light spin-1 gauge boson U, axionlike spin-0 particles, or light dark matter particles. Spacetime may be extended to new fermionic coordinates as in superspace (x^μ, θ), or extra space dimensions as $(x^\mu, x^5, x^6, \ldots)$. The two notions are closely connected as extended supersymmetric theories are naturally formulated with extra (compact) space dimensions, but this will not be discussed here, although extra dimensions may well play a crucial role for supersymmetry breaking.

OPEN QUESTIONS IN THE STANDARD MODEL

The Standard Model describes strong, electromagnetic and weak interactions of quarks and leptons, invariant under a $SU(3) \times SU(2) \times U(1)$ gauge group. It involves

$$\left\{ \begin{array}{ll} \textit{spin-1 gauge bosons:} & \textit{gluons, } W^+, W^-, Z, \textit{photon,} \\ \textit{spin-}\frac{1}{2} \textit{ fermions:} & \textit{quarks and leptons.} \end{array} \right. \tag{1}$$

One also expects a spin-0 Brout-Englert-Higgs boson, still unobserved, in connection with the spontaneous breaking of the electroweak symmetry [2]- [5].

While remarkably successful, this model leaves many questions unanswered. To mention a few, do fundamental Higgs fields actually exist? Many physicists were long reluctant to accept fundamental spin-0 fields, still unknown in Nature, viewing them as a trick to mimic or simulate the effects of the spontaneous breaking of the electroweak symmetry. But supersymmetric theories, in which spin-0 fields are related to spin-$\frac{1}{2}$ ones and thus of equal dignity (with spin-0 mass parameters equal to spin-$\frac{1}{2}$ ones as long as supersymmetry is unbroken), are the natural place for such fundamental spin-0 fields and particles.

CP1241, *Invisible Universe: Proceedings of the Conference,* edited by J.-M. Alimi and A. Füzfa

What is the origin of the Brout-Englert-Higgs potential

$$V(\varphi) = \lambda\,|\varphi|^4 - \mu^2\,|\varphi|^2 \ ? \tag{2}$$

The mass of the corresponding spin-0 boson is $m_h = \mu\,\sqrt{2} = v\,\sqrt{2\lambda}$. μ is related to λ through $v = 2^{-1/4}\,G_F^{-1/2} \simeq$ 246 GeV, but what fixes the value of the quartic coupling λ ? Is there only one doublet of spin-0 fields as in the standard model, or is this sector more complicated ? Do other new particles exist, maybe also new forces ? After LEP, we think we know all sequential quarks and leptons. The interrogation about new particles is essential, especially in view of the increasing evidence for non-baryonic dark matter, and the need to know its nature.

Other interrogations deal with the role of gravity, related to the curvature of spacetime through general relativity. Can it be more closely connected with particle physics ? Can one get a consistent theory of quantum gravity ? What about the cosmological constant Λ associated with an intrinsic curvature of empty spacetime, or induced by its energy density and related pressure, usually infinite in quantum field theory ? How could its value be so small ?

Can interactions be unified ? In grand-unified theories, according to which $SU(3) \times SU(2) \times U(1) \subset SU(5)$, for example, spin-1 gluons are related to the electroweak bosons $W^{\pm}$, Z and photon and to additional gauge bosons, and spin-$\frac{1}{2}$ quarks and leptons are also related. But this brings in a whole new set of questions, about the Higgs potential and gauge symmetry breaking, the many new coupling constants introduced, the origin of the hierarchy of mass scales, a problematic relation between quark and lepton masses, etc.. Asking whether it may be possible to relate particles of different spins leads to supersymmetry.

CAN SUPERSYMMETRY APPLY TO THE FUNDAMENTAL LAWS OF NATURE ?

According to common knowledge, supersymmetry relates (or should relate) bosons with fermions:

$$\underbrace{\text{bosons}}_{\text{integer spin}} \quad \stackrel{\text{SUSY}}{\longleftrightarrow} \quad \underbrace{\text{fermions}}_{\text{half-integer spin}} \ . \tag{3}$$

Can this idea be of any help in understanding the real world of fundamental particles and interactions ? In particular, can one relate the bosons, messengers of interactions, to the fermions, constituants of matter, as tried with the photon and the neutrino, the W^- and the electron [6], and arrive to some sort of

$$\text{Unification} \quad \text{FORCES} \ \longleftrightarrow \ \text{MATTER} \ ? \tag{4}$$

This would be very attractive, but unfortunately things don't work out that way !

The supersymmetry algebra:

$$\left\{ \begin{array}{lcl} \{\, Q, \bar{Q} \,\} & = & -\,2\,\gamma_\mu\,P^\mu \\ [\, Q, P^\mu \,] & = & 0 \end{array} \right. \tag{5}$$

was introduced in the years 1970-73 [7]-[9] with various motivations including: is this algebra at origin of parity non-conservation ? or is the neutrino a Goldstone particle ? Note that the supersymmetric Volkov-Akulov model [8] involves only fermions, in contrast with the common idea that the supersymmetry algebra must relate fundamental fermions to fundamental bosons, and automatically requires the existence of superpartners.

But Nature is obviously not supersymmetric, it seems. Unbroken supersymmetry would imply, for supersymmetry acting linearly in the usual way, equal masses for bosons and fermions, which is not the case.

Special features of spontaneous supersymmetry breaking

We are used to spontaneously broken symmetries, which could be the case for supersymmetry, allowing for bosons and fermions to have different masses. But spontaneous supersymmetry breaking did not seem possible ! A special feature of supersymmetry is that the Hamiltonian, which determines which state will be the stable vacuum, appears in the anticommutation relations (5), indicating that a supersymmetric vacuum state, which verifies $Q|0> = 0$ and thus $H|0> = 0$, should have vanishing energy (while non-supersymmetric states have positive energy in global supersymmetry) and thus be stable.

In spite of that, spontaneous supersymmetry breaking turned out to be possible, but it is very constrained. Instead of simply trying to make a supersymmetric state unstable, one has to arrange for such a state to be totally absent, as it would otherwise be stable! One can indeed find such special situations for which supersymmetry does get spontaneously broken, using in global supersymmetry one or the other of two mechanisms relying on a $U(1)$ gauge group [10] or on an appropriate set of chiral superfields including at least a gauge singlet one [11]. One always has to fight the tendency for all fields to reach if possible a value at which the potential would vanish, in which case supersymmetry would be conserved.

This usually requires, especially with the latter mechanism relying on non-vanishing v.e.v.'s for F-components of chiral superfields, not using the most general superpotential $\mathscr{W}$ compatible with supersymmetry and gauge symmetries; but using very carefully chosen superpotentials, as obtained with the help of additional symmetries such as an R symmetry acting chirally on the supersymmetry generator [11].

Where is the spin-$\frac{1}{2}$ Goldstone fermion of supersymmetry ("goldstino")?

Spontaneous breaking of global supersymmetry should lead to a massless spin-$\frac{1}{2}$ Goldstone fermion. But where is this Goldstone fermion of supersymmetry, which cannot be identified with a neutrino? Why has not it been observed? The present understanding is that within supergravity theories [12, 13], in which supersymmetry is realized locally, this spin-$\frac{1}{2}$ "goldstino" must be eliminated in favor of additional degrees of freedom for the spin-$\frac{3}{2}$ gravitino, which acquires a mass.

Still there is a caveat in considering that the goldstino disappears in this way: if the gravitino is very light its $\pm\frac{1}{2}$ polarisation states would interact with *enhanced gravitational strength*, proportionally to $\kappa/m_{3/2} \propto 1/(m_{3/2}\, m_{\rm Planck})$, much more strongly than for the $\pm\frac{3}{2}$ states which still interact with normal gravitational strength (proportionally to $\kappa = \sqrt{8\pi G_N} \simeq 4\ 10^{-19}\ \text{GeV}^{-1}$) [14, 15]. A very light gravitino would thus still behave very much as the spin-$\frac{1}{2}$ goldstino of spontaneously broken global supersymmetry, according to the equivalence theorem of supersymmetry.

This may have observable effects, e.g. through decays of supersymmetric particles, like

$$\textit{neutralino} \to \gamma + \textit{unobserved gravitino}, \tag{6}$$

allowing the lightest neutralino to decay radiatively, if the gravitino mass $m_{3/2}$ is small enough for the decay to actually occur (with a rate proportional to $G_N/m_{3/2}^2$ and thus a lifetime proportional to $m_{3/2}^2\, m_{\rm Planck}^2$), i.e. if the supersymmetry-breaking scale parameter is not too large [1]. This light gravitino is then usually the lightest supersymmetric particle, stable from R-parity conservation. This is what is expected today in the so-called "GMSB" models characterized by such a light gravitino, with the production of photons + missing energy-momentum carried away by gravitinos.

Relating bosons with fermions does not seem possible

Independently of the question of supersymmetry breaking, one has to find out which bosons and fermions could be related. One may try as a warm-up exercise the tentative associations

$$\left\{\begin{array}{ccc} \textit{photon} & \stackrel{?}{\longleftrightarrow} & \textit{neutrino} \\ W^\pm & \stackrel{?}{\longleftrightarrow} & e^\pm \\ \textit{gluons} & \stackrel{?}{\longleftrightarrow} & \textit{quarks} \\ & \dots & \end{array}\right. \tag{7}$$

---

[1] The gravitino mass may be expressed as $m_{3/2} = \kappa d/\sqrt{6} = \kappa F/\sqrt{3}$, with $\Lambda_{ss} = \sqrt{F} = (\frac{3}{8\pi})^{1/4}\sqrt{m_{3/2}\, m_{\rm Planck}}$ often called the supersymmetry-breaking scale.

But we have no chance to realize in this way systematic associations of known fundamental bosons and fermions, especially as we know more fundamental fermionic field degrees of freedom, describing quarks and leptons, than bosonic ones. In addition these fields have also different gauge and internal (B and L) quantum numbers.

Still the exercise of trying to relate the photon with a "neutrino" and the W^- with an "electron" and "heavy electron", together with a charged Higgs boson H^-, turned out to be fruitful [6]. With two doublet Higgs superfields now called

$$H_1 = \begin{pmatrix} H_1^0 \\ H_1^- \end{pmatrix} \quad \text{and} \quad H_2 = \begin{pmatrix} H_2^+ \\ H_2^0 \end{pmatrix}, \tag{8}$$

as required to get *two* massive Dirac charginos (or winos) from charged gaugino ($\lambda^- = \widetilde{W}^-_{L+R}$) and higgsino ($\tilde{h}^-_{1L}$ and $(\tilde{h}^+_{2L})^c$) components, it led us to the gauge and Higgs sector of supersymmetric extensions of the standard model, the would-be "lepton" candidates being reinterpreted as charginos and neutralinos. With only one doublet superfield say H_1, we could only construct, from $\widetilde{W}^-_{L+R}$ and $\tilde{h}^-_{1L}$, one-and-a-half Dirac chargino, getting stuck with a massless charged chiral fermion. These two doublet Higgs superfields H_1 and H_2 allow for the generation of charged-lepton and down-quark masses (for H_1), and up-quark masses (for H_2), respectively [1].

What to do with the Majorana fermions of supersymmetric theories ?

In supersymmetry we have to deal with the systematic appearance of *self-conjugate Majorana fermions*, while Nature seems to know only Dirac fermions. How can we obtain the usual Dirac fermions, and how can we attribute them conserved quantum numbers like B and L? This problem gets more acute as baryon number B and lepton number L are carried by fundamental fermions only, quarks and leptons, not by bosons. This gets impossible to realize in a supersymmetric theory where bosons and fermions are related. It also seemed to make supersymmetry irrelevant to the real world.

The solution consists first in accepting the existence of Majorana fermions as belonging to a new class of particles. The photon will not be associated with ν_e, ν_μ or ν_τ, but with its own new "neutrino", a "*photonic neutrino*" called in 1977 the *photino*; and similarly for the gluons associated with *gluinos*, etc. [1, 14]. The Majorana fermions of supersymmetry are thus identified as gluinos and neutralinos, or combine into charginos. At the same time one introduces new bosons carrying baryon and lepton numbers, now known as *squarks and sleptons* – although this does not necessarily guarantee yet that B and L will remain conserved at least to a sufficiently good approximation.

Supersymmetry thus does not relate directly known bosons and fermions. All particles should be associated with new superpartners, according to

$$\begin{cases} \text{known bosons} & \longleftrightarrow \quad \text{new fermions,} \\ \text{known fermions} & \longleftrightarrow \quad \text{new bosons.} \end{cases} \tag{9}$$

This was long mocked as a sign of irrelevance of supersymmetry. But times have changed, and the same feature gets now frequently viewed as an "obvious" consequence of the supersymmetry algebra.

SUPERPARTNERS AND R-PARITY, NEUTRALINOS, MISSING-ENERGY, AND DARK MATTER

Even after accepting the introduction of superpartners one still has to face another potential problem. With so many spin-0 particles, many of them, squarks and sleptons, carrying B and L, how can we get interactions mediated essentially by spin-1 gauge bosons, avoiding unwanted squark and slepton exchanges that would lead to B and L violations ?

To be able to account for B and L conservation laws the superpotential $\mathcal{W}$ of supersymmetric extensions of the standard model should be taken as *an even function of quark and lepton superfields* [1]. This is of course related with the introduction in these theories of a continuous R-symmetry or simply of its discrete version known as R-parity. These related symmetries actually forbid direct exchanges of (R-odd) squarks and sleptons between ordinary (R-even) quarks and leptons, which would lead to unwanted processes with B and L non-conservation, such as proton decay at a much too high rate. The only allowed direct exchanges of spin-0 particles between quarks and leptons are those of spin-0 Higgs bosons. These are very weakly coupled to light quarks and leptons proportionally to their masses, as represented in Fig. 1 for the charged spin-0 boson now known as $H^{\pm}$.

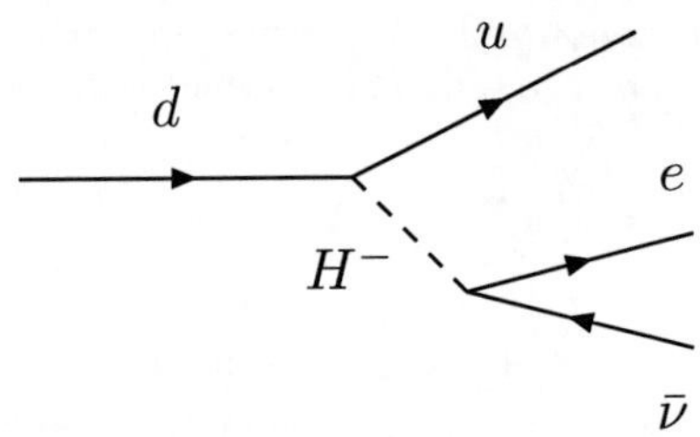

FIGURE 1. *Charged Higgs boson contribution to β-decay amplitudes; direct exchanges of squarks and sleptons ($\tilde{q}, \tilde{l}$) between quarks and leptons do not occur with a superpotential $\mathcal{W}$ even in quark and lepton superfields, as expressed with R-parity [1].*

These properties may be elegantly expressed in terms of a discrete symmetry named R-parity, first obtained as a discrete Z_2 remnant of the continuous $U(1)_R$ R symmetry acting chirally on the supersymmetry generator according to

$$Q \overset{R}{\to} e^{-\gamma_5 \alpha}\, Q \ , \tag{10}$$

and surviving electroweak breaking, with the two Higgs doublets h_1 and h_2 defined as having $R = 0$ [1, 6]. R-parity is then defined as $R_p = (-1)^R$, which is simply $+1$ for ordinary (R-even) particles including Higgs bosons, and -1 for their (R-odd) superpartners. This may be reexpressed in terms of B and L through [16]

$$R_p = (-1)^{2S}\,(-1)^{3B+L} \ . \tag{11}$$

This discrete symmetry requires that supersymmetric particles, with $R_p = -1$, be produced in pairs, with their decays producing again particles of $R_p = -1$ (or an odd number of them).

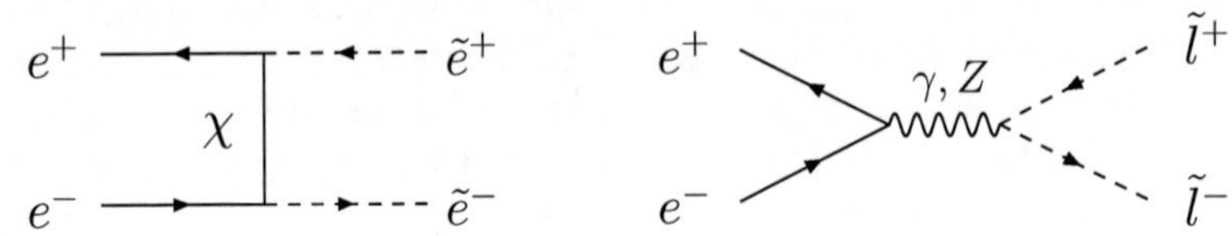

FIGURE 2. *Slepton pair production in e^+e^- annihilations, induced by γ, Z or neutralino exchanges.*

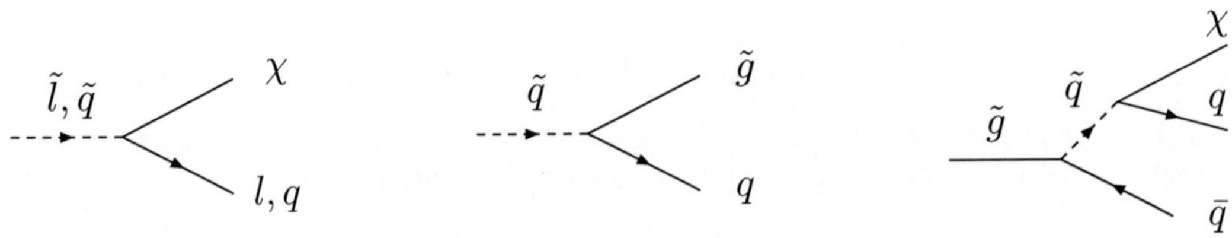

FIGURE 3. *Superpartner decays, ultimately leading to unobserved LSP's in the final state [1]. $\tilde{l}$, $\tilde{q}$ and $\tilde{g}$ denote sleptons, squarks and gluinos, and χ a neutralino (mixing of $\tilde{W}_3$, $\tilde{W}'$, $\tilde{h}_1^\circ$ and $\tilde{h}_2^\circ$, ...).*

The "lightest supersymmetric particle" (LSP) must then be stable, and is usually a neutralino. The pair-production of supersymmetric particles should ultimately lead to two unobserved neutralinos, the famous "missing energy-momentum" signature [1, 16] often used to search for supersymmetry [17]. These neutralinos have interactions of roughly weak-interaction strength through squark, slepton or Z ... exchanges [15] as illustrated in Fig. 4, depending on squark and slepton masses and more specifically on the details of the mass spectrum.

Once they acquire a sizeable-enough mass as from gravity-induced gaugino mass terms m_1 and m_2 [18] (with m_2 also allowing for both charginos to be heavier than m_W [19] as is now required), these weakly-interacting neutralinos [15] (or WIMPs, for "weakly-interacting massive particles"), stable from R-parity conservation [16], can annihilate sufficiently to satisfy cosmological constraints with a relic density evaluated as in [20], becoming natural candidates for the non-baryonic dark matter of the Universe [21, 22].

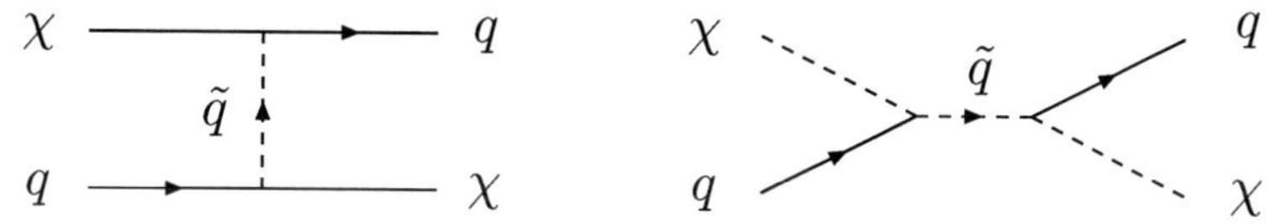

FIGURE 4. *Squark-exchange contributions to χ-nucleon scattering amplitudes* [15].

Among other possibilities (including axions, ...), dark matter may also be constituted by light gravitinos [23] interacting with ordinary matter with enhanced gravitational strength (proportionally to $G_N/m_{3/2}^2$) but still much more weakly than neutrinos [15] so that they decouple much earlier. This results in a present number density well below that of neutrinos, making such light gravitinos non-baryonic dark matter candidates with a mass $\sim$ keV, significantly higher than considered at the time for neutrino candidates. Light gravitinos would behave, in particle physics experiments, very much as very-weakly-interacting spin-$\frac{1}{2}$ goldstinos. They could be produced, for example, in the decay of the lightest neutralino (then the "next-to lightest" supersymmetric particle, or NLSP) into γ + gravitino, as indicated in (6).

BACK TO SUPERSYMMETRY BREAKING

Once we accept that supersymmetry does not relate directly known bosons and fermions, we have to account for the fact that superpartners have not been observed, which requires them to be sufficiently heavy. If a few GeV's or $\gtrsim 15$ GeV could be sufficient in the late seventies or early eighties, at the time of the PETRA and PEP experiments, we know now, from LEP and Fermilab experiments, that most superpartners, excepted possibly for some of the neutralinos, should be heavier than roughly about one hundred GeV's, up to several hundreds for squarks and gluinos [17].

Squark and slepton masses

To discuss how squarks may be heavy we have to study the mass (or mass^2) splittings between squarks and quarks, generated in global supersymmetry from the auxiliary $<D>$'s of neutral gauge superfields [10] and $<F>$'s of chiral ones [11], most notably the neutral Higgs superfields H_1^0 and H_2^0 directly coupled to quark superfields, in connection with the supercurrent conservation equation [14]. This results in the mass sum rule [24]

$$\overline{\Sigma}\, m^2(\text{squarks}) \;=\; \overline{\Sigma}\, m^2(\text{quarks}) \quad \left\{ \begin{array}{l} \text{in global supersymmetry,} \\ \text{up to radiative corrections,} \\ \text{with } SU(3)\times SU(2)\times U(1) \ \ (\text{or } SU(5)) \text{ gauge group.} \end{array} \right. \tag{12}$$

And in particular, for the first quark family,

$$m^2(\tilde{u}_1)+m^2(\tilde{u}_2)+m^2(\tilde{d}_1)+m^2(\tilde{d}_2) \;=\; 2\,(m_u^2+m_d^2)\ . \tag{13}$$

Whenever applicable as specified in (12) it expresses that the average squark mass^2 within each family should be the same as for spin-$\frac{1}{2}$ quarks, which is not tenable at least for the superpartners of light quarks. Indeed this would imply that one at least among the two $\tilde{u}$ and the two $\tilde{d}$ squarks should have a very small or actually a negative mass^2, which is not possible as it would lead to a spontaneous breaking of the charge and color symmetries [24] [2].

Making all four $\tilde{u}$ and $\tilde{d}$ squarks heavy thus requires bypassing the obstruction illustrated by the tree-level sum rule (13), corresponding to the goldstino of spontaneously broken global supersymmetry being a linear combination of the neutral electroweak gauginos $\tilde{W}_3$ and $\tilde{W}'$ with the Majorana components of neutral chiral superfields, especially the neutral higgsinos $\tilde{h}_1^\circ$ and $\tilde{h}_2^\circ$ (and possible singlinos, ...). This may be done using supergravity, or radiative corrections, which corresponds in general to the so-called (somewhat improperly) gravity-mediated or gauge-mediated

[2] Simply considering that the average mass^2 should be the same for all bosons and fermions [25], $\overline{\Sigma} m_B^2 = \overline{\Sigma} m_F^2$, in the absence of an axial $U(1)_A$ gauge symmetry which can invalidate this relation [1], does not in general allow to conclude directly that an extra contribution to the sum rule is needed, as squark (and slepton) masses could be in balance with other large fermion masses.

supersymmetry-breaking models [22]. One may also extend the gauge group to include at least an extra $U(1)$ with non-vanishing axial couplings, an axial $U(1)_A$ in the simplest case, to take advantage of its $<D>$ contribution to the squark mass^2 to make all squarks heavy (or at least to generate an additional contribution to squark mass^2), as in the initial proposal. This one corresponds to the USSM with a $SU(3) \times SU(2) \times U(1) \times$ extra-$U(1)$ gauge group, as naturally suggested by the 2-Higgs doublet structure of supersymmetric extensions of the standard model, allowing for the possibility of "rotating" independently the two doublets h_1 and h_2, as discussed in Sec. . Or one may consider combinations of these mechanisms.

How to generate gluino masses

If squarks and sleptons can be made heavy we still have to face the question of gluino masses [26]. Even if supersymmetry is spontaneously broken gluino masses cannot be generated spontaneously at the classical level, as gluinos are coupled to colored spin-0 fields which cannot be translated, or the color gauge symmetry would be broken. Could radiative corrections generate gluino masses ? But these are also forbidden by the continuous $U(1)_R$ R-symmetry if it is present, which acts chirally on the supersymmetry generator as in (10), so that $\tilde{g} \to e^{\gamma_5 \alpha} \tilde{g}$.

In the absence of such a continuous R-symmetry acting chirally on gluinos, reduced to its R-parity subgroup, gluino couplings to a new set of massive "messenger" quarks vectorially coupled to gauge fields, and sensitive to the source of supersymmetry-breaking from $<D>$'s or $<F>$'s, may indeed induce gluino masses from radiative corrections, possibly with the help of considering a metastable vacuum state [26]. This requires, however, the messenger masses to be quite large, otherwise the resulting radiatively-generated gluino masses would be too small (especially in comparison with the several hundred GeV's now required), raising in particular the question of the origin of such a large new mass scale.

Another commonly used possibility relies on the fact that ($N = 1$) supergravity theories necessitate in any case a breaking of the continuous R symmetry down to R-parity so that the gravitino can acquire a mass $m_{3/2}$, allowing for the $SU(3), SU(2)$ and $U(1)$ gauginos to acquire mass terms m_3, m_2 and m_1 as well [14, 18]. One may also consider obtaining massive gluinos without breaking the continuous $U(1)_R$ symmetry by introducing a second Majorana octet of "paragluinos" so as to turn gluinos into Dirac particles, which leads us naturally to extended supersymmetry.

Indeed beyond $N = 1$, $N = 2$ supersymmetric theories, or theories with extra compact dimensions, lead to the consideration of *Dirac gluinos*, and associated scalar gluons [27], with the possibility of generating their masses from momenta along the compact dimensions. The $4d$ mass scale of superpartners may then be obtained from the compactification scale through formulas like

$$m_{\text{gluinos}} = m_2 = m_1 = m_{3/2} = \frac{\pi}{L} = \frac{1}{2R}, \ \dots, \tag{14}$$

up to radiative corrections, L denoting the size of a compact dimension, or R the corresponding compactification "radius" [28]. This leads to the possibility that this compactification scale be relatively "low" (as compared to the Planck or GUT scales), and that the discovery of superpartners in the TeV mass range be followed by the discovery of series of heavy copies for all particles, corresponding to the Kaluza-Klein excitations of the extra dimensions [28].

In view of the variety of mechanims one may invoke to break supersymmetry, one can parametrize the possible results using soft supersymmetry-breaking terms (which break explicitly the supersymmetry) [29]. The price to pay is the introduction of many arbitrary parameters, leading to the feeling that supersymmetric theories are not predictive. This is not true as we know quite well what particles to expect and how they should interact, but it is difficult to predict their masses. We refer the reader to [22] for a review of the various versions of supersymmetric extensions of the standard model, and possible ways to break supersymmetry. We note that the "Minimal Supersymmetric Standard Model" or MSSM, where supersymmetry is broken explicitly through soft supersymmetry-breaking terms, is in fact *not supersymmetric*, as long as the mechanism generating these terms from spontaneous breaking is not specified, and the corresponding particles involved in this mechanism are ignored.

THE SUPERSYMMETRIC STANDARD MODEL

The required ingredients for a supersymmetric extension of the Standard Model, minimal or not, are [1]

$$\left\{ \begin{array}{l} 1)\ SU(3)\times SU(2)\times U(1) \text{ gauge superfields } \ [\ \times \textit{possible extra } U(1)\ ], \\ 2)\ \text{chiral quark and lepton superfields}, \\ 3)\ \text{two doublet Higgs superfields } H_1 \text{ and } H_2, \\ 4)\ \text{trilinear superpotential for } q \text{ and } l \text{ masses}. \end{array} \right. \tag{15}$$

The superpotential, taken to be an even function of quark and lepton superfields i.e. invariant under R-parity, includes the terms

$$\mathscr{W} \ = \ h_e\, H_1 . \bar{E}\, L \ + \ h_d\, H_1 . \bar{D}\, Q \ - \ h_u\, H_2 . \bar{U}\, Q \quad [\, + \,\mu\, H_1 H_2\,]\ . \tag{16}$$

H_1 and H_2 are separately responsible for charged-lepton and down-quark masses, and up-quark masses, respectively, with $m_e = h_e v_1/\sqrt{2}$, $m_d = h_d v_1/\sqrt{2}$, $m_u = h_u v_2/\sqrt{2}$, and $\tan\beta = v_2/v_1$. The minimal content of the Supersymmetric Standard Model is shown in Table 1.

TABLE 1. Minimal particle content of the Supersymmetric Standard Model.

Spin 1	Spin 1/2	Spin 0
gluons g photon γ	gluinos $\tilde{g}$ photino $\tilde{\gamma}$	
$W^{\pm}$ Z	winos $\widetilde{W}^{\pm}_{1,2}$ zinos $\widetilde{Z}_{1,2}$ higgsino $\tilde{h}^{\circ}$	$H^{\pm}$ H $h,\ A$ } Higgs bosons
	leptons l quarks q	sleptons $\tilde{l}$ squarks $\tilde{q}$

The neutral gaugino fields $\tilde{W}_3$ and $\tilde{W}'$ mix with the higginos $\tilde{h}_1^{\circ}, \tilde{h}_2^{\circ}$ into four Majorana neutralinos. We get, more generally,

$$\{W_3, W';\, h_1^{\circ}, h_2^{\circ};\, ...\} \ \stackrel{\text{SUSY}}{\longleftrightarrow} \ \underbrace{\{\tilde{W}_3, \tilde{W}';\, \tilde{h}_1^{\circ}, \tilde{h}_2^{\circ};\, ...\}}_{\textit{neutralinos}}\ . \tag{17}$$

The association

$$\textit{neutral gauge } (\gamma, Z, ...) \textit{ and Higgs bosons } (h, H, A,\ ...) \ \longleftrightarrow \ \textit{dark matter} \tag{18}$$

serves as a substitute for (4), now relating the mediators of electroweak forces (or mass generation) to *dark matter*, instead of normal matter. The neutralino sector, and thus the LSP which serves as a natural dark matter candidate, may also involve an extra higgsino (*singlino*) described by with a chiral singlet superfield S coupled to H_1 and H_2 through a $\lambda\ H_1 H_2 S$ superpotential as in the N/nMSSM, or *another gaugino* associated with an extra $U(1)$ gauge symmetry as in the USSM (cf. Secs. and).

In any case the quartic Higgs interactions of the two doublets h_1 and h_2 appear as *electroweak gauge interactions*, with a well specified potential [6]

$$V_{\text{quartic}} \ = \ \frac{g^2+g'^2}{8}\ (h_1^{\dagger} h_1 - h_2^{\dagger} h_2)^2 \ + \ \frac{g^2}{2}\ |h_1^{\dagger} h_2|^2\ . \tag{19}$$

This quartic Higgs potential of the MSSM appears also in all its extensions. *Quartic Higgs couplings are now fixed by the electroweak gauge couplings g and g'*, to $(g^2+g'^2)/8$ and $g^2/2$.

A quartic coupling $\lambda_{\text{sm}} = (g^2+g'^2)/8$ corresponds in the standard model to a Higgs boson mass $m_h = v\sqrt{2\,\lambda_{\text{sm}}} = m_Z$. The quartic coupling for the neutral component $\varphi^{\circ}_{\text{sm}} = h_1^{\circ} \cos\beta + h_2^{\circ *} \sin\beta$, which acquires a v.e.v. $v/\sqrt{2}$ and is responsible for electroweak breaking, is directly obtained from (19) as

$$\lambda_{\text{sm}} \ = \ \frac{g^2+g'^2}{8}\ (\cos^2\beta - \sin^2\beta)^2\ . \tag{20}$$

It is reduced by $\cos^2 2\beta$ compared to $(g^2+g'^2)/8$. This translates in all such models, and most notably in the MSSM, to an upper bound on the lightest Higgs mass m_h equal to m_Z reduced by $|\cos 2\beta|$,

$$m_h \leq m_Z\,|\cos 2\beta|\;, \tag{21}$$

or, including radiative corrections [22, 30],

$$m_h \leq m_Z\,|\cos 2\beta| + \underbrace{\text{rad. corr.}}_{\textit{should be large !}}\;. \tag{22}$$

The tree-level constraint (21) is rather problematic in view of present experimental results [17]. This necessitates to rely in the MSSM on large radiative corrections to make the lightest Higgs boson sufficiently heavy. Taking all constraints into account only a rather restricted region of the parameter space survives, generally requiring rather large squark masses $\approx$ TeV scale. This also creates (or recreates) a "little" hierarchy problem, since one has to explain how to obtain a relatively small m_Z^2 from a linear combination involving significantly larger quantities of typically $\approx$ TeV^2 scale.

BEYOND MSSM: THE N/NMSSM, WITH A SINGLET AND A TRILINEAR COUPLING $\lambda H_1 H_2 S$

From the beginning the μ parameter introduced as a direct $\mu\,H_1 H_2$ superpotential mass term for the doublet superfields H_1 and H_2 appeared a source of difficulties. In particular the corresponding direct mass terms $\mu^2\,(h_1^\dagger h_1 + h_2^\dagger h_2)$, leading in the presence of a $\xi\,D'$ term for the weak hypercharge $U(1)$ to Higgs mass^2 parameters $(\mu^2 \pm \frac{\xi g'}{2})$, were an obstacle for a satisfactory electroweak breaking, with both doublets h_1 and h_2 acquiring non-vanishing v.e.v.'s $v_1/\sqrt{2}$ and $v_2/\sqrt{2}$ already at the classical level.

Furthermore in any case, with a quartic coupling for neutral Higgs interactions at most equal to $(g^2+g'^2)/8$ as shown by (19,20), even adding as in the MSSM a non-diagonal mass term proportional to $\Re\,(h_1 h_2)$ (as may be generated from a small or very small coupling to a singlet S with a linear σS superpotential term), the potential can never be steep enough to make all neutral Higgs bosons heavier than m_Z, classically, as this seems desirable now.

This μ parameter was thus immediately promoted, in 1974, to a dynamical superfield variable $\mu(x,\theta)$. I.e. $\mu\;H_1H_2$ was replaced by a trilinear superpotential coupling $\lambda\;H_1H_2S$ with an extra singlet chiral superfield S, leading to the superpotential [6]

$$\mathscr{W} = \lambda\,H_1\,H_2 S + f(S)\;, \tag{23}$$

with

$$f(S) = \frac{\kappa}{3}\,S^3 + \frac{\mu_S}{2}\,S^2 + \sigma S \tag{24}$$

for the general NMSSM. This was further reduced to the linear term $f(S) = \sigma S$ of the nMSSM (in modern terminology) with the help of a continuous R symmetry of the superpotential, under which

$$H_{1,2}(x,\theta) \to H_{1,2}(x,\theta\, e^{-i\alpha}),\;\; S(x,\theta) \to e^{2i\alpha}\, S(x,\theta\, e^{-i\alpha})\,, \tag{25}$$

so that

$$\mathscr{W}(x,\theta) \to e^{2i\alpha}\,\mathscr{W}(x,\theta\, e^{-i\alpha}) \tag{26}$$

as required for its F-component to be R-invariant. The extra trilinear and bilinear terms $\frac{\kappa}{3}S^3$ and $\frac{\mu_S}{2}S^2$ are excluded from the superpotential (23,24) by this continuous R-symmetry.

With the dynamical change $\mu \to \lambda S$, the doublet mass terms in the potential of h_1, h_2 get replaced by $\lambda^2\,|s|^2$ $(h_1^\dagger h_1 + h_2^\dagger h_2)$, with the Higgs potential (19) replaced as in [6] (where λ^2 was called $h^2/2$, with $\tan\delta = v'/v''$ instead of $\tan\beta = v_2/v_1$) by:

$$V = \frac{g^2+g'^2}{8}\;(h_1^\dagger h_1 - h_2^\dagger h_2)^2 + \frac{g^2}{2}\;|h_1^\dagger h_2|^2 + \left|\lambda\, h_1 h_2 + \frac{\partial f(s)}{\partial s}\right|^2 + \lambda^2\,|s|^2\;(h_1^\dagger h_1 + h_2^\dagger h_2) + \dots\;. \tag{27}$$

The crucial point is that this introduces an extra quartic coupling

$$V^{\lambda}_{\text{quartic}} \;=\; \lambda^2\,|h_1 h_2|^2 \;+\; \lambda^2\,|s|^2\,(h_1^\dagger h_1 + h_2^\dagger h_2)\ , \tag{28}$$

in addition to the $(g^2+g'^2)/8$ and $g^2/2$ terms in (19), essential to make all Higgs bosons sufficiently heavy in the N/nMSSM, provided λ is large enough. In particular the quartic coupling associated with the "standard" neutral Higgs combination ($\varphi^\circ_{\rm sm} = h_1^\circ\cos\beta + h_2^{\circ *}\sin\beta$) responsible for electroweak breaking, with $<\varphi^\circ_{\rm sm}> \;=\; v/\sqrt{2}$, is immediately identified as

$$\lambda_{\rm sm} \;=\; \frac{g^2+g'^2}{8}\ \cos^2 2\beta + \frac{\lambda^2}{4}\ \sin^2 2\beta\ . \tag{29}$$

The potential (27) is written as a function of $\varphi^\circ_{\rm sm}$, very much as in the standard model, as

$$V(\varphi^\circ_{\rm sm},...) \;=\; \lambda_{\rm sm}\,(\,|\varphi^\circ_{\rm sm}|^2 - \tfrac{v^2}{2}\,)^2 + ... \;=\; \frac{1}{2}\,(2\,\lambda_{\rm sm}\,v^2)\,(\,\sqrt{2}\,\Re\,\varphi^\circ_{\rm sm} - v)^2 + ...\ .\ , \tag{30}$$

It shows that neutral Higgs bosons cannot all have mass^2 larger than $2\,\mu^2_{\rm sm} = 2\,\lambda_{\rm sm}\,v^2$, providing directly from (29) an upper bound on the lightest Higgs mass $\mu_{\rm sm}\sqrt{2} = \sqrt{2\,\lambda_{\rm sm}\,v^2}$, valid in all such models independently of their details (affecting only terms of dimension ≤ 3 in the potential), and most notably in the N/nMSSM:

$$m_h^2 \;\leq\; m_Z^2\cos^2 2\beta + \frac{\lambda^2 v^2}{2}\ \sin^2 2\beta \quad [+\ \text{rad. corr.}]\ . \tag{31}$$

This does not require evaluating explicitly as usual the spin-0 boson mass matrix [31]. This upper bound is verified as an equality when $(\sqrt{2}\,\Re\,\varphi^\circ_{\rm sm} - v)$ (corresponding to the physical standard model Higgs boson field) is actually associated with a mass eigenstate, otherwise the inequality is strict.

For $\lambda^2 \leq (g^2+g'^2)/2$ one still gets $m_h \leq m_Z$ (up to radiative corrections) as all neutral doublet combinations have quartic coupling $\leq (g^2+g'^2)/8$. For $\lambda^2 \geq (g^2+g'^2)/2$ however, $\lambda^2 v^2/2 \;\geq\; m_Z^2$ and (31) reads

$$m_h^2 \;\leq\; m_Z^2 + \underbrace{\left(\frac{\lambda^2 v^2}{2} - m_Z^2\right)}_{\geq\,0}\ \sin^2 2\beta \quad [+\ \text{rad. corr.}]\ . \tag{32}$$

allowing for all Higgs bosons to be sufficiently heavy already at the classical level.

A large trilinear coupling λ can make all Higgs bosons heavy already at the classical level by providing from (28) *a steep enough contribution to the quartic potential closing the quartically-flat valley corresponding to the direction* $|h_1^\circ|^2 \approx |h_2^\circ|^2$. A simple illustration of the mass inequality (31) was provided long ago with the nMSSM superpotential $\mathscr{W} = \lambda\ H_1 H_2 S + \sigma S$ [6]. With supersymmetry unbroken for neutral particles in the absence of gravity-induced or soft supersymmetry-breaking terms the mass spectrum includes

$$\begin{cases} \text{1 neutral Higgs of mass } m_Z\,, & \textit{in Z massive gauge multiplet,} \\[2mm] \text{4 neutral Higgses of mass } \dfrac{\lambda\, v}{\sqrt{2}}\,, & \end{cases} \tag{33}$$

(with $\lambda\,v/\sqrt{2}$ noted $hv/2$ in [6]), independently of $\tan\beta$, which remains an arbitrary parameter. For $\lambda^2 \leq (g^2 + g'^2)/2$ the lightest neutral bosons have mass $\lambda\,v/\sqrt{2} \leq m_Z$, in agreement with (31). For $\lambda^2 \geq (g^2+g'^2)/2$ on the other hand, the lightest one has mass m_Z (again in agreement with (31)) and the four others $\lambda\,v/\sqrt{2}\ \geq m_Z$.

The limiting case $\lambda = \sqrt{(g^2+g'^2)/2}$ actually corresponds to *an unbroken effective* $N=2$ *supersymmetry in the neutral Z-and-Higgs subsector*. This one includes within a massive gauge multiplet of $N=2$ supersymmetry the Z boson, 5 neutral spin-0 bosons of mass m_Z, and 4 Majorana inos also of mass m_Z [27]. The latter are built from the two neutral higgsinos $\tilde h_1^\circ$ and $\tilde h_2^\circ$, the gaugino $\lambda_Z = \cos\theta\ \tilde W_3 - \sin\theta\ \tilde W'$, and the extra singlino described by S playing the role of a second gaugino ζ_Z.

With the superpotential restricted to $\lambda\,H_1 H_2 S + \sigma S$ from the use of a continuous $U(1)_R$ R symmetry acting as in (25), and the super-Yukawa coupling λ fixed to $\lambda^2 = (g^2+g'^2)/2$ by the second supersymmetry generator acting effectively in this sector, the quartic neutral Higgs potential, which depends only on $|h_1^\circ|^2 + |h_2^\circ|^2$ and $|s|^2$, has an $O(4) \approx SU(2) \times SU(2)$ symmetry. It reads

$$V_{\text{quartic}}^{(N=2)}(h_1^\circ, h_2^\circ, s) = \frac{g^2+g'^2}{8} \, (|h_1^\circ|^2 + |h_2^\circ|^2)^2 + \frac{g^2+g'^2}{2} \, |s|^2 \, (|h_1^\circ|^2 + |h_2^\circ|^2) \, , \tag{34}$$

very much as for a purely electroweak interaction. It depends only on $g^2 + g'^2$, not on any arbitrary superpotential coupling.

Additional quadratic terms proportional to $|h_1^\circ|^2 - |h_2^\circ|^2$ and $\Re(h_1^\circ h_2^\circ)$ can be obtained in a supersymmetric way, from the weak-hypercharge $\xi D'$ term and the linear σS superpotential term, respectively. They are in practice effectively equivalent in this sector as they may be related by a $SU(2)$ symmetry [27], and responsible for spontaneous electroweak breaking as usual with $< h_1^\circ > = v \cos\beta/\sqrt{2}$, $< h_2^\circ > = v \sin\beta/\sqrt{2}$. All five neutral spin-0 particles, h, H, A and the two extra ones described by S, acquire the same mass $m_Z = \sqrt{(g^2+g'^2)}\, v/2$, independently of $\tan\beta$, in agreement with the unbroken effective $N = 2$ supersymmetry then present in this sector.

In the presence of additional soft supersymmetry-breaking terms the $\mu H_1 H_2$ term replaced in (23) by the trilinear coupling $\lambda \, H_1 H_2 S$ may be effectively regenerated through a translation of the singlet s, leading to an effective

$$\mu_{\text{eff}} = \lambda < s > \, , \tag{35}$$

which appears in particular in the expression of the chargino (wino) mass matrix

$$\mathcal{M} = \begin{pmatrix} m_2 & m_W \sqrt{2} \sin\beta \\ m_W \sqrt{2} \cos\beta & \mu_{\text{eff}} \end{pmatrix} . \tag{36}$$

Having both the direct gaugino mass term m_2 and the effective higgsino mass term μ_{eff} different from zero is essential to allow for both charginos (winos) to be heavier than m_W [18, 19], as is now necessary [17].

THE $U(1)_A$ LIMIT OF THE N/NMSSM, AND THE USSM

Supersymmetric theories require two doublets for electroweak breaking and quark and lepton mass generation. This allows for the possibility of rotating them independently, according to an extra-$U(1)$ symmetry U_A acting as $h_1 \to e^{i\alpha} \, h_1$, $h_2^c \to e^{-i\alpha} \, h_2^c$ or equivalently $h_2 \to e^{i\alpha} \, h_2$ [32], i.e. in terms of superfields

$$H_1 \overset{U_A}{\to} e^{i\alpha} \, H_1 \, , \quad H_2 \overset{U_A}{\to} e^{i\alpha} \, H_2 \, . \tag{37}$$

This symmetry was identified and extended to

$$S \overset{U_A}{\to} e^{-2i\alpha} \, S \tag{38}$$

as soon as the singlet S was introduced [6], so that the trilinear superpotential $\lambda \, H_1 H_2 S$ be invariant, while S itself is not. The linear superpotential term σS, which breaks explicitly the extra $U(1)_A$ symmetry as shown by (38), allows us to avoid the appearance in [6] of a classically massless "axion" ($A = \sqrt{2}$ Im $(h_1^\circ \sin\beta + h_2^\circ \cos\beta)$), that would otherwise be present as seen in the spectrum (33) with $\lambda = 0$, including a massless "axion", a massless "modulus", and two decoupled spin-0 states described by S.

The extra $U(1)_A$ symmetry acts axially on quark and lepton superfields according to

$$\{L, \bar{E}; Q, \bar{D}, \bar{U}\} \overset{U_A}{\to} e^{-i\frac{\alpha}{2}} \, \{L, \bar{E}; Q, \bar{D}, \bar{U}\} \tag{39}$$

so that the trilinear superpotential (16) responsible for quark and lepton masses be $U(1)_A$-invariant. If imposed to all superpotential interactions it requires the superpotential contribution $f(S)$ to $\mathcal{W}$ to vanish, i.e. $\kappa = \mu_S = \sigma = 0$ in (23, 24). This leads to the $U(1)_A$-invariant limit of the N/nMSSM, which favors the appearance of a massless or light (axionlike) pseudoscalar a, associated with the spontaneous or almost-spontaneous breaking of the extra $U(1)_A$ symmetry (in connection with the soft-supersymmetry breaking terms which may be added to the Lagrangian density).

One may also gauge the extra $U(1)_A$ symmetry as initially proposed in [1], under the assumption that the corresponding anomalies have been suitably cancelled using e.g. mirror or exotic fermions, or other mechanisms. This leads to USSM theories with an extra-$U(1)$ neutral gauge boson Z' (also called U if it is light). This new boson, which did not show up in neutral-current phenomenology nor direct production at Fermilab, should normally have a large mass $\gtrsim$ TeV if the new gauge coupling (g") is comparable to g or g', as for an extra $U(1)$, $U(1)_A$ in the simplest case, taken as a subgroup of $E(6)$.

But the additional Z may also be light if the extra $U(1)$ gauge coupling is small [33], leading to an interesting new phenomenology, with the possibility of a new interaction involving also dark matter particles, mediated by a light U boson that could manifest at MeV or GeV energies (cf. Sec.).

In all such theories beyond the MSSM, N/n/USSM, ... , the lightest neutralino, candidate for non-baryonic dark matter, is a combination of the neutral electroweak gauginos and higgsinos $\tilde{W}_3, \tilde{W}'$ and $\tilde{h}_1^\circ, \tilde{h}_2^\circ$ as usual, with the singlino described by S and possibly an extra-$U(1)$ gaugino if present.

LIGHT DARK MATTER, ...

Relatively light dark matter particles are normally excluded, as their relic abundance would be too large. This would not be the case, however, if they can annihilate through a new interaction significantly stronger than weak interactions, at least at lower energies. A new light U boson associated with the gauging of an extra-$U(1)$ symmetry [33], with rather small couplings at least to ordinary particles, can indeed induce sufficient annihilations of light dark matter particles in the early Universe as represented in Fig. 5 a, and lead to an interesting new phenomenology in both particle and astroparticle physics [34]. One may also consider spin-0 dark matter particles annihilating through exchanges of new heavy fermions such as mirror fermions, as represented in Fig. 5 b, partly inspired from $N=2$ supersymmetry or theories with extra dimensions.

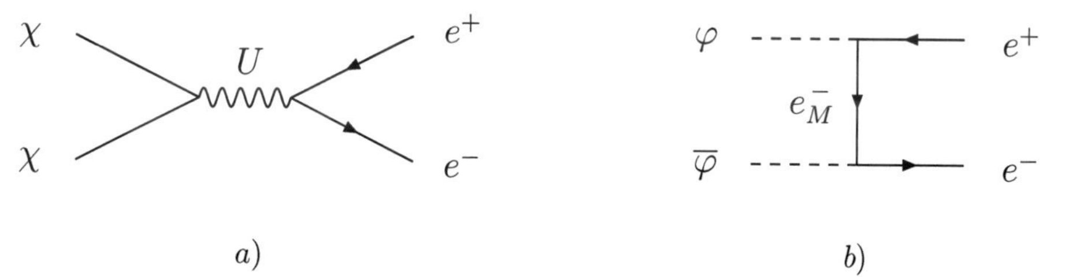

FIGURE 5. *a) Light dark matter annihilation induced by U exchanges; b) scalar dark matter annhilation induced by heavy-fermion exchanges.*

CONCLUSIONS

Supersymmetric extensions of the standard model lead us to search for gluinos, charginos and neutralinos, squarks and leptons, new charged and neutral Higgs bosons, and we are expecting that their discovery could occur at LHC. R-parity is responsible for the stability of the lightest supersymmetric particle, usually a weakly-interacting neutralino, candidate for the non-baryonic dark matter of the Universe. This construction has many appealing consequences, allowing for the possibility of producing dark matter at particle colliders, and paving the way to extra dimensions through extended supersymmetry. Another frontier can be explored at lower energies, in search for light weakly or very-weakly coupled new particles.

REFERENCES

1. P. Fayet, *Phys. Lett.* B **64**, 159 (1976); B **69**, 489 (1977).
2. F. Englert and R. Brout, *Phys. Rev. Lett.* **13**, 321 (1964).
3. P. Higgs, *Phys. Lett.* **12**, 132 (1964); *Phys. Rev. Lett.* **13**, 508 (1964).
4. G. Guralnik, C. Hagen and T. Kibble, *Phys. Rev. Lett.* **13**, 585 (1964).

5. T. Kibble, *Phys. Rev.* **155**, 1554 (1967).
6. P. Fayet, *Nucl. Phys.* B **90**, 104 (1975).
7. Yu Gol'fand and E. Likhtman, *ZhETF Pis. Red.* **13**, 452 (1971) [*JETP Lett.* 13, **323** (1971)].
8. D. Volkov and V. Akulov, *Phys. Lett.* B **46**, 109 (1973).
9. J. Wess and B. Zumino, *Nucl. Phys.* B **70**, 39 (1974); *Phys. Lett.* B **49**, 52 (1974); *Nucl. Phys.* B **78**, 1 (1974).
10. P. Fayet and J. Iliopoulos, *Phys. Lett.* B **51**, 461 (1974).
11. P. Fayet, *Phys. Lett.* B **58**, 67 (1975); L. O'Raifeartaigh, *Nucl. Phys.* B **96**, (1975) 331.
12. S. Ferrara, D. Freedman and P. van Nieuwenhuizen, *Phys. Rev.* D **13**, 3214 (1976).
13. S. Deser and B. Zumino, *Phys. Lett.* B **62**, 335 (1976).
14. P. Fayet, *Phys. Lett.* B **70**, 461 (1977).
15. P. Fayet, *Phys. Lett.* B **86**, 272 (1979).
16. G. Farrar and P. Fayet, *Phys. Lett.* B **76**, 575 (1978).
17. C. Amsler *et al.*, Particle Data Group, *Phys. Lett.* B **667**, 1 (2008), and 2009 update.
18. E. Cremmer, P. Fayet and L. Girardello, *Phys. Lett.* B **122**, 41 (1983).
19. P. Fayet, *Phys. Lett.* B **125**, 178 (1983).
20. B. Lee and S. Weinberg, *Phys. Rev. Lett.* **39**, 165 (1977).
21. H. Goldberg, *Phys. Rev. Lett.* **50**, 1419 (1983); J. Ellis *et al.*, *Nucl. Phys.* B **238**, 453 (1984); G. Bertone, D. Hooper and J. Silk, *Phys. Rept.* **405**, 279 (2005).
22. S. Martin, *A Supersymmetry Primer*, hep-ph/9709356 (2008), and references therein.
23. P. Fayet, Proc. 16th Renc. Moriond, Les Arcs, v.1, 347 (1981); H. Pagels and J. Primack, *Phys. Rev. Lett.* **48**, 223 (1982).
24. P. Fayet, *Phys. Lett.* B **84**, 416 (1979).
25. S. Ferrara, L. Girardello and F. Palumbo, *Phys. Rev.* D **20**, 403 (1979).
26. P. Fayet, *Phys. Lett.* B **78**, 417 (1978).
27. P. Fayet, *Nucl. Phys.* B **113**, 135 (1976); B **149**, 137 (1979).
28. P. Fayet, *Phys. Lett.* B **159**, 121 (1985); *Nucl. Phys.* B **263**, 649 (1986); Proc. 2nd Nobel Symp. on El. Part. Phys., Marstrand (1986), Physica Scripta T **15**, 46 (1987).
29. L. Girardello and M. Grisaru, *Nucl. Phys.* B **194**, 65 (1982).
30. I. Aitchison, *Supersymmetry in Particle Physics. An Elementary Introduction.* SLAC-R-865 (2007) and Cambridge Univ. Pr. (2007).
31. M. Drees, *Int. J. Mod. Phys.* A **4**, 3635 (1989); J. Ellis *et al.*, *Phys. Rev.* D **39**, 844 (1989).
32. P. Fayet, *Nucl. Phys.* B **78**, 14 (1974); R. Peccei and H. Quinn, Phys. Rev. D **16**, 1791 (1977).
33. P. Fayet, *Phys. Lett.* B **95**, 285 (1980); Nucl. Phys. B **187**, 184 (1981).
34. C. Bœhm and P. Fayet, Nucl. Phys. B **683**, 219 (2004); P. Fayet, *Phys. Rev.* D **70**, 023514 (2004); D **75**, 115017 (2007).

Experimental particle physics aspects of dark matter and dark matter candidates

NEUTRINO SIGNATURE OF INERT DOUBLET DARK MATTER (ABSTRACT)

Sarah Andreas[1,2,3]

[1]*Institut für Theoretische Physik E, RWTH Aachen University, D-52056 Aachen, Germany*

[2]*Service de Physique Théorique, Université Libre de Bruxelles, B-1050 Brussels, Belgium*

[3]*Deutsches Elektronen-Synchrotron DESY, Notkestraße 85, D-22607 Hamburg, Germany, Currently at DESY, email: Sarah.Andreas@desy.de*

In the framework of the Inert Doublet Model and extensions, the signature of neutrinos from dark matter annihilation in the Earth, the Sun and at the Galactic centre is presented. The model contains an extra Higgs doublet, a neutral component of which is chosen as dark matter candidate. There are three distinct mass ranges for which consistency both with WMAP abundance and direct searches can be obtained: a low (4 - 8 GeV), a middle (60 - 70 GeV) and a high (500 - 1500 GeV) WIMP mass range. The first case is of interest as we showed that the model can at the same time give the correct WMAP abundance and account for the positive DAMA results without contradicting other direct searches. We present how capture in the Sun can further constrain this scenario using Super-Kamiokande data. Indirect detection through neutrinos is challenging for the middle and high mass ranges. For the former, the presence of the so-called 'iron resonance' gives rise to larger neutrino fluxes for WIMP masses around 60 - 70 GeV since capture by the Earth is enhanced. The addition of light right-handed Majorana neutrinos to the particle content of the model further increases the signal since it opens a direct annihilation channel into mono-energetic neutrinos. Neutrinos from the Galactic centre might be detected for heavy WIMPs if the dark matter density at the Galactic centre is substantially boosted.

Keywords: dark matter, astronomical observations of theories beyond the SM, neutrino experiments
PACS: 95.35.+d, 12.60.Fr, 95.85.Ry

RELIC DENSITY AND FUTURE COLLIDERS: INVERSE PROBLEM(S) (ABSTRACT)

Alexandre Arbey[1] **and Farvah Mahmoudi**[2]

[1]*Université de Lyon, Lyon, F-69000, France; Université Lyon 1, Villeurbanne, F-69622, France; Centre de Recherche Astrophysique de Lyon, Observatoire de Lyon, 9 avenue Charles André, Saint-Genis Laval cedex, F-69561, France; CNRS, UMR 5574; Ecole Normale Supérieure de Lyon, Lyon, France.*

[2]*Laboratoire de Physique Corpusculaire de Clermont-Ferrand (LPC), Université Blaise Pascal, CNRS/IN2P3, 63177 Aubière Cedex, France.*

Relic density calculations are often used to constrain particle physics models, and in particular supersymmetry. We will show that the presence of additional energy or entropy before the Big-Bang nucleosynthesis can however

CP1241, *Invisible Universe: Proceedings of the Conference,* edited by J.-M. Alimi and A. Füzfa

completely change the relic density constraints on the SUSY parameter space. Therefore one should be extremely careful when using the relic density to constrain supersymmetry as it could give misleading results, especially if combined with the future collider data. Alternatively, we will also show that combining the discoveries of the future colliders with relic density calculations can shed light on the inaccessible pre-BBN dark time physics. Finally we will present SuperIso Relic, a new relic density calculator code in Supersymmetry, which incorporates alternative cosmological models, and is publicly available.

Keywords: Dark matter; Relic density; Supersymmetry; Very early Universe
PACS: 11.30.Pb, 12.60.Jv, 95.35.+d, 95.36.+x

GALACTIC DARK MATTER HALO MADE OF SPIN-ZERO BOSONS (ABSTRACT)

A. Bernal[1], J. Barranco[1], D. Alic[2,3] and C. Palenzuela[1,4]

[1]*Max-Planck-Institut für Gravitationsphysik, Albert Einstein Institut, 14476 Golm, Germany.*

[2]*Department of Physics, Universitat de les Illes Balears, Cra. Valldemossa, Spain.*

[3]*Currently at Max-Planck-Institut für Gravitationsphysik, Albert Einstein Institut, 14476 Golm, Germany.*

[4]*Currently at Canadian Institute of Theoretical Astrophysics, Toronto, Ontario, Canada.*

Scalar Field Dark Matter models, in which the dark matter particle is a spin-0 boson, are becoming a serious alternative to the Cold Dark Matter paradigm. In the present work, we have constructed gravitational-bounded configurations made of massive spin zero bosons that can be used as models for galactic dark matter halos. These configurations are constructed in the context of Einstein's General Relativity and contrary to previous studies, where the bosons populate only the ground state, different excited states are coexisting simultaneously. We show that these "mixed states" are stable under radial perturbations and that they provide more realistic rotation curves than single configurations.

Keywords: Dark matter, galactic halo, rotation curves
PACS: 04.40.-b,04.40.Dg,95.35.+d

DARK MATTER SOLUTIONS TO THE COSMIC RAY LEPTON PUZZLE (ABSTRACT)

Pierre Brun

CEA, Irfu, Service de Physique des Particules, Centre de Saclay, F-91191 Gif sur Yvette, France

Recent measurements of cosmic ray leptons by PAMELA, ATIC, HESS and Fermi revealed interesting excesses. Many authors suggested particle Dark Matter (DM) annihilations could be at the origin of these effects. In this paper, we critically assess this interpretation by reviewing some results questioning the naturalness and robustness of such an interpretation. Natural values for the DM particle parameters lead to a poor leptons production so that models often require signal enhancement effects that we constrain here. Considering DM annihilations are likely to produce antiprotons as well, we use the PAMELA antiproton to proton ratio measurements to constrain possible exotic contribution. We also consider the possibility of an enhancement due to a nearby clump of DM. This scenario appears unlikely when compared to the state-of- the-art cosmological N-body simulations. We conclude that the bulk of the observed signals most likely has no link with DM and is rather a new, yet unconsidered source of background for searches in these channels.

Keywords: Particle dark matter, cosmic ray leptons, enhancement factors
PACS: 95.35.+d

EARLY UNIVERSE COSMOLOGY WITH MIRROR DARK MATTER (ABSTRACT)

P. Ciarcelluti

IFPA, Département AGO, Université de Liège, B-4000 Liège, Belgium

Mirror matter is a stable self-collisional dark matter candidate. If exact mirror parity is a conserved symmetry of nature, there could exist a parallel hidden (mirror) sector of the Universe which has the same kind of particles and the same physical laws of our (visible) sector. The two sectors interact each other predominantly via gravity, therefore mirror matter is naturally "dark". Here I briefly review the cosmological signatures of mirror dark matter, as Big Bang nucleosynthesis, primordial structure formation and evolution, cosmic microwave background and large scale structure power spectra, together with its compatibility with the interpretation of the DAMA annual modulation signal in terms of photonÐmirror-photon kinetic mixing. Summarizing the present status of research and comparing theoretical results with observations/experiments, it emerges that mirror matter is not just a viable, but a promising dark matter candidate.

Keywords: dark matter, structure formation, Big Bang nucleosynthesis, cosmic microwave background, large scale structure
PACS: 26.35.+c, 95.30.Cq, 95.35.+d, 98.65.Dx, 98.70.Vc, 98.80.-k

EDELWEISS-II DARK MATTER SEARCH : STATUS AND FIRST RESULTS (ABSTRACT)

Maryvonne De Jésus[1] for the EDELWEISS Collaboration

[1]*IPNL, Université de Lyon, Université Lyon 1, CNRS/IN2P3, 4 rue E. Fermi, 69622 Villeurbanne, France*

EDELWEISS is a direct dark matter search experiment operating in the low radioactivity environment of the Modane UndergroundLaboratory. The experiment runs at very low temperature massive Ge detectors in order to identify nuclear recoils induced by elastic scattering of WIMPs from the galactic halo. The commissioning of the second phase of the experiment, involving more than 7 kg of Ge detector, has been completed in 2007. These ionization/heat detectors allow an event by event electronic/nuclear recoil discrimination. Preliminary results of a new type of bolometer with interdigitized electrodes for the charge collection with active rejection are presented.

Keywords: EDELWEISS, EURECA, Dark Matter, WIMPS, ID Detectors, Cryogenic.
PACS: 95.35.+d, 14.80.Ly, 98.80.Es, 29.40.W, 95.85.Ry, 07.57.K

REGULAR BLACK HOLE REMNANTS (ABSTRACT)

Irina Dymnikova[1,2]

[1] *Department of Math. and Comp. Science, University of Warmia and Mazury, 10-561 Olsztyn, Poland*

[2] *A.F. Ioffe Physico-Technical Institute, Politekhnicheskaya 26, 194021 St.Petersburg, Russia*

The Einstein equations admit the class of regular solutions asymptotically de Sitter at the origin ($\Lambda\delta_k^i$ as $r \to 0$) and at infinity ($\lambda\delta_k^i$ as $r \to \infty$). In a certain range of mass M they describe a regular cosmological black holes with three horizons, a cosmological horizon r_c, a black hole horizon $r_b < r_c$ and an internal horizon $r_a < r_b$. Generic asymptotic behavior of a metric function defines generic dynamics of evaporation: It involves a phase transition where a specific heat is broken and changes its sign. Thermodynamical preference for a final product of the Hawking evaporation is a thermodynamically stable double-horizon ($r_a = r_b$) remnant with the positive specific heat.

Keywords: regular black holes, dark matter
PACS: 98.80.Es; 04.70.Bw; 04.20.Dw

MODELLING ULTRA-FINE STRUCTURE IN DARK MATTER HALOS (ABSTRACT)

Daniele S. M. Fantin, Michael R. Merrifield and Anne M. Green

School of Physics & Astronomy, University of Nottingham, University Park, Nottingham, NG7 2RD, United Kingdom.

Dark matter plays a fundamental role in theories of formation and evolution of galaxies, so every attempt to model galaxy formation and evolution has to consider the presence of dark halos. Moreover, mergers and accretions appear to be fundamental driving mechanisms in determining the present day properties of galaxies. Our project studies the ultra-fine distribution of dark matter in the Milky Way and the implications for the next generation of dark matter particle detectors. We have developed a model for halo mergers in a Milky Way-like galaxy. The signals expected in the lab-based dark matter detection experiments depend on the phase-space distribution on submilliparsec scales: with our numerical technique it is possible to resolve structures, produced by mergers of subhalos with a larger parent halo, inaccessible to conventional N-body simulations. When applied in a cosmological context, this method becomes a powerful instrument to reproduce and analyse the complete multiple merger history of a Milky Way-like system. Preliminary results suggest that fine-grained features from mergers that occurred a few Gyrs ago survive until now. These features may raise also the possibility of using them to unravel the complete merger history of the Milky Way.

Keywords: methods: numerical - Galaxy: evolution - Galaxy: halo - Galaxy: dynamics and kinematics - Galaxy: solar neighbourhood - dark matter
PACS: 95.35.+d; 98.35.G

SELF-ANNIHILATING DARK MATTER AND THE CMB: REIONIZING THE UNIVERSE AND CONSTRAINING CROSS SECTIONS (ABSTRACT)

Fabio Iocco[1,2]

Institut de Physique Théorique, CNRS, URA 2306 & CEA/Saclay, F-91191 Gif-sur-Yvette, France

Institut d'Astrophysique de Paris, UMR 7095-CNRS Paris, Université Pierre et Marie Curie, 98bis bd Arago, F-75014, Paris, France.

I summarize the recent advances in determining the effects of self-ahhinilating WIMP dark matter on the modification of the recombination history, at times earlier than the formation of astrophysical objects. Depending on mass and self-annihilation cross section, WIMP DM can reproduce sizable amounts of the total free electron abundance at $z \gtrsim 6$; as known, this affects the CMB temperature and polarization correlation spectra, and can be used to place stringent bounds in the particle mass vs cross-section plane. WMAP5 data already strongly disfavor the region capable to explain the recent positron anomaly in terms of DM annihilation, whereas in principle the Planck mission has the potential to see a signal produced by a candidate in that region, or from WIMPs with thermal annihilation cross-sections $\langle \sigma v \rangle \sim 3 \times 10^{-26} \mathrm{cm}^3/\mathrm{s}$ and masses with values $m_\chi \lesssim 50$ GeV/c^2.

Keywords: dark matter, reionization, CMB
PACS: 26.35.+c, 98.80.Cq, 98.80.Ft

THE PUZZLES OF DARK MATTER SEARCHES (ABSTRACT)

Maxim Yu. Khlopov[1,2]

[1] *Centre for Cosmoparticle Physics "Cosmion" 115409 Moscow, Russia; Moscow Engineering Physics Institute (National Nuclear Research University), 115409 Moscow, Russia*

[2] *APC laboratory 10, rue Alice Domon et Léonie Duquet 75205 Paris Cedex 13, France*

Positive results of dark matter searches in DAMA/NaI and DAMA/LIBRA experiments, being put together with negative results of other groups, imply nontrivial particle physics solutions for cosmological dark matter. Stable particles with charge -2 bind with primordial helium in O-helium "atoms" (OHe), representing a specific Warmer than Cold nuclear-interacting form of dark matter. Slowed down in the terrestrial matter, OHe is elusive for direct methods of underground Dark matter detection like those used in CDMS experiment, but its reactions with nuclei can lead to annual variations of energy release in the interval of energy 2-6 keV in DAMA/NaI and DAMA/LIBRA experiments. Schrodinger equation for system of nucleus and OHe is considered and reduced to an equation of relative motion in a spherically symmetrical potential well, formed by the Yukawa tail of nuclear scalar isoscalar attraction potential, acting on He beyond the nucleus, and dipole Coulomb repulsion between the nucleus and OHe at distances from the nuclear surface, smaller than the size of OHe. The values of coupling strength and mass of meson, mediating scalar isoscalar nuclear potential, are rather uncertain. Within these uncertainties we find a narrow window of these parameters, at which the sodium and/or iodine nuclei have a few keV binding energy with OHe. The concentration of OHe in the matter of underground detectors is adjusted to the incoming flux of cosmic O-helium at the timescale less than few minutes. Therefore the rate of radiative capture of Na and/or I by OHe should experience annual modulations. Transitions to more energetic levels of Na+OHe (I+OHe) system imply tunneling through dipole Coulomb barrier that leads to suppression of annual modulation of events with MeV-tens MeV energy release in the correspondence with the results of DAMA experiments. The proposed explanation inevitably leads to prediction of abundance of anomalous Na (and/or I) corresponding to the signal, observed by DAMA. At nuclear parameters, reproducing DAMA results, the energy release predicted for detectors with chemical content other than NaI differ in the most cases from the one in DAMA detector. In particular, it is shown that in the case of CDMS the energy of

OHe-germanium bound state is beyond the range of 2-6 keV and its formation should not lead to ionization in the energy interval of DAMA signal.

Keywords: elementary particles, nuclear reactions, nucleosynthesis, abundances, dark matter, early universe, large-scale structure of universe
PACS: 12.60.Cn,98.90.+s,12.60.Nz,14.60.Hi,26.35.+c,36.90.+f,03.65.Ge

COSMIC RAY POSITRON EXCESS: IS THE DARK MATTER SOLUTION A GOOD BET? (ABSTRACT)

Julien Lavalle

Università di Torino & INFN, Dipartimento di Fisica Teorica, Via Giuria 1, 10125 Torino – Italia

The recent observation by the PAMELA satellite of a rising positron fraction up to ? 100 GeV has triggered a considerable amount of putative interpretations in terms of dark matter (DM) annihilation or decay. Here, we make a critical reassessment of such a possibility, recalling the elementary conditions with respect to the standard astrophysical background that would make it likely, showing that they are not fulfilled. Likewise, we argue that, as now well accepted, DM would need somewhat contrived properties to contribute significantly to the observed positron signal, even when including e.g. clumpiness effects. This means that most of natural DM candidates arising in particle physics beyond the standard model are not expected to be observed in the cosmic antimatter spectrum, unfortunately. However, this does not prevent them from remaining excellent DM candidates, this only points towards the crucial need of developing much more complex detection strategies (multimessenger, multiwavelength, multiscale searches)

Keywords: Dark matter; Galactic cosmic rays
PACS:95.35.+d;12.60.-i;95.30.Cq;96.50.S-

DARK MATTER SEARCHES WITH IMAGING ATMOSPHERIC CHERENKOV TELESCOPES (ABSTRACT)

E. Moulin

CEA Saclay, Irfu, Service de Physique des Particules, F-91991 Gif-sur-Yvette Cedex, France

The annihilation of WIMPs is required to reproduce the correct dark matter relic abundance in the Universe. This process occurs in dense regions. Amongst the plausible targets to look for dark matter are the Galactic Center, the center of galaxy clusters, dwarf satellite galaxies and substructures in Galactic haloes. WIMP annihilations produce high energy gamma-rays in the final state. which may be detected by imaging atmospheric Cherenkov telescopes (IACTs). I will review on the recent results from observations of ongoing IACTs.

Keywords: Gamma-rays : observations, Dark matter
PACS:98.70.Rz, 95.35.+d

RADIO SIGNAL CONSTRAINTS ON GALACTIC DARK MATTER ANNIHILATION (ABSTRACT)

E. Borriello[1], A. Cuoco[2], G. Mangano[3], G. Miele[1,3] and C. A. Moura[3,4]

[1] *Università "Federico II", Dipartimento di Scienze Fisiche, Napoli, Italy.*

[2] *Department of Physics and Astronomy, University of Aarhus, Ny Munkegade, Bygn. 1520 8000 Aarhus Denmark.*

[3] *INFN Sezione di Napoli, Complesso Universitario di Monte S. Angelo, Via Cintia, I-80126 Napoli, Italy.*

[4] *Instituto de Física "Gleb Wataghin" - UNICAMP, 13083-970, Campinas, São Paulo, Brazil.*

Annihilation of Dark Matter usually produces together with gamma rays comparable amounts of electrons and positrons. The e^+e^- gyrating in the galactic magnetic field then produce secondary synchrotron radiation which thus provides an indirect mean to constrain the DM signal itself. To this purpose, we calculate the radio emission from the galactic halo as well as from its expected substructures and we then compare it with the measured diffuse radio background. We employ a multi-frequency approach using data in the relevant frequency range 100 MHz–100 GHz, as well as the WMAP Haze data at 23 GHz. The derived constraints are of the order $\langle \sigma_A v \rangle = 10^{-24}\,\mathrm{cm^3 s^{-1}}$ for a DM mass $m_\chi = 100\,\mathrm{GeV}$ even though sensibly depending on the astrophysical uncertainties.

Keywords: Dark Matter Annihilation, Synchrotron Radiation, Radio Signal
PACS: 95.30.Gv, 95.35.+d, 95.85.Bh, 98.70.Vc

ARE NON-RELATIVISTIC NEUTRINOS THE DARK MATTER PARTICLES? (ABSTRACT)

Theo M. Nieuwenhuizen

Institute for Theoretical Physics, University of Amsterdam, Valckenierstraat 65, 1018 XE Amsterdam, The Netherlands

The dark matter of a spherical, relaxed galaxy cluster is modeled by isothermal, non-interacting fermions; the galaxies and X-ray gas by isothermal classical distributions. A fit to lensing data of the cluster Abell 1689 works well and yields a mass of a few eV. This low value casts doubt on the existence of a Cold Dark Matter particle.

The best case is the neutrino, for which in the cluster all 12 left- and righthanded modes are available. The fit gives an average mass $1.45(h/0.70)^{1/2}$ eV, with 2% error, while neutrino oscillations bring deviations of order meV. A neutrino mass between 0.2 and 2 eV will be searched in the Katrin experiment in 2012. The ideal value is $m_\nu = Y_e m_e = 1.4998$ eV, where $Y_e = 2^{3/4} G_F{}^{1/2} m_e$ is the Yukawa coupling of the electron. It occurs for reduced Hubble constant $h = 0.744$ with 4% error, right on top of and slightly sharper than the presently best supernova value of Riess et al. 2009, $h = 0.742$ with 4.8% error.

In the cluster the neutrinos have a temperature of 0.045 K and a de Broglie length of 0.20 mm. They establish a quantum structure of several million light years across, the largest known in the Universe. The virial α-particle temperature of $9.9 \pm 1.1\,\mathrm{keV}/k_B$ coincides with the average one of X-rays, while also the gas profile comes out well.

Active neutrinos alone with the 1.45 eV mass give some 9.5 % dark matter, more than allowed by the cold dark matter papradigm. A dark matter fraction of some 19%, $\Omega_\nu = (h/0.70)^{-3/2} 0.189$ (4), occurs for 12 degrees of freedom, i. e., for 3 families of left plus right handed neutrinos. The sterile modes may be produced in the early universe if there is a small Majorana mass matrix of order meV, on top of the Dirac matrix with ~ 1.45 eV masses.

The neutrinos are free-streaming in the early universe and play no role during the decoupling. But now they are not homogeneous anymore. They condense on the Abell 1689 cluster fairly late, at redshift $z \sim 6-8$, a prediction testable in future observations. Thereby the spead up the intracluster gas to virial speeds of 10 keV, which causes reionization without assistance of heavy stars. Within the analysis, the baryons are poor tracers of the dark matter density.

This work is described in Theo M. Nieuwenhuizen, *Do non-relativistic neutrinos constitute the dark matter?* Europhysics Letters 86, 59001 (2009). This text of this paper is an update of this work. Structure formation is presently believed to need cold dark matter. However, hydrodynamics alone may explain baryonic clustering without this trigger. Th. M. Nieuwenhuizen, C. H. Gibson and R. E. Schild, *Gravitational hydrodynamics of large scale structure formation*, Europhysics Letters 2009, to appear.

Keywords: Dark matter, Galaxy groups, clusters, and superclusters, Non-standard-model neutrinos, right-handed neutrinos, etc.
PACS: 95.35.+d, 98.65.-r, 14.60.St

THE ARDM - A TON - SCALE LIQUID ARGON EXPERIMENT FOR DIRECT DARK MATTER DETECTION (ABSTRACT)

Polina Otyugova, *on behalf of ArDM collaboration*

Physik-Institut der Universität Zürich, Winterthurerstrasse 190, CH-8057 Zürich, Switzerland

The ArDM is a ton-scale double phase detector for the direct search of the Weakly Interacting Massive Particle (WIMPs) as Dark Matter candidates. The detector is based on a liquid Argon (LAr) target. The present goal is to assemble, fully characterize the detector on the surface and then operate it at an underground facility. The scintillation light and ionization charge produced by recoiling nuclei in WIMP-Ar collision can be measured independently. The discrimination of the WIMP induced nuclear recoils from the electron/gamma background is done using the pulse discrimination technique and the ratio between the produced light and charge. The experiment and the last results from the detector commissioning are presented.

Keywords: Dark Matter, liquid argon
PACS: 29.40

MULTI-MESSENGER CONSTRAINTS ON THE ANNIHILATING DARK MATTER INTERPRETATION OF THE POSITRON EXCESS (ABSTRACT)

Miguel Pato[1], Lidia Pieri[2] and Gianfranco Bertone[3]

Institut d'Astrophysique de Paris, UMR7095-CNRS, 98bis bd Arago, 75014, Paris, France.

[1] *email=pato@iap.fr*

[2] *email=pieri@iap.fr*

[3] *email=bertone@iap.fr*

The rise in the energy spectrum of the positron ratio, observed by the PAMELA satellite above 10 GeV, and other cosmic ray measurements, have been interpreted as a possible signature of Dark Matter annihilation in the Galaxy. However, the large number of free parameters, and the large astrophysical uncertainties, make it difficult to draw conclusive statements about the viability of this scenario. Here, we perform a multi-wavelength, multi-messenger analysis, that combines in a consistent way the constraints arising from different astrophysical observations. We show that if standard assumptions are made for the distribution of Dark Matter (we build models on the recent Via Lactea II and Aquarius simulations) and the propagation of cosmic rays, current Dark Matter models cannot explain the observed positron flux without exceeding the observed fluxes of antiprotons or gamma-ray and radio photons. To visualize the multi-messenger constraints, we introduce "star plots", a graphical method that shows in the same plot theoretical predictions and observational constraints for different messengers and wavelengths.

Keywords: dark matter, N-body simulations
PACS: 95.35.+d, 98.35.Gi, 98.35.Jk

ON THE CIRCULAR POLARISATION OF LIGHT FROM AXION-PHOTON MIXING (ABSTRACT)

A. Payez, J.R. Cudell and D. Hutsemékers

University of Liège, Allée du 6 Août 17, 4000 Liège, Belgium

From the analysis of measurements of the linear polarisation of visible light coming from quasars, the existence of large-scale coherent orientations of quasar polarisation vectors in some regions of the sky has been reported. Here, we show that this can be explained by the mixing of the incoming photons with nearly massless pseudoscalar (axion-like) particles in extragalactic magnetic fields. We present a new treatment in terms of wave packets and discuss its implications for the circular polarisation.

Keywords: axion, polarisation, wave packets, active galactic nuclei, large-scale structure of the Universe
PACS: 14.80.Mz, 95.30.Gv, 98.54.Aj, 98.65.Dx

NEUTRALINO DARK MATTER IN MSSM MODELS WITH NON-UNIVERSAL HIGGS MASSES (ABSTRACT)

Pearl Sandick

Theory Group and Texas Cosmology Center, The University of Texas at Austin, TX 78712, USA

We consider the Minimal Supersymmetric Standard Model (MSSM) with varying amounts of non-universality in the soft supersymmetry-breaking contributions to the Higgs scalar masses. In addition to the constrained MSSM (CMSSM) in which these are universal with the soft supersymmetry-breaking contributions to the squark and slepton masses at the input GUT scale, we consider scenarios in which both the Higgs masses are non-universal by the same amount (NUHM1), and scenarios in which they are independently non-universal (NUHM2). As the lightest neutralino is a dark matter candidate, we demand that the relic density of neutralinos not be in conflict with measurements by WMAP and others, and examine the viable regions of parameter space. Prospects for direct detection of neutralino dark matter via elastic scattering in these scenarios are discussed.

Keywords: Supersymmetry, Dark Matter
PACS: 11.30.Pb, 95.35.+d

THE XENON100 DARK MATTER EXPERIMENT (ABSTRACT)

E. Tziaferi (on behalf of the XENON100 collaboration)

Physics Institute, University of Zürich, Winterthurerstr. 190, Zürich, Switzerland

The XENON100 experiment is searching for WIMPs, which are particles that may consist dark matter. It is located in the underground laboratory of Gran Sasso (LNGS) in Italy at a depth of ~3600 m.w.e..The experiment description, its performance and the expected background based on Monte Carlo simulations and material screening along with the projected sensitivities of the experiment are presented. In addition, a brief description of the upgrade XENON100 detector is given.

Keywords: Dark Matter, Liquid Noble Gas Detector, Background Radiation
PACS: 95.35+d, 95.55Vj, 98.70.Vc

ANNIHILATION LINES FROM DARK MATTER WITH THE FERMI-LAT (ABSTRACT)

Tomi Ylinen, on behalf of the Fermi LAT Collaboration

The Oskar Klein Centre for Cosmoparticle Physics & Department of Physics, Royal Institute of Technology (KTH), AlbaNova, SE-106 91 Stockholm, Sweden School of Pure and Applied Natural Sciences, University of Kalmar, SE-391 82 Kalmar, SwedenPhysics Institute, University of Zürich, Winterthurerstr. 190, Zürich, Switzerland

Dark matter is today one of the most intriguing but so far unresolved issues in physics. Many extensions of the Standard Model of particle physics predict a stable Weakly Interacting Massive Particle (WIMP) that may annihilate directly into two gamma-rays. If the WIMPs are non-relativistic, the gamma-rays from this channel will have an energy equal to the mass of the WIMP. The signature caused by this annihilation is a spectral line, smeared out only by the energy resolution of the detector. The signal would be a "smoking gun" for dark matter, since no other astrophysical source should be able to produce it. We present here the preliminary results from the search for a dark matter line on a limited data set from the Fermi Large Area Telescope (LAT), the main instrument onboard the Fermi Gamma-ray Space Telescope, which was successfully launched on June 11, 2008. The Fermi-LAT is a pair-conversion detector for gamma-rays with an energy range from 20 MeV to 300 GeV and has an unprecedented resolution and sensitivity.

Keywords: dark matter, annihilation, lines, Fermi, LAT
PACS: 95.35.+d, 95.85.Pw, 98.70.Vc

GRAVITATION AND COSMOLOGY

A PHYSICAL INTERPRETATION OF GRAVITATIONAL FIELD EQUATIONS

T. Padmanabhan

IUCAA, Pune University Campus,
Ganeshkhind, Pune 411007, INDIA.
email: nabhan@iucaa.ernet.in

Abstract. It is possible to provide a thermodynamic interpretation for the field equations in any diffeomorphism invariant theory of gravity. This insight, in turn, leads us to the possibility of deriving the gravitational field equations from another variational principle without using the metric as a dynamical variable. I review this approach and discuss its implications.

Keywords: gravity, field equations, black hole entropy, horizon, Lanczos-Lovelock models, Rindler frame
PACS: 04.20.-q,04.20.Fy,04.60.-m,04.62.+v,04.70.Dy

1. MOTIVATION AND SUMMARY

An unsatisfactory feature of all theories of gravity is that the field equations do not have any direct physical interpretation. The lack of an elegant principle which can lead to the dynamics of gravity ("how matter tells spacetime to curve") is quite striking when we compare this situation with the kinematics of gravity ("how spacetime makes the matter move"). The latter can be determined through the principle of equivalence by demanding that all freely falling observers, at all events in spacetime, must find that the equations of motion for matter reduce to their special relativistic form. Our first aim will be to remedy this and provide a physical interpretation to field equations describing gravity in any diffeomorphism invariant theory. This, in turn, will lead us to the possibility of *deriving* the gravitational field equations from a thermodynamic variational principle without using the metric as a dynamical variable.

The alternative interpretation is based on the thermodynamics of horizons. Several recent investigations have shown that there is indeed a deep connection between gravitational dynamics and horizon thermodynamics (for a review, see [1]). For example, studies have shown that:

- Gravitational field equations in *a wide variety of theories*, when evaluated on a horizon, reduce to a thermodynamic identity $TdS = dE + PdV$. This result, first pointed out in ref.[2], has now been demonstrated [3] in several cases like the stationary axisymmetric horizons and evolving spherically symmetric horizons in Einstein gravity, static spherically symmetric horizons and dynamical apparent horizons in Lovelock gravity, and three dimensional BTZ black hole horizons, FRW cosmological models in various gravity theories and even [4] in the case Horava-Lifshitz Gravity. It is not possible to understand, in the conventional approach, why the field equations should encode information about horizon thermodynamics.
- Gravitational action functionals in a wide class of theories have a a surface term and a bulk term. In the conventional approach, we *ignore* the surface term completely (or cancel it with a counter-term) and obtain the field equation from the bulk term in the action. Any solution to the field equation obtained by this procedure is logically independent of the nature of the surface term. But when the *surface term* (which was ignored) is evaluated at the horizon that arises in any given solution, it gives the entropy of the horizon! (Again, this result extends far beyond Einstein's theory to situations in which the entropy is not just proportional to horizon area.) This is possible only because there is a specific holographic relationship [5, 6] between the surface term and the bulk term which, however, is an unexplained feature in the conventional approach to gravitational dynamics. Since the surface term has the thermodynamic interpretation as the entropy of horizons, and is related holographically to the bulk term, we are again led to an indirect connection between spacetime dynamics and horizon thermodynamics.

Based on these features — *which have no explanation in the conventional approach* — one can argue that there is a conceptual reason why we need to relate horizon thermodynamics with gravitational dynamics (in a wide class of

CP1241, *Invisible Universe: Proceedings of the Conference,* edited by J.-M. Alimi and A. Füzfa

theories far more general than just Einstein gravity) and revise our perspective towards spacetime. To set the stage for this future discussion, we begin by recalling the implications of the existence of temperature for horizons.

In the study of normal macroscopic systems — like, for example, a solid or a gas — one can *deduce* the existence of microstructure just from the fact that the object can be heated. This was the insight of Boltzmann which led him to suggest that heat is essentially a form of motion of the microscopic constituents of matter. That is, the existence of temperature is sufficient for us to infer the existence of microstructure even without any direct experimental evidence.

The non-zero temperature of a horizon shows that we can actually heat up a spacetime, just as one can heat up a solid or a gas. An unorthodox way of doing this would be to take some amount of matter and arrange it to collapse and form a black hole. The Hawking radiation [7] emitted by the black hole can be used to heat up, say, a pan of water just as though the pan was kept inside a microwave oven. In fact the same result can be achieved by just accelerating through the inertial vacuum carrying the pan of water which will eventually be heated to a temperature proportional to the acceleration [8]. These processes show that the temperatures of all horizons are as real as any other temperature [9]. Since they give rise to a class of hot spacetimes, it follows *à la* Boltzmann that the spacetimes should possess microstructure.

In the case of a solid or gas, we know the nature of this microstructure from atomic and molecular physics. Hence, in principle, we can work out the thermodynamics of these systems from the underlying statistical mechanics. This is not possible in the case of spacetime because we have no clue about its microstructure. However, one of the remarkable features of thermodynamics — in contrast to statistical mechanics — is that the thermodynamic description is fairly insensitive to the details of the microstructure and can be developed as a fairly broad frame work. For example, a thermodynamic identity like $TdS = dE + PdV$ has a universal validity and the information about a *given* system is only encoded in the form of the entropy functional $S(E,V)$. In the case of normal materials, this entropy arises because of our coarse graining over microscopic degrees of freedom which are not tracked in the dynamical evolution. In the case of spacetime, the existence of horizons for a particular class of observers makes it mandatory that these observers integrate out degrees of freedom hidden by the horizon.

To make this notion clearer, let us start from the principle of equivalence which allows one to construct local inertial frames (LIF) with coordinates X^i, around any event in an arbitrary curved spacetime. Given the LIF, we can now construct a local Rindler frame (LRF) by boosting along one of the directions with an acceleration κ, thereby locally transforming the metric to a form given by:

$$ds^2 = -dT^2 + dX^2 + d\mathbf{x}_\perp^2 = -\kappa^2 x^2 dt^2 + dx^2 + dL\mathbf{x}_\perp^2 = -2\kappa l\, dt^2 + \frac{dl^2}{2\kappa l} + d\mathbf{x}_\perp^2 \tag{1}$$

where $T = x\sinh(\kappa t); X = x\cosh(\kappa t)$ and $l = (1/2)\kappa x^2$. The observers at rest (with $x =$ constant) in the LRF will perceive the $X = T$ null surface as a horizon $\mathcal{H}$ (fig 1). These local Rindler observers and the freely falling inertial observers will attribute different thermodynamical properties to matter in the spacetime. For example, they will attribute different temperatures and entropies to the vacuum state as well as excited states of matter fields. When some matter with energy δE moves close to the horizon — say, within a few Planck lengths because it formally takes infinite Rindler time for matter to actually cross $\mathcal{H}$ — the local Rindler observer will consider it to have transfered an entropy $\delta S = (2\pi/\kappa)\delta E$ to the horizon degrees of freedom. We will show (in Sec. 3) that, when the metric satisfies the field equations of any diffeomorphism invariant theory, this transfer of entropy can be given [10] a geometrical interpretation as the change in the entropy of the horizon.

This result allows us to associate an entropy functional with the null surfaces which the local Rindler observers perceive as horizons. We can now demand that the sum of the horizon entropy and the entropy of matter that flows across the horizons (both as perceived by the local Rindler observers), should be an extremum for all observers in the spacetime. This leads to a constraint on the geometry of spacetime which can be stated, in $D = 4$, as

$$(G_{ab} - 8\pi T_{ab})n^a n^b = 0 \tag{2}$$

for all null vectors n^a in the spacetime. The general solution to Eq. (2) is given by $G_{ab} = 8\pi T_{ab} + \rho_0 g_{ab}$ where ρ_0 has to be a constant because of the conditions $\nabla_a G^{ab} = 0 = \nabla_a T^{ab}$. Hence the thermodynamic principle leads uniquely to Einstein's equation with a cosmological constant in 4-dimensions. Notice, however, that Eq. (2) has a new symmetry which the standard Einstein's theory does not posses; viz., it is invariant under the transformation $T_{ab} \to T_{ab} + \lambda g_{ab}$. This has important implications for the cosmological constant problem which we will discuss in Sec. 4.3. In $D > 4$, the same entropy maximization leads to a more general class of theories called Lanczos-Lovelock models (see Sec. 4.2).

The constraint on the background geometry in Eq. (2) arises from our demand that the thermodynamic extremum principle should hold for *all* local Rindler observers. This is identical to the manner in which freely falling observers are used to determine how gravitational field influences matter. Demanding the validity of special relativistic laws for the matter variables, as determined by all the freely falling observers, allows us to determine the influence of gravity on matter. In a similar manner, demanding the maximization of entropy of horizons (plus matter), as measured by all local Rindler observers, leads to the dynamical equations of gravity.

In this approach, both the entropy of horizons as well as the entropy of matter flowing across the horizon will be observer dependent thereby introducing a new level of observer dependence in thermodynamics. In particular, observers in different states of motion will have different regions of spacetime accessible to them; for example, an observer falling into a black hole will not perceive its horizon in the same manner as an observer who is orbiting around it. Therefore we are forced to accept that the notion of entropy is an observer dependent concept. (At a conceptual level this is no different from the fact different freely falling observers will measure physical quantities differently; but in this case, standard rules of special relativity allow us to translate the results between the observers. We do not yet have a similar set of rules for quantum field theory in noninertial frames.) All these features suggest a deep relationship between quantum theory, thermodynamics and gravity which forms the main theme of this article.

The rest of the article is organized as follows: In the next section, I briefly review some of the background results needed for our discussion. Section 3 describes an interpretation of gravitational field equations in a general, diffeomorphism invariant, theory of gravity. Using these results, it is possible to introduce an entropy maximization principle from which one can obtain the same field equations without using the metric as a dynamical variable. This is done in Sec. 4 and the last section summarizes the results.

2. REVIEW OF SOME STANDARD RESULTS

We begin by summarizing some of the standard results we will need later on. In particular, we will briefly review: (i) the field equations in a general class of theories of gravity as well as (ii) the relation between Noether charge and gravitational entropy of horizons.

2.1 A general class of theories of gravity

Consider a theory for gravity described by the metric g_{ab} coupled to matter with some degrees of freedom generally denoted by q_A. We will take the action describing such a theory in $D-$dimensions to be

$$A = \int d^D x\sqrt{-g}\left[L(R^{ab}_{cd}, g^{ab}) + L_{matt}(g^{ab}, q_A)\right] \tag{3}$$

where L is some scalar built from metric and curvature and L_{matt} is the matter Lagrangian depending on the metric and matter variables q_A. (We have assumed that L does not involve derivatives of curvature tensor, to simplify the discussion.) Varying g^{ab} in this action with suitable boundary conditions, we will get the equations of motion (see e.g. Refs. [6, 11]):

$$2E_{ab} - T_{ab} = 0 \tag{4}$$

where

$$E_{ab} = P_a^{\ cde} R_{bcde} - 2\nabla^c\nabla^d P_{acdb} - \frac{1}{2} L g_{ab}; \quad P^{abcd} \equiv \frac{\partial L}{\partial R_{abcd}} \tag{5}$$

When $\nabla_a P^{abcd} = 0$ these field equations describe a class of theories, called the Lanczos-Lovelock theories which have very interesting geometrical features [12]. In this case, we have

$$E_{ab} = P_a^{\ cde} R_{bcde} - \frac{1}{2} L g_{ab}; \quad P^{abcd} \equiv \frac{\partial L}{\partial R_{abcd}} \tag{6}$$

The crucial difference between Eq. (5) and Eq. (6) is that, the E_{ab} in Eq. (6) contains no derivatives of the metric higher than second order thereby leading to field equations which are second order in the metric. In contrast, Eq. (5) can contain up to fourth order derivatives of the metric.

The Lagrangians which lead to the constraint $\nabla_a(\partial L/\partial R_{abcd}) = 0$ and hence to Eq. (6) are, of course, quite special. They can be written as a sum of terms, each involving products of curvature tensors with the m–th term being a product of m curvature tensors. The general Lanczos-Lovelock Lagrangian has the form,

$$L = \sum_{m=1}^{K} c_m L_{(m)} \; ; \; L_{(m)} = \frac{1}{16\pi} 2^{-m} \delta^{a_1 a_2 \ldots a_{2m}}_{b_1 b_2 \ldots b_{2m}} R^{b_1 b_2}_{a_1 a_2} \cdots R^{b_{2m-1} b_{2m}}_{a_{2m-1} a_{2m}} \,, \tag{7}$$

where the c_m are arbitrary constants and $L_{(m)}$ is the m-th order Lanczos-Lovelock Lagrangian. The $m = 1$ term is proportional to $\delta^{ab}_{cd} R^{cd}_{ab} \propto R$ and leads to Einstein's theory. It is conventional to take $c_1 = 1$ so that the $L_{(1)}$, reduces to $R/16\pi$. The normalizations for $m > 1$ are somewhat arbitrary for individual $L_{(m)}$ since the c_ms are unspecified at this stage. The $m = 2$ term gives rise to what is known as Gauss-Bonnet theory. Because of the determinant tensor, it is obvious that in any given dimension D we can only have K terms where $2K \leq D$. It follows that, if $D = 4$, then only the $m = 1, 2$ are non-zero. Of these, the Gauss-Bonnet term (corresponding to $m = 2$) gives, on variation of the action, a vanishing bulk contribution in $D = 4$. (In dimensions $D = 5$ to 8, one can have both the Einstein-Hilbert term and the Gauss-Bonnet term and so on.)

2.2 Noether charge and the horizon entropy

Since our aim is to provide a thermodynamic interpretation of the field equations in Eq. (4), we first need an expression for horizon entropy in this theory. This has been already obtained by Wald [13] but we will introduce it in a manner appropriate for our purpose.

In any generally covariant theory, the infinitesimal coordinate transformations $x^a \to x^a + \xi^a$ lead to conservation of a Noether current that can be obtained as follows: The variation of the gravitational Lagrangian resulting from arbitrary variations of δg^{ab} generically has to a surface term and hence can be expressed in the form,

$$\delta(L\sqrt{-g}) = \sqrt{-g}\left(E_{ab}\delta g^{ab} + \nabla_a \delta v^a\right). \tag{8}$$

When the variations in δg^{ab} arise through the diffeomorphism $x^a \to x^a + \xi^a$ we have, $\delta(L\sqrt{-g}) = -\sqrt{-g}\nabla_a(L\xi^a)$, with $\delta g^{ab} = (\nabla^a \xi^b + \nabla^b \xi^a)$. Substituting these in Eq. (8) and using the (generalized) Bianchi identity $\nabla_a E^{ab} = 0$, we obtain the conservation law $\nabla_a J^a = 0$, for the current,

$$J^a \equiv \left(L\xi^a + \delta_\xi v^a + 2E^{ab}\xi_b\right) \tag{9}$$

where $\delta_\xi v^a$ represents the boundary term which arises for the specific variation of the metric in the form $\delta g^{ab} = (\nabla^a \xi^b + \nabla^b \xi^a)$. It is also convenient to introduce the antisymmetric tensor J^{ab} by $J^a = \nabla_b J^{ab}$. Using the known expression for $\delta_\xi v^a$ in Eq. (9), it is possible to write an explicit expression for the current J^a for any diffeomorphism invariant theory. For the general class of theories we are considering, the J^{ab} and J^a can be expressed [11] in the form

$$J^{ab} = 2P^{abcd}\nabla_c \xi_d - 4\xi_d \left(\nabla_c P^{abcd}\right) \tag{10}$$

$$J^a = -2\nabla_b \left(P^{adbc} + P^{acbd}\right)\nabla_c \xi_d + 2P^{abcd}\nabla_b \nabla_c \xi_d - 4\xi_d \nabla_b \nabla_c P^{abcd} \tag{11}$$

where $P_{abcd} \equiv (\partial L/\partial R^{abcd})$. These expressions simplify significantly at any event $\mathscr{P}$ where ξ^a behaves like an (approximate) Killing vector and satisfies the conditions

$$\nabla_{(a}\xi_{b)} = 0; \quad \nabla_a \nabla_b \xi_c = R_{cbad}\xi^d \tag{12}$$

(which a true Killing vector will satisfy everywhere). Then one can easily prove that $\delta_\xi v^a = 0$ at the event $\mathscr{P}$; the expression for Noether current simplifies considerably and is given by

$$J^a \equiv \left(L\xi^a + 2E^{ab}\xi_b\right). \tag{13}$$

By considering physical processes involving horizons, it can be shown that the first law of black hole dynamics, for example, is consistent with the identification of the following expression as the horizon entropy:

$$S_{\text{Noether}} = \beta \int d^{D-1}\Sigma_a J^a = \frac{1}{2}\beta \int d^{D-2}\Sigma_{ab} J^{ab} \tag{14}$$

where $\beta^{-1} = 2\pi/\kappa$ is the temperature of the horizon. In the final expression in Eq. (14) the integral is over any surface with $(D-2)$ dimension which is a spacelike cross-section of the Killing horizon on which the norm of ξ^a vanishes. As an example, consider the special case of Einstein gravity for which Eq. (10) reduces to

$$J^{ab} = \frac{1}{16\pi}\left(\nabla^a \xi^b - \nabla^b \xi^a\right) \tag{15}$$

If ξ^a be the timelike Killing vector in the spacetime describing a Schwarzschild black hole, we can compute the Noether charge Q as an integral of J^{ab} over any two surface which is a spacelike cross-section of the Killing horizon on which the norm of ξ^a vanishes. The area element on the horizon can be taken to be $d\Sigma_{ab} = (l_a\xi_b - l_b\xi_a)\sqrt{\sigma}d^{D-2}x$ with l_a being an auxiliary vector field satisfying the condition $l_a\xi^a = -1$. Then the integral in Eq. (14) reduces to

$$S_{Noether} = -\beta\frac{1}{8\pi}\int \sqrt{\sigma}d^{D-2}x(l_a\xi_b)\nabla^b\xi^a = \beta\frac{1}{8\pi}\kappa\int\sqrt{\sigma}d^{D-2}x = \frac{1}{4}A_H \tag{16}$$

where A_H is the horizon area. (We have used the relations $\xi^a\nabla_a\xi^b = \kappa\xi^b$, $l_a\xi^a = -1$, $\beta\kappa = 2\pi$ and the fact that ξ^a is a Killing vector.) This result, of course, agrees with the standard one.

The expression for entropy in Eq. (14) allows one to interpret $\beta_{loc}J^a$, where β_{loc} is the appropriately redshifted local temperature near the horizon, as an entropy density associated with the horizon. This is the interpretation which we will exploit in what follows.

3. THE MEANING OF GRAVITATIONAL FIELD EQUATIONS

With this background, we are now in a position to provide a thermodynamic interpretation for the gravitational field equations in the local rindler frames (LRFs) around any event. We will do this in a manner analogous to the way we use the freely falling observers to determine the kinematics of gravity. At every event in spacetime, we will introduce local Rindler observers and use the horizon thermodynamics perceived by these Rindler observers to constrain the background geometry. We shall begin by making the notion of local Rindler observers and their coordinate systems well defined.

3.1 Local Rindler frame and its horizon

Let us choose any event $\mathscr{P}$ and introduce a local inertial frame (LIF) around it with Riemann normal coordinates $X^a = (T, \mathbf{X})$ such that $\mathscr{P}$ has the coordinates $X^a = 0$ in the LIF. Let k^a be a future directed null vector at $\mathscr{P}$ and we align the coordinates of LIF such that k^a lies in the $X-T$ plane at $\mathscr{P}$ (Fig. 1). We next transform from the LIF to a local Rindler frame (LRF) coordinates x^a by accelerating along the X-axis with an acceleration κ by the usual transformation. The metric near the origin now reduces to the form in Eq. (1) where $(t, x, \mathbf{x}_\perp)$ are the coordinates of LRF. Let ξ^a be the approximate Killing vector corresponding to translation in the Rindler time such that the vanishing of $\xi^a\xi_a \equiv -N^2$ characterizes the location of the local horizon $\mathscr{H}$ in LRF. Usually, we shall do all the computation on a timelike surface infinitesimally away from $\mathscr{H}$ with $N=$ constant, usually called a "stretched horizon". Let the timelike unit normal to the stretched horizon be r_a.

This LRF (with metric in Eq. (1)) and its local horizon $\mathscr{H}$ will exist within a region of size $L \ll \mathscr{R}^{-1/2}$ (where $\mathscr{R}$ is a typical component of curvature tensor of the background spacetime) as long as $\kappa^{-1} \ll \mathscr{R}^{-1/2}$. This condition can always be satisfied by taking a sufficiently large κ. Essentially, the introduction of the LRF uses the fact that we have two length scales in the problem at any event. First is the length scale $\mathscr{R}^{-1/2}$ associated with the curvature components of the background metric over which we have no control. The second is the length scale κ^{-1} associated with the accelerated trajectory which we can choose. Hence we can always ensure that $\kappa^{-1} \ll \mathscr{R}^{-1/2}$.

In fact, this is clearly seen in the Euclidean sector in which the horizon maps to the origin (Fig. 2). The locally flat frame in the Euclidean sector will exist in a region of radius $\mathscr{R}^{-1/2}$ while the trajectory of a uniformly accelerated

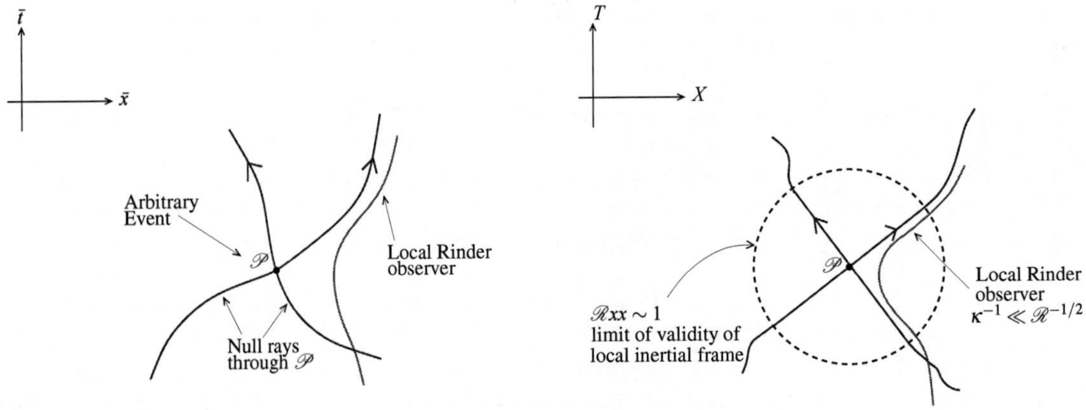

FIGURE 1. The left frame illustrates schematically the light rays near an event $\mathscr{P}$ in the $\bar{t}-\bar{x}$ plane of an arbitrary spacetime. The right frame shows the same neighbourhood of $\mathscr{P}$ in the locally inertial frame at $\mathscr{P}$ in Riemann normal coordinates (T,X). The light rays now become 45 degree lines and the trajectory of the local Rindler observer becomes a hyperbola very close to $T = \pm X$ lines which act as a local horizon to the Rindler observer.

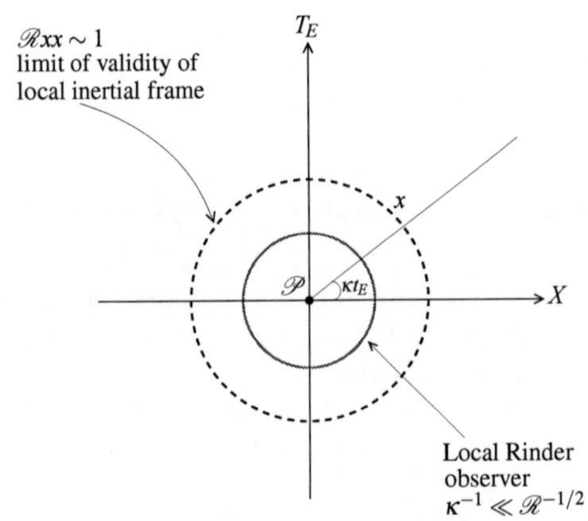

FIGURE 2. The region around $\mathscr{P}$ shown in figure 1 is represented in the Euclidean sector obtained by analytically continuing to imaginary values of T by $T_E = iT$. The horizons $T = \pm X$ collapse to the origin and the hyperbolic trajectory of the Rindler observer becomes a circle of radius κ^{-1} around the origin. The Rindler coordinates (t,x) become — on analytic continuation to $t_E = it$ — the polar coordinates $(r = x, \theta = \kappa t_E)$ near the origin.

observer will be a circle of radius κ^{-1}. Hence one can always keep the latter inside the former. In the Euclidean sector the Rindler observer's trajectory is a circle of radius κ^{-1} which can be made arbitrarily close to the origin. Suppose the observer's trajectory has the usual form $X = \kappa^{-1}\cosh\kappa t; T = \kappa^{-1}\sinh\kappa t$ which is maintained for a time interval of the order of $2\pi/\kappa$. Then, the trajectory will complete a full circle *in the Euclidean sector* irrespective of what happens later. When we work in the limit of $\kappa \to \infty$, this becomes arbitrarily local in both space *and* time. The metric in Eq. (1) is just the metric of the locally flat region in polar coordinates.

More generally, one can choose a trajectory $x^i(\tau)$ such that its acceleration $a^j = u^i\nabla_i u^j$ (where u^i is the time-like four velocity) satisfies the condition $a^j a_j = -\kappa^2$. In a suitably chosen LIF this trajectory will reduce to the standard hyperbola of a uniformly accelerated observer. This construction also defines local Rindler horizons around any event. Further, the local temperature on the stretched horizon will be $\kappa/2\pi N$ so that $\beta_{\rm loc} = \beta N$ with $\beta \equiv 2\pi/\kappa$.

3.2 Thermodynamic interpretation of gravitational field equations

The local Rindler observers will view the thermodynamics of matter around them very differently from the freely falling observers (We will say more about this later on in Sec. 3.3). In particular, they will attribute a loss of entropy $\delta S = (2\pi/\kappa)\delta E$ when matter with an amount of energy δE gets close to the horizon (within a few Planck lengths, say). In the Rindler frame the appropriate energy-momentum density is $T^a_b\xi^b$. (It is the integral of $T^a_b\xi^b d\Sigma_a$ that gives the Rindler Hamiltonian H_R, which leads to evolution in Rindler time t and appears in the thermal density matrix $\rho = \exp -\beta H_R$.) A local Rindler observer, moving along the orbits of the Killing vector field ξ^a with four velocity $u^a = \xi^a/N$, will associate an energy $\delta E = u^a(T_{ab}\xi^b)dV_{\rm prop}$ with a proper volume $dV_{\rm prop}$. If this energy gets transfered across the horizon, the corresponding entropy transfer will be $\delta S_{\rm matter} = \beta_{\rm loc}\delta E$ where $\beta_{\rm loc} = \beta N = (2\pi/\kappa)N$ is the local (redshifted) temperature of the horizon and N is the lapse function. Since $\beta_{\rm loc}u^a = (\beta N)(\xi^a/N) = \beta\xi^a$, we find that

$$\delta S_{\rm matter} = \beta\xi^a\xi^b T_{ab}\, dV_{\rm prop} \tag{17}$$

Consider now the gravitational entropy associated with the local horizon. From the discussion of Noether charge as horizon entropy [see Eq. (14)], we know that $\beta_{\rm loc}J^a$, associated with the (approximate) Killing vector ξ^a, can be thought of as local entropy current. Therefore, $\delta S = \beta_{\rm loc}u_aJ^a dV_{\rm prop}$ can be interpreted as the gravitational entropy associated with a volume $dV_{\rm prop}$ as measured by an observer with four-velocity u^a. (The conservation of J^a ensures that there is no irreversible entropy production or dissipation in the spacetime.) Since ξ^a satisfies Eq. (12) locally, it follows that $\delta_\xi v = 0$ giving the current to be $J^a = \left(L\xi^a + 2E^a_b\xi_b\right)$. For observers moving along the orbits of the Killing vector ξ^a with $u^a = \xi^a/N$ we get

$$\delta S_{\rm grav} = \beta N u_a J^a dV_{\rm prop} = \beta[\xi^j\xi^a(2E_{aj}) + L(\xi_j\xi^j)]\, dV_{\rm prop} \tag{18}$$

As one approaches the horizon, $\xi^a\xi_a \to 0$ making the second term vanish and we find that

$$\delta S_{\rm grav} = \beta[\xi^j\xi^a(2E_{aj})]\, dV_{\rm prop} \tag{19}$$

In the same limit ξ^j will become proportional to the original null vector k^j we started with (viz., ξ^i goes to $\kappa\lambda k^i$ where λ is the affine parameter associated with the null vector k^a we started with) keeping everything finite. We now see that the condition $\delta S_{\rm grav} = \delta S_{\rm matter}$ leads to the result

$$[2E^{ab} - T^{ab}]k_a k_b = 0 \tag{20}$$

Since the original null vector k_a was arbitrary, this equation should hold for all null vectors for all events in the spacetime. This is equivalent to $2E^{ab} - T^{ab} = \lambda g^{ab}$ with some constant λ. (The constancy of λ follows from the conditions $\nabla_a E^{ab} = 0$, $\nabla_a T^{ab} = 0$.)

This provides a purely thermodynamical interpretation of the field equations of any diffeomorphism invariant theory of gravity. Note that Eq. (20) is not quite the same as the standard equation in Eq. (4) because Eq. (20) has an extra symmetry which standard gravitational field equations do not have: These equations are invariant under the shift $T^{ab} \to T^{ab} + \mu g^{ab}$ with some constant μ. (This symmetry has important implications [14] for cosmological constant problem which we will discuss in Sec. 4.3.) While the properties of LRF are relevant conceptually to define the intermediate notions (local Killing vector, horizon temperature ....), the essential result is independent of these notions. *Just as we introduce local inertial frames to decide how gravity couples to matter, we use local Rindler frames to interpret the physical content of the field equations.*

In the above interpretation we used the local, conserved, current J^a. To understand why this is to be expected for consistency, recall that we have crucially used the "democracy of all observers" in demanding entropy balance to hold for all observers. The mathematical content of this active version of general covariance is captured by the diffeomorphism invariance of the underlying theory which determines the dynamics of the spacetime. Because the diffeomorphism invariance of the theory forced us to treat all observers at an equal footing, the diffeomorphism invariance must also lead to the conserved current J^a. We have already seen that this is the case.

The expression for the Noether current is not unique in the sense that one can add to it any term of the form $Q^a = \nabla_b Q^{ab}$ where Q^{ab} is an anti-symmetric tensor thereby ensuring $\nabla_a Q^a = 0$. This ambiguity has been extensively discussed in the literature but for providing the thermodynamic interpretation to the field equations, we have ignored this ambiguity and used the expression in Eq. (9). There are several reasons why the ambiguity is irrelevant for our purpose. First, in a truly thermodynamic approach, one specifies the system by specifying a thermodynamic potential, say, the entropy functional. In a local description, this translates into specifying the entropy current which determines the theory. So it is perfectly acceptable to make a specific choice for J^a consistent with the symmetries of the problem. Second, we shall often be interested in theories in which the equations of motion are no higher than second order. In these (so called Lanczos-Lovelock models) it is not natural to add any extra term to the Noether current such that it is linear in ξ^a as we approach the horizon with a coefficient determined entirely from metric and curvature. Finally, we shall obtain in Sec. 4 the field equation from maximizing an entropy functional where this ambiguity will not arise.

Once we realize that the real physical meaning of the field equations is contained in Eq. (20), it is possible to re-interpret these equations in several alternative ways all of which have the same physical content. We shall mention two of them.

Consider an observer who sees some matter energy flux crossing the horizon. Let r_a be the spacelike unit normal to the stretched horizon Σ, pointing in the direction of increasing N. If ξ^a is the approximate, timelike Killing vector corresponding to Rindler time, then the energy flux through a patch of stretched horizon with normal r_a will be $T^b_a\xi^a r_b$ and the associated entropy flux will be $\beta_{loc}T^b_a\xi^a r_b$ where $\beta_{loc}^{-1} = \beta^{-1}/N$ is the local temperature. We require this entropy flux must match the entropy change of the locally perceived horizon. The gravitational entropy current is given by $\beta_{loc}J^a$, such that $\beta_{loc}(r_aJ^a)$ gives the corresponding gravitational entropy flux. So we require $\beta_{loc}r_aJ^a = \beta_{loc}T^a_b r_a\xi^b$ to hold at all events. The product r_aJ^a for the vector r^a, which satisfies $\xi^a r_a = 0$ on the stretched horizon is $r_aJ^a = 2E^b_a\xi^a r_b$. Hence we get $\beta_{loc}r_aJ^a = 2E^b_a\xi^a r_b = \beta_{loc}T^b_a\xi^a r_b$. As $N \to 0$ and the stretched horizon approaches the local horizon and Nr^i approaches ξ^i (which in turn is proportional to k^i) so that $\beta_{loc}r_a = \beta N r_a \to \beta\xi_a$. So, as we approach the horizon we obtain Eq. (20).

There is another way of interpreting this result which will be useful for further generalizations. Instead of allowing matter to flow across the horizon, one could have equally well considered a virtual, infinitesimal (Planck scale), displacement of the $\mathscr{H}$ normal to itself engulfing some matter. We only need to consider infinitesimal displacements because the entropy of the matter is not 'lost' until it crosses the horizon; that is, until when the matter is at an infinitesimal distance (a few Planck lengths) from the horizon. Hence an infinitesimal displacement of $\mathscr{H}$ normal to itself will engulf some matter. Some entropy will be again lost to the outside observers unless displacing a piece of local Rindler horizon itself costs some entropy. We, therefore, expect the entropy balance condition derived earlier to ensure this and indeed it does. An infinitesimal displacement of a local patch of the stretched horizon in the direction of r_a, by an infinitesimal proper distance ε, will change the proper volume by $dV_{prop} = \varepsilon\sqrt{\sigma}d^{D-2}x$ where σ_{ab} is the metric in the transverse space. The flux of energy through the surface will be $T^a_b\xi^b r_a$ and the corresponding entropy flux can be obtained by multiplying the energy flux by $\beta_{\rm loc}$. Hence the 'loss' of matter entropy to the outside observer because the virtual displacement of the horizon has engulfed some matter is $\delta S_m = \beta_{\rm loc}\delta E = \beta_{\rm loc}T^j_a\xi^a r_j dV_{prop}$. To find the change in the gravitational entropy, we again use the Noether current J^a corresponding to the local Killing vector ξ^a. Multiplying by r^a and $\beta_{\rm loc} = \beta N$, we get

$$\beta_{\rm loc}r_aJ^a = \beta_{\rm loc}\xi^a r_b(2E^b_a) + \beta N(r_a\xi^a)L \tag{21}$$

As the stretched horizon approaches the true horizon, $Nr^a \to \xi^a$ and $\beta\xi^a\xi_a L \to 0$ making the last term vanish. So the condition $\delta S_{\rm grav} = \delta S_m$ leads to

$$\delta S_{\rm grav} = \beta(2E_{aj})\xi^a\xi^j dV_{prop} = \delta S_m = \beta T_{aj}\xi^a\xi^j dV_{prop} \tag{22}$$

which is again the same as Eq. (20). Obviously, there are many other equivalent ways of presenting this result.

3.3 Aside: Observer dependence of horizons and entropy

As an aside, we shall comment on some new conceptual issues brought about by the existence of horizons and entropy which are relevant in this context. We recall that the mathematical formulation leading to the association of temperature with any horizon is fairly universal and it does not distinguish between different horizons, like for example Rindler horizon in flat space or a Schwarzschild black hole event horizon or a de Sitter horizon [9]. Assuming that temperature and entropy arise for fundamentally the same reason, it would be extremely unnatural *not* to associate entropy with all horizons.

This feature, however, brings in a new layer of observer dependent thermodynamics into the theory which — though it need not come as a surprise — has to be tackled. We know that while an inertial observer will attribute zero temperature and zero entropy to the inertial vacuum, a Rindler observer will attribute a finite temperature and non-zero (formally divergent; 'entanglement') entropy to the same vacuum state. *So entropy is indeed an observer dependent concept.* When one does quantum field theory in curved spacetime, it is not only that particles become an observer dependent notion so do the temperature and entropy. This notion can be made more quantitative as follows:

Consider an excited state of a quantum field with energy δE above the ground state as specified in an inertial frame. When we integrate out the unobservable modes for the Rindler observer in this state, we will get a density matrix ρ_1 and the corresponding entropy will be $S_1 = -\text{Tr}\,(\rho_1 \ln \rho_1)$. The inertial vacuum state has the density matrix ρ_0 and the entropy $S_0 = -\text{Tr}\,(\rho_0 \ln \rho_0)$. The difference $\delta S = S_1 - S_0$ is finite and represents the entropy attributed to the excited state by the Rindler observer. (This is finite though S_1 and S_0 can be divergent.) In the limit of $\kappa \to \infty$, which would correspond to a Rindler observer who is very close to the horizon, we can show that

$$\delta S = \beta \delta E = \frac{2\pi}{\kappa} \delta E \tag{23}$$

To prove this, note that if we write $\rho_1 = \rho_0 + \delta\rho$, then in the limit of $\kappa \to \infty$ we can concentrate on states for which $\delta\rho/\rho_0 \ll 1$. Then we have

$$\begin{aligned} -\delta S &= \text{Tr}\,(\rho_1 \ln \rho_1) - \text{Tr}\,(\rho_0 \ln \rho_0) \simeq \text{Tr}\,(\delta\rho \ln \rho_0) \\ &= \text{Tr}\,(\delta\rho(-\beta H_R)) = -\beta \text{Tr}\,((\rho_1 - \rho_0) H_R) \equiv -\beta \delta E \end{aligned} \tag{24}$$

where we have used the facts Tr $\delta\rho \approx 0$ and $\rho_0 = Z^{-1} \exp(-\beta H_R)$ where H_R is the Hamiltonian for the system in the Rindler frame. The last line defines the δE in terms of the difference in the expectation values of the Hamiltonian in the two states. This is the amount of entropy a Rindler observer would claim to be lost when the matter disappears into the horizon. (This result can be explicitly proved for, say, one particle excited states of the field [15].)

One might have naively thought that the expression for entropy of matter crossing the horizon should consist of its energy δE and *its own* temperature T_{matter} rather than the horizon temperature T_h. But the correct expression is $\delta S = \delta E/T_h$; the horizon acts as a system with some internal degrees of freedom and temperature T_h *as far as Rindler observer is concerned* so that when one adds an energy δE to it, the entropy change is $\delta S = (\delta E/T_h)$. Obviously, a Rindler observer (or an observer at rest just outside a black hole horizon) will attribute all these entropy changes to the horizon she perceives while an inertial observer (or an observer falling through the Schwarzschild horizon) will see none of these phenomena. This requires us to accept the fact that many thermodynamic phenomena needs to be now thought of as specifically observer dependent. For example, if we drop some hot matter into a Schwarzschild black hole, then, when it gets to a few Planck lengths away from the horizon it would interact with the microscopic horizon degrees of freedom *as far as an outside observer is concerned.* A freely falling observer through the horizon will have a completely different picture. We have learnt to live with this dichotomy as far as elementary kinematics goes; we now need to do the same as regards thermodynamics and quantum processes.

As far as an outside observer is concerned, matter takes an infinite amount of coordinate time to cross the horizon. However, this is irrelevant from a practical point of view. We have, for example, considerable evidence of very different nature to suggest Planck length acts as lower bound to the length scales that can be operationally defined and that no measurements can be ultra sharp at Planck scales [16]. So one cannot really talk about the location of the event horizon ignoring fluctuations of this order. Hence from the operational point of view, we only need to consider matter reaching within few Planck lengths of the horizon to talk about entropy loss which is what we have done in our discussion. In fact, physical processes very close to the horizon must play an important role in order to provide a complete picture of the issues we are discussing. There is already some evidence [17] that the infinite redshift induced by the horizon plays a crucial role in this though a mathematically rigorous model is lacking.

4. GRAVITY: THE INSIDE STORY

4.1 An entropy extremum principle for gravitational field equations

The last interpretation of the field equations (see Eq. (22)) given in Sec. 3.2 is similar to switching from a passive point of view to an active point of view. Instead of allowing matter to fall into the horizon, we made a virtual displacement of the horizon surface to engulf the matter when it is infinitesimally close to the horizon. But for the theory to be consistent, this displacement of the horizon surface degrees of freedom should cost some entropy. By determining the functional form of this entropy density, we should be able to obtain the field equations of gravity through an extremum principle. Recall that thermodynamics relies entirely on the form of the entropy functional to make predictions. Hence, if we can determine the form of entropy functional for gravity (S_{grav}) in terms of the normals to the null surfaces, then it seems natural to demand that the dynamics should follow from the extremum prescription $\delta[S_{grav}+S_{matter}]=0$ for *all null surfaces in the spacetime* where S_{matter} is the relevant matter entropy.

The form of S_{matter} and $S_{\rm grav}$ can be determined as follows. Let us begin with S_{matter} which is easy to ascertain from the previous discussion. If T_{ab} is the matter energy-momentum tensor in a general $D(\geq 4)$ dimensional spacetime then an expression for matter entropy *relevant for our purpose* can be taken to be

$$S_{\rm matt}=\int_{\mathscr{V}} d^Dx\sqrt{-g}T_{ab}n^a n^b \tag{25}$$

where n^a is a null vector field. From our Eq. (17) we see that the entropy density associated with proper 3-volume is $\beta(T_{ab}\xi^a\xi^b)dV_{prop}$ where — on the horizon — the vector ξ^a becomes proportional to a null vector n^a. If we now use the Rindler coordinates in Eq. (1) in which $\sqrt{-g}=1$ and interpret the factor β as arising from an integration of dt in the range $(0,\beta)$ we find that the entropy density associated with a proper four volume is $(T_{ab}n^a n^b)$. This suggests treating Eq. (25) as the matter entropy. For example, if T_{ab} is due to an ideal fluid at rest in the LIF then $T_{ab}n^a n^b$ will contribute $(\rho+P)$, which — by Gibbs-Duhem relation — is just $T_{local}s$ where s is the entropy density and $T_{local}^{-1}=\beta N$ is the properly redshifted temperature with $\beta=2\pi/\kappa$ being the periodicity of the Euclidean time coordinate. Then

$$\int dS=\int\sqrt{h}d^3x\, s=\int\sqrt{h}d^3x\beta_{\rm loc}(\rho+P)=\int\sqrt{h}Nd^3x\beta(\rho+P)=\int_0^\beta dt\int d^3x\sqrt{-g}T^{ab}n_a n_b \tag{26}$$

which matches with Eq. (25) in the appropriate limit.

This argument works for any matter source, not necessarily the ones with which we conventionally associate an entropy. What is really relevant is only the *energy* flux close to the horizon from which one can obtain an entropy flux. We *do* have the notion of *energy* flux across a surface with normal r^a being $T_{ab}\xi^b r^a$ which holds for *any* source T^{ab}. Given some energy flux δE in the Rindler frame, there is an associated entropy flux loss $\delta S=\beta\delta E$ as given by Eq. (23). It is *this* entropy which is given by Eq. (23) and Eq. (25). The only non-trivial feature in Eq. (25) is the integration range for time which is limited to $(0,\beta)$. This is done by considering the integrals in the Euclidean sector and rotating back to the Lorentzian sector but the same result can be obtained working entirely in the Euclidean sector. (There is an ambiguity in the overall scaling of n^a since if n^a is null so is $f(x)n^a$ for all $f(x)$; we will comment on this ambiguity, which turns out to be irrelevant, later on.)

Next, let us consider the expression for $S_{\rm grav}$. We will first describe the simplest possible choice and then consider a more general expression. The simplest choice is to postulate S_{grav} to be a quadratic expression [18] in the derivatives of the normal:

$$S_{grav}=-4\int_{\mathscr{V}} d^Dx\sqrt{-g}P_{ab}{}^{cd}\nabla_c n^a\nabla_d n^b \tag{27}$$

where the explicit form of $P_{ab}{}^{cd}$ is ascertained below. Given Eq. (25) and Eq. (27) the expression for the total entropy becomes:

$$S[n^a]=-\int_{\mathscr{V}} d^Dx\sqrt{-g}\left(4P_{ab}{}^{cd}\nabla_c n^a\nabla_d n^b-T_{ab}n^a n^b\right), \tag{28}$$

The field equations of gravity are to be determined by extremizing this entropy functional. However, there is one crucial conceptual difference between the extremum principle introduced here and the conventional one. Usually, given a set of dynamical variables n_a and a functional $S[n_a]$, the extremum principle will give a set of equations *for* the dynamical variable n_a. Here the situation is completely different. We expect the variational principle to hold for *all* null vectors n^a thereby leading to a constraint on the *background metric.* Obviously, the functional in Eq. (28) must be rather special to accomplish this and one needs to impose restrictions on $P_{ab}{}^{cd}$ (and T_{ab}, though that condition turns out to be trivial) to achieve this.

It turns out — as we shall see below — that two conditions are sufficient to ensure this. First, the tensor P_{abcd} should have the same algebraic symmetries as the Riemann tensor R_{abcd} of the D-dimensional spacetime. This condition can be ensured if we define $P_a{}^{bcd}$ as

$$P_a{}^{bcd}=\frac{\partial L}{\partial R^a{}_{bcd}} \tag{29}$$

where $L=L(R^a{}_{bcd},g^{ik})$ is some scalar. The motivation for this choice arises from the fact that this approach leads to the same field equations as the one with L as gravitational Lagrangian in the conventional approach (which explains the choice of the symbol L). Second, we will postulate the condition:

$$\nabla_a P^{abcd} = 0. \tag{30}$$

as well as $\nabla_a T^{ab} = 0$ which is anyway satisfied by any matter energy-momentum tensor. One possible motivation for Eq. (30) is to ensure that the field equations are no higher than second order in the metric. (If we think of gravity as an emergent phenomenon like elasticity, then n^a is like the displacement field in elasticity. The standard entropy functional [19] used in theory of elasticity has the form in Eq. (30) with coefficients being elastic constants. Here the coefficients are P^{abcd} and the condition in Eq. (30) may be interpreted as saying the 'elastic constants of spacetime solid' are actually 'constants' [20].) This is, however, not a crucial condition and in fact we will see below how this condition in Eq. (30) can be relaxed.

4.2 The field equations

Varying the normal vector field n^a after adding a Lagrange multiplier function $\lambda(x)$ for imposing the condition $n_a \delta n^a = 0$, we get

$$-\delta S = 2\int_{\mathscr{V}} d^D x\sqrt{-g}\left[4P_{ab}{}^{cd}\nabla_c n^a\left(\nabla_d \delta n^b\right) - T_{ab}n^a\delta n^b - \lambda(x) g_{ab} n^a \delta n^b\right] \tag{31}$$

where we have used the symmetries of $P_{ab}{}^{cd}$ and T_{ab}. An integration by parts and the condition $\nabla_d P_{ab}{}^{cd} = 0$, leads to

$$-\delta S = 2\int_{\mathscr{V}} d^D x\sqrt{-g}\left[-4P_{ab}{}^{cd}\left(\nabla_d\nabla_c n^a\right) - \left(T_{ab} + \lambda g_{ab}\right)n^a\right]\delta n^b + 8\int_{\partial\mathscr{V}} d^{D-1}x\sqrt{h}\left[k_d P_{ab}{}^{cd}\left(\nabla_c n^a\right)\right]\delta n^b, \tag{32}$$

where k^a is the D-vector field normal to the boundary $\partial\mathscr{V}$ and h is the determinant of the induced metric on $\partial\mathscr{V}$. As usual, in order for the variational principle to be well defined, we require that the variation δn^a of the vector field should vanish on the boundary. The second term in Eq. (32) therefore vanishes, and the condition that $S[n^a]$ be an extremum for arbitrary variations of n^a then becomes

$$2P_{ab}{}^{cd}\left(\nabla_c\nabla_d - \nabla_d\nabla_c\right)n^a - \left(T_{ab} + \lambda g_{ab}\right)n^a = 0, \tag{33}$$

where we used the antisymmetry of $P_{ab}{}^{cd}$ in its upper two indices to write the first term. The definition of the Riemann tensor in terms of the commutator of covariant derivatives reduces the above expression to

$$\left(2P_b{}^{ijk}R^a{}_{ijk} - T^a_b + \lambda\delta^a_b\right)n_a = 0, \tag{34}$$

and we see that the equations of motion *do not contain* derivatives with respect to n^a which is, of course, the crucial point. This peculiar feature arose because of the symmetry requirements we imposed on the tensor $P_{ab}{}^{cd}$. We need the condition in Eq. (34) holds for *arbitrary* null vector fields n^a. One can easily show[18] that this requires

$$16\pi\left[P_b{}^{ijk}R^a{}_{ijk} - \frac{1}{2}\delta^a_b L\right] = 8\pi T^a_b + \Lambda\delta^a_b \tag{35}$$

where Λ is an arbitrary integration constant. It is also easy to see that Eq. (35) is equivalent to Eq. (20) with E_{ab} given by Eq. (6). One crucial difference between Eq. (4) [along with Eq. (6)] and Eq. (35) is the introduction of the cosmological constant Λ; we will discuss this later on. We mentioned earlier that the expression in Eq. (28) depends on the overall scaling of n^a which is arbitrary, since $f(x)n^a$ is a null vector if n^a is null. But since the arbitrary variation of n^a (with the constraint $n_a n^a = 0$) includes scaling variations of the type $\delta n^a = \varepsilon(x) n^a$, it is clear that this ambiguity is irrelevant for determining the equations of motion.

To summarize, we have proved the following. Suppose we start with the Lagrangian in Eq. (3), define a P^{abcd} by Eq. (29) ensuring that it satisfies Eq. (30). Varying the metric with this action will lead to field equations in Eq. (6). We have now proved that we will get the *same* field equations (but with a cosmological constant) if we start with the expression in Eq. (28), maximize it with respect to n^a and demand that it holds for all n^a.

This result might appear a little mysterious at first sight, but the following alternative description will make clear why this works. Note that, using the constraints on P^{abcd} we can prove the identity

$$\begin{aligned}4P_{ab}{}^{cd}\nabla_c n^a\nabla_d n^b &= 4\nabla_c[P_{ab}{}^{cd}n^a\nabla_d n^b]-4n^aP_{ab}{}^{cd}\nabla_c\nabla_d n^b=4\nabla_c[P_{ab}{}^{cd}n^a\nabla_d n^b]-2n^aP_{ab}{}^{cd}\nabla_{[c}\nabla_{d]}n^b\\ &= 4\nabla_c[P_{ab}{}^{cd}n^a\nabla_d n^b]-2n^aP_{ab}{}^{cd}R^b{}_{icd}n^i=4\nabla_c[P_{ab}{}^{cd}n^a\nabla_d n^b]+2n^aE_{ai}n^i \end{aligned} \tag{36}$$

where the first equality uses Eq. (30), the second equality uses the antisymmetry of $P_{ab}{}^{cd}$ in c and d, the third equality uses the standard identity for commutator of covariant derivatives and the last one is based on Eq. (5) when $n_a n^a=0$ and Eq. (30) hold. Using this in the expression for S in Eq. (28) and integrating the four-divergence term, we can write

$$S[n^a]=-\int_{\partial\mathscr{V}}d^{D-1}xk_c\sqrt{h}(4P_{ab}{}^{cd}n^a\nabla_d n^b)-\int_{\mathscr{V}}d^Dx\sqrt{-g}(2E_{ab}-T_{ab})n^an^b \tag{37}$$

So, when we vary $S[n^a]$ (ignoring the surface term) we are effectively varying $(2E_{ab}-T_{ab})n^an^b$ with respect to n_a and demanding that it holds for all n_a. There is an ambiguity of adding a term of the form $\lambda(x)g_{ab}$ in the integrand of the second term in Eq. (37) leading to the final equation $(2E_{ab}=T_{ab}+\lambda(x)g_{ab})$ but the Bianchi identity $\nabla_aE^{ab}=0$ along with $\nabla_aT^{ab}=0$ will make $\lambda(x)$ actually a constant.

It is now clear how we can find an S for any theory, even if Eq. (29) does not hold. This can be achieved by starting from the expression $(2E_{ab}-T_{ab})n^an^b$ as the entropy density, using Eq. (5) for E_{ab} and integrating by parts (see the discussion after Eq. (14) in ref. [10]). In this case, we get for $S_{\rm grav}$ the expression:

$$S_{\rm grav}=-4\int_V d^Dx\sqrt{-g}\left[P^{abcd}\nabla_c n_a\nabla_d n_b+(\nabla_dP^{abcd})n_b\nabla_c n_a+(\nabla_c\nabla_dP^{abcd})n_an_b\right] \tag{38}$$

Varying this with respect to n^a will then lead to the correct equations of motion and — incidentally — the same surface term.

While one could indeed work with this more general expression, there are four reasons to prefer the imposition of the condition in Eq. (29). First, it is clear from Eq. (5) that when L depends on the curvature tensor and the metric, E_{ab} can depend up to the fourth derivative of the metric if Eq. (29) is not satisfied. But when we impose Eq. (29) then we are led to field equations which have, at most, second derivatives of the metric tensor which is a desirable feature. Second, as we shall see below, with that condition we can actually determine the form of L; it turns out that in $D=4$, it uniquely selects Einstein's theory, which is probably a nice feature. In higher dimensions, it picks out a very geometrical extension of Einstein's theory in the form of Lanczos-Lovelock theories. Third, it is difficult to imagine why the terms in Eq. (38) should occur with very specific coefficients. In fact, it is not clear why we cannot have derivatives of R_{abcd} in L, if the derivatives of P_{abcd} can occur in the expression for entropy. Finally, if we take the idea of elastic constants being constants, then one is led to Eq. (29). None of these rigorously exclude the possibility in Eq. (38) and in fact this model has been explored recently [21].

So far we have not fixed P^{abcd} so we have not fixed the theory. In a complete theory, the explicit form of P^{abcd} will be determined by the long wavelength limit of the microscopic theory just as the elastic constants can — in principle — be determined from the microscopic theory of the lattice. In the absence of such a theory, we need to determine P^{abcd} by general considerations which is possible when P^{abcd} satisfies Eq. (29). Since this condition is identically satisfied by Lanczos-Lovelock models which are known to be unique, our problem can be completely solved by taking the P^{abcd} as a series in the powers of derivatives of the metric as:

$$P^{abcd}(g_{ij},R_{ijkl})=c_1\overset{(1)}{P}{}^{abcd}(g_{ij})+c_2\overset{(2)}{P}{}^{abcd}(g_{ij},R_{ijkl})+\cdots, \tag{39}$$

where $c_1,c_2,\cdots$ are coupling constants with the m th order term derived from the Lanczos-Lovelock Lagrangian:

$$\overset{(m)}{P}{}_{ab}^{cd}\propto\delta^{cda_3\ldots a_{2m}}_{abb_3\ldots b_{2m}}R^{b_3b_4}_{a_3a_4}\cdots R^{b_{2m-1}b_{2m}}_{a_{2m-1}a_{2m}}=\frac{\partial L_{(m)}}{\partial R^{ab}_{cd}}. \tag{40}$$

where $\delta^{cda_3\ldots a_{2m}}_{abb_3\ldots b_{2m}}$ is the alternating tensor. The lowest order term depends only on the metric with no derivatives. The next term depends (in addition to metric) linearly on curvature tensor and the next one will be quadratic in curvature etc. The lowest order term in Eq. (39) (which leads to Einstein's theory) is

$$\overset{(1)}{P}{}^{ab}_{cd}=\frac{1}{16\pi}\frac{1}{2}\delta^{ab}_{cd}=\frac{1}{32\pi}(\delta^a_c\delta^b_d-\delta^a_d\delta^b_c). \tag{41}$$

To the lowest order, when we use Eq. (41) for ${P_b}^{ijk}$, the Eq. (35) reduces to Einstein's equations. The corresponding gravitational entropy functional [1] is:

$$S_{\rm GR}[n^a] = \int_{\mathscr{V}} \frac{d^Dx}{8\pi}\left(\nabla_a n^b \nabla_b n^a - (\nabla_c n^c)^2\right) \tag{42}$$

The next order term (which arises from the Gauss-Bonnet Lagrangian) in Eq. (7) is:

$$\overset{(2)}{P}{}^{ab}_{\ \ cd} = \frac{1}{16\pi}\frac{1}{2}\delta^{aba_3a_4}_{cdb_3b_4}R^{b_3b_4}_{a_3a_4} = \frac{1}{8\pi}\left(R^{ab}_{cd} - G^a_c\delta^b_d + G^b_c\delta^a_d + R^a_d\delta^b_c - R^b_d\delta^a_c\right) \tag{43}$$

and similarly for all the higher orders terms. None of them can contribute in $D=4$ so we get Einstein's theory as the unique choice if we assume $D=4$. If we assume that P^{abcd} is to be built *only* from the metric, then this choice is unique in all D.

While the matter term in the functional in Eq. (28) has a natural interpretation in terms of entropy transfered across the horizon, the interpretation of the gravitational part needs to be made explicit. The interpretation of $S_{\rm grav}$ as entropy arises from the following two facts. First, we see from the identity Eq. (36) that this term differs from $2E_{ij}n^in^j$ by a total divergence. On the other hand, we have seen earlier that the term $2E_{ij}n^in^j$ can be related to the gravitational entropy of the horizon through the Noether current which suggests the identification. In fact, Eq. (37) shows that when the equations of motion holds *the total entropy of a bulk region is entirely on its boundary*. Further if we evaluate this boundary term

$$-S|_{\rm on-shell} = 4\int_{\partial\mathscr{V}} d^{D-1}x\, k_a\sqrt{h}\left(P^{abcd}n_c\nabla_b n_d\right) \tag{44}$$

(where we have manipulated a few indices using the symmetries of P^{abcd}) in the case of a *stationary* horizon which can be locally approximated as Rindler spacetime, one gets exactly the Wald entropy of the horizon [18].

4.3 Cosmological constant

The approach outlined above has important implications for the cosmological constant problem, which we shall now briefly mention [22]. In the conventional approach, we start with an action principle which depends on matter degrees of freedom and the metric and vary (i) the matter degrees of freedom to obtain the equations of motion for matter and (ii) the metric g^{ab} to obtain the field equations of gravity. The equations of motion for *matter* remain invariant if one adds a constant, say, $-\rho_0$ to the matter Lagrangian. However, gravity breaks this symmetry which the matter sector has and ρ_0 appears as a cosmological constant term in the field equations of gravity. If we interpret the evidence for dark energy in the universe (see ref. [23]; for a critical look at data, see ref. [24] and references therein) as due to the cosmological constant, then its value has to be fine-tuned to satisfy the observational constraints. It is not clear why a particular parameter in the low energy sector has to be fine-tuned in such a manner.

In the alternative perspective described here, the functional in Eq. (28) is clearly invariant under the shift $L_m \to L_m - \rho_0$ or equivalently, $T_{ab} \to T_{ab} + \rho_0 g_{ab}$, since it only introduces a term $-\rho_0 n_a n^a = 0$ for any null vector n_a. In other words, one *cannot* introduce the cosmological constant as a low energy parameter in the action in this approach. We saw, however, that the cosmological constant can reappear as an *an integration constant* when the equations are solved. The integration constants which appear in a particular solution have a completely different conceptual status compared to the parameters which appear in the action describing the theory. It is much less troublesome to choose a fine-tuned value for a particular integration constant in the theory if observations require us to do so. From this point of view, the cosmological constant problem is considerably less severe when we view gravity from the alternative perspective.

This extra symmetry under the shift $T_{ab} \to T_{ab} + \rho_0 g_{ab}$ arises because we are not treating metric as a dynamical variable in an action principle.[2] In fact one can state a stronger result [25]. Consider any model of gravity satisfying

[1] Interestingly, the integrand in S_{GR} has the $Tr(K^2) - (TrK)^2$ structure. If we think of the $D=4$ spacetime being embedded in a sufficiently large k-dimensional *flat* spacetime we can obtain the same structure using the Gauss-Codazzi equations relating the (zero) curvature of k-dimensional space with the curvature of spacetime.

[2] It is sometimes claimed that a spin-2 graviton in the linear limit *has to* couple to T_{ab} in a universal manner, in which case, one will have the graviton coupling to the cosmological constant. In our approach, the linearized field equations for the spin-2 graviton field $h_{ab} = g_{ab} - \eta_{ab}$, in a suitable gauge, will be $(\Box h_{ab} - T_{ab})n^an^b = 0$ for all null vectors n^a. This equation is still invariant under $T_{ab} \to T_{ab} + \rho_0 g_{ab}$ showing that the graviton does *not* couple to cosmological constant.

the following three conditions: (1) The metric is varied in a local action to obtain the equations of motion. (2) We demand full general covariance of the equations of motion. (3) The equations of motion for matter sector are invariant under the addition of a constant to the matter Lagrangian. Then, we can prove a 'no-go' theorem that the cosmological constant problem cannot be solved in such model. That is, we cannot solve cosmological constant problem unless we drop one of these three demands. Of these, we do not want to sacrifice general covariance encoded in (2); neither do we have a handle on low energy matter Lagrangian so we cannot avoid (3). So the only hope we have is to introduce an approach in which gravitational field equations are obtained from varying some degrees of freedom other than g_{ab} in a maximization principle. This suggests that the so called cosmological constant problem has its roots in our misunderstanding of the nature of gravity.

Our approach at present is not yet developed far enough to predict the value of the cosmological constant .But providing a mechanism in which the *bulk cosmological constant decouples from gravity* is a major step forward. It was always thought that some unknown symmetry should make the cosmological constant (almost) vanish and weak (quantum gravitational) effects which break this symmetry could lead to its small value. *Our approach provides a model which has such symmetry.* The small value of the observed cosmological constant has to arise from non-perturbative quantum gravitational effects at the next order, for which we do not yet have a fully satisfactory model. (See, however, Ref. [26].)

5. CONCLUSIONS

It is useful to distinguish clearly (i) the mathematical results which can be rigorously proved from (ii) interpretational ideas which might evolve when our understanding of these issues deepen.

(i) From a purely algebraic point of view, without bringing in any physical interpretation or motivation, we can prove the following mathematical results:

- Consider a functional of null vector fields $n^a(x)$ in an arbitrary spacetime given by Eq. (28) [or, more generally, by Eq. (38)]. Demanding that this functional is an extremum for all null vectors n^a leads to the field equations for the background geometry given by $(2E_{ab} - T_{ab})n^a n^b = 0$ where E_{ab} is given by Eq. (6) [or, more generally, by Eq. (5)]. Thus field equations in a wide class of theories of gravity can be obtained from an extremum principle without varying the metric as a dynamical variable.
- These field equations are invariant under the transformation $T_{ab} \to T_{ab} + \rho_0 g_{ab}$, which relates to the freedom of introducing a cosmological constant as an integration constant in the theory. Further, this symmetry forbids the inclusion of a cosmological constant term in the variational principle by hand as a low energy parameter. *That is, we have found a symmetry which makes the bulk cosmological constant decouple from the gravity.* When linearized around flat spacetime, the graviton inherits this symmetry and does not couple to the cosmological constant.
- On-shell, the functional in Eq. (28) [or, more generally, by Eq. (38)] contributes only on the boundary of the region. When the boundary is a horizon, this terms gives precisely the Wald entropy of the theory.

It is remarkable that one can derive not only Einstein's theory uniquely in $D = 4$ but even Lanczos-Lovelock theory in $D > 4$ from an extremum principle involving the null normals *without varying g_{ab} in an action functional!*.

(ii) As regards interpretation of these results, we see from Eq. (36) that in the case of Einstein's theory, we have a Lagrangian $n^a(\nabla_{[a}\nabla_{b]})n^b$ for a vector field n^a (except for a surface term) that becomes vacuous in flat spacetime in which covariant derivatives become partial derivatives. There is no dynamics in n^a (in the usual sense) but they do play a crucial role. This raises the question as to the physical meaning of these null vectors and the interpretation of our approach. As described above, this interpretation is essentially thermodynamical and the physical picture is made of the following qualitative ingredients:

- Assume that the spacetime is endowed with certain microscopic degrees of freedom capable of exhibiting thermal phenomena. This is just the Boltzmann paradigm: *If one can heat it, it must have microstructure!*; and one can heat up a spacetime.
- Whenever a class of observers perceive a horizon, they are "heating up the spacetime" and the degrees of freedom close to a horizon participate in a very *observer dependent* thermodynamics. Matter which flows close to the horizon (say, within a few Planck lengths of the horizon) transfers energy to these microscopic, near-horizon, degrees of freedom *as far as the observer who sees the horizon is concerned.* Just as entropy of a normal system

at temperature T will change by $\delta E/T$ when we transfer to it an energy δE, here also an entropy change will occur. (A freely falling observer in the same neighbourhood, of course, will deny all these!)

- We proved that when the field equations of gravity hold, one can interpret this entropy change in a purely geometrical manner involving the Noether current. From this point of view, the normals n^a to local patches of null surfaces are related to the (unknown) degrees of freedom that can participate in the thermal phenomena involving the horizon.
- Just as demanding the validity of special relativistic laws with respect to all freely falling observers leads to the kinematics of gravity, demanding the local entropy balance in terms of the thermodynamic variables as perceived by local Rindler observers leads to the field equations of gravity in the form $(2E_{ab} - T_{ab})n^a n^b = 0$.

As stressed in earlier sections, this involves a new layer of observer dependent thermodynamics. At a conceptual level, this may be welcome when we note that every key progress in physics involved realizing that something we thought as absolute is not absolute. With special relativity it was the flow of time and with general relativity it was the concept of global inertial frames and when we brought in quantum fields in curved spacetime it was the notion of particles and temperature. We now know that the temperature attributed to even vacuum state depends on the observer. It seems necessary to integrate the entire thermodynamic machinery (involving what we usually consider to be the 'real' temperature) with this notion of LRFs having their own (observer dependent) temperature. There is scope for further work in this direction.

ACKNOWLEDGMENTS

I thank Jean-Michel Alimi for organizing a stimulating conference and providing excellent hospitality.

REFERENCES

1. T. Padmanabhan, *A Dialogue on the Nature of Gravity* [arXiv:0910.0839]; *Gen. Rel. Grav.*, **40**, 2031-2036 (2008); *Phys. Reports*, **406**, 49 (2005) [gr-qc/0311036]; AIP Conference Proceedings, **989** 114 (2007) [arXiv:0706.1654].
2. T Padmanabhan, *Class. Quan. Grav.*, **19**, 5387, (2002) [gr-qc/0204019]
3. For a small sample, see e.g., D. Kothawala et al., *Phys. Lett.* **B 652**, 338 (2007) [arXiv:gr-qc/0701002]; A. Paranjape et al., *Phys. Rev.* **D 74**, 104015 (2006) [arXiv:hep-th/0607240]; Dawood Kothawala, T. Padmanabhan, *Phys. Rev.*, **D 79**, 104020 (2009)[arXiv:0904.0215]; R. G. Cai, et al., *Phys. Rev.* **D 78**, 124012 (2008) [arXiv:0810.2610]; *Phys. Rev.* **D 75**, 084003 (2007) [arXiv:hep-th/0609128]; M. Akbar and R. G. Cai, *Phys. Lett.* ,**B 635**, 7 (2006) [arXiv:hep-th/0602156]; *Phys. Lett.* **B 648**, 243 (2007) [arXiv:gr-qc/0612089]; Y. Gong and A. Wang,Phys. Rev. Lett. **99**, 211301 (2007) [arXiv:0704.0793 [hep-th]]; S. F. Wu, G. H. Yang and P. M. Zhang, [arXiv:0710.5394]; S. F. Wu, B. Wang and G. H. Yang, Nucl. Phys. B **799**, 330 (2008) [arXiv:0711.1209]; S. F. Wu, B. Wang, G. H. Yang and P. M. Zhang, [arXiv:0801.2688].
4. R. G. Cai and N. Ohta, *Horizon Thermodynamics and Gravitational Field Equations in Horava-Lifshitz Gravity,* [arXiv:0910.2307].
5. T. Padmanabhan, *Gen.Rel.Grav.,* **34** 2029 (2002) [gr-qc/0205090]; *Brazilian Jour.Phys.* (Special Issue) **35**, 362 (2005) [gr-qc/0412068]; *Mod.Phys.Letts.* **A 17**, 1147 (2002) [hep-th/0205278]; AIP Conference Proceedings **861**, 858, [astro-ph/0603114].
6. A. Mukhopadhyay, T. Padmanabhan, *Phys.Rev.*,**D 74**, 124023 (2006) [hep-th/0608120].
7. S. W. Hawking, *Commun. Math. Phys.* **43** (1975) 199–220.
8. P. C. W. Davies, *J. Phys.* **A8** (1975) 609–616; W. G. Unruh, Phys. Rev. D14 (1976) 870.
9. T. Padmanabhan, *Mod. Phys. Lett.* **A 17** (2002) 923 [gr-qc/0202078].
10. T. Padmanabhan, *Entropy density of spacetime and thermodynamic interpretation of field equations of gravity in any diffeomorphism invariant theory*, [arXiv:0903.1254].
11. Nathalie Deruelle et al., *Class. Quant. Grav.* **21**, (2004), 1971, [gr-qc/0310098]; G.L. Cardoso et al., [arXiv:hep-th/9904005v2]; T. Padmanabhan, *Gravitation: Foundations and Frontiers*, (Cambridge University Press, 2010), in press.
12. C Lanczos, *Z. Phys.* **73**, 147, (1932); *Annals Math.* **39**, 842, (1938); D Lovelock, *J. Math. Phys.*, **12**, 498 (1971).
13. R. M. Wald, *Phys. Rev.* **D 48** (1993) 3427–3431.
14. T. Padmanabhan, *Adv. Sci. Lett.*, **2**, 174 (2009) [arXiv:0807.2356];
15. S. Kolekar, D. Kothawala, T. Padmanabhan, work in progress (2009).
16. H. S. Snyder *Phys. Rev.*, **71**, 38 (1947); B. S. DeWitt, *Phys. Rev. Lett.*, **13**, 114 (1964); T. Yoneya *Prog. Theor. Phys.*, **56**, 1310 (1976); T. Padmanabhan *Ann. Phys.* (N.Y.), **165**, 38 (1985); *Class. Quantum Grav.* **4**, L107 (1987); T. Padmanabhan *Phys. Rev. Lett.* **78**, 1854 (1997) [hep-th/9608182]; *Phys. Rev.* **D 57**, 6206 (1998); K.Srinivasan et al., *Phys. Rev.* **D 58** 044009 (1998) [gr-qc/9710104]; X.Calmet et al.,Phys.Rev.Lett.**93**:211101,(2004); *Int.J.Mod.Phys.* **D14** (2005) 2195; M. Fontanini et al. *Phys.Lett.* **B 633**, 627 (2006) [hep-th/0509090]. For a review, see L.J. Garay, *Int. J. Mod. Phys.* **A10**, 145 (1995).

17. T. Padmanabhan, *Phys. Rev. Lett.* **81** (1998) 4297[hep-th/9801015]; *Phys. Rev.* **D 59** (1999) 124012 [hep-th/9801138].
18. T. Padmanabhan, *Gen.Rel.Grav.*, **40**, 529 (2008) [arXiv:0705.2533]; T. Padmanabhan, A. Paranjape, *Phys.Rev.*,**D75**, 064004, (2007)[gr-qc/0701003].
19. L. D. Landau and E. M. Lifshitz, *Theory of Elasticity*, Pergamon Press; 2 edition (1981).
20. T. Padmanabhan, *Int.Jour.Mod.Phys.*,**D 13**, 2293 (2004) [gr-qc/0408051]; *Int.J.Mod.Phys.*, **D14**,2263 (2005) [gr-qc/0510015].
21. Shao-Feng Wu et.al., (2009), [arXiv:0909.1367v2].
22. For a review, see e.g., T. Padmanabhan, *Physics Reports* **380**, 235 (2003) [hep-th/0212290].
23. S.J. Perlmutter et al., *Astrophys. J.* **517**,565 (1999); A.G. Reiss et al., *Astron. J.* (1998), **116**,1009; J. L. Tonry et al., *Ap. J*, (2003), **594**, 1; B. J. Barris, *Astrophys.J.*, **602** (2004), 571; A. G.Reiss et al., *Astrophys.J.* **607**, (2004), 665.
24. H.K.Jassal et al., *Phys.Rev.* **D 72**, 103503 (2005) [astro-ph/0506748]; [astro-ph/0601389]; T. Padmanabhan,T. Roy Choudhury, *MNRAS* **344**, 823 (2003) [astro-ph/0212573]; T. Roy Choudhury, T. Padmanabhan, *Astron.Astrophys.* **429**, 807 (2005), [astro-ph/0311622]; S. Nesseris, L.Perivolaropoulos, *JCAP* **0702**,025 (2007); Y. Wang, P. Mukherjee, *Phys. Rev.* **D 76**, 103533 (2007).
25. T. Padmanabhan, *Gen.Rel.Grav.*, **38**, 1547(2006); Current Science, **88**,1057, (2005), [astro-ph/0411044]; *Int.J.Mod.Phys.*, **D 15**, 2029 (2006) [gr-qc/0609012].
26. T. Padmanabhan *Class.Quan.Grav.*, **22**, L107, (2005) [hep-th/0406060]. For earlier attempts in similar spirit, see T. Padmanabhan, *Class.Quan.Grav.* **19**, L167 (2002), [gr-qc/0204020]; D. Sorkin, *Int.J.Theor.Phys.* **36**, 2759 (1997); for related work, see Volovik, G. E., [gr-qc/0405012]; J. V. Lindesay et al., [astro-ph/0412477]; Y. S. Myung, [hep-th/0412224]; J.D.Barrow, [gr-qc/0612128]; E.Elizalde et al., [hep-th/0502082].

The Big bang and the Quantum

Abhay Ashtekar

Institute for Gravitation and the Cosmos, Penn State, University Park, PA 16802, U.S.A.

Abstract.
General relativity predicts that space-time comes to an end and physics comes to a halt at the big-bang. Recent developments in loop quantum cosmology have shown that these predictions cannot be trusted. Quantum geometry effects can resolve singularities, thereby opening new vistas. Examples are: The big bang is replaced by a quantum bounce; the 'horizon problem' disappears; immediately after the big bounce, there is a super-inflationary phase with its own phenomenological ramifications; and, in presence of a standard inflation potential, initial conditions are naturally set for a long, slow roll inflation independently of what happens in the pre-big bang branch. As in my talk at the conference, I will first discuss the foundational issues and then the implications of the new Planck scale physics near the Big Bang.

PACS: 04.60.Kz,04.60Pp,98.80Qc,03.65.Sq

INTRODUCTION

At this conference we heard of the spectacular progress that has occurred in observational cosmology in recent years. We also learned about the upcoming missions that are poised to provide new data to further constrain or even rule out leading theoretical models. These advances have been and continue to be the engines that drive contemporary cosmology. They have brought to forefront the astonishing success of the simple Friedmann, Lemaître, Robertson, Walker (FLRW) models, and perturbations thereof. Indeed, it appears that the rich data that we now have and are likely to accumulate in the near future would be adequately described by these simple applications of general relativity and quantum field theory on the resulting cosmological backgrounds.

However these theories are conceptually incomplete. They assume that the universe begins with a big bang at which matter densities and space-time curvature become infinite. With inflationary scenarios there were initial hopes that perhaps the big-bang singularity could be avoided because the inflaton fails to satisfy the strong energy condition often used in the singularity theorems of general relativity. However, Borde, Guth and Vilenkin [1] soon established that this hope was misplaced. Inflation can be eternal in the future but, if we go back in time using Einstein's equations, one again finds that the space-time ends and physics simply comes to a halt at the big bang. But all our experience with fundamental physics suggests that this cannot be the situation in the real world. This must be a prediction of a theory that has been pushed well beyond the domain of its validity. To know what *really* happened near the putative big bang, we must work with a genuine unification of general relativity and quantum physics, an unification which does not pre-suppose that space-time is a smooth continuum, and which can encompass the rich non-linear structure of strong gravity that lies beyond the scope of perturbation theory.

But the burden on this desired unification is heavy. In the context of cosmology there is a long list of fundamental questions that must be satisfactorily addressed by such a theory. Here is an illustrative list encompassing some of the contemporary issues.

- If general relativity is transcended, how close to the putative Big Bang does a smooth space-time of Einstein's make sense? In particular, can one show from first principles that this approximation is valid at the onset of inflation?
- Is the Big-Bang singularity naturally resolved by the quantum version of Einstein's equations? Or, is some external input such as a new principle or a boundary condition at the Big Bang essential? An outstanding example of such an external input is the Hartle-Hawking proposal [2].
- Is the quantum evolution across the 'singularity' deterministic? One needs a fully non-perturbative framework to answer this question in the affirmative. In the Pre-Big-Bang [3] and Ekpyrotic/Cyclic [4, 5] scenarios, for example, so far the answer is in the negative because these theories pre-suppose the space-time continuum of general relativity and this approximation fails at the big-bang.

CP1241, *Invisible Universe: Proceedings of the Conference,* edited by J.-M. Alimi and A. Füzfa

• If the singularity is resolved, what is on the 'other side'? Is there just a 'quantum foam', far removed from any classical space-time, or, is there another large, classical universe?

Such questions are fundamental and must be faced squarely because the resolution of classical singularities can profoundly shift the paradigm underlying contemporary cosmology. This in itself makes it imperative that we understand the quantum nature of the big bang. But there could also be another rich pay-off: the new paradigm may enable us to address open issues that are observationally significant. For example, if there is a classical pre-big-bang branch to the universe, the horizon problem would disappear and the observed large scale homogeneity could be simply a consequence of the fact even the most distant parts of the universe would have been in causal contact in the past. If there is a pre-big-bang branch, we would not have to specify the initial conditions for perturbations on the singularity, where the applicability of current theories is least reliable. There would be more natural ways of specifying these conditions which, in turn, may well lead to small, potentially observable deviations from current predictions. Finally, our experience with general relativity itself suggests that, once the physics of the Planck regime is well understood, we may be handed with novel predictions of central importance to the next generation of astrophysicists and cosmologists.

Loop quantum gravity (LQG) [6, 7, 8] is well suited to embark on this mission because it does not pre-suppose a classical space-time –it is background independent. At a fundamental level, everything, *including geometry,* is described in the paradigm of quantum physics [9, 10, 11, 12]. Classical space-times emerge only on coarse graining of semi-classical quantum states. Finally, since the approach is fully non-perturbative, it is well suited for the strong field regime near the putative big bang.

In this chapter I will discuss loop quantum cosmology (LQC), the application of the principles of LQG to cosmology [13, 14]. Initial ideas appeared in [15, 16] and were developed in detail for a variety of cosmological models in [17, 18, 19, 27, 20, 21, 22, 23, 24, 25, 26, 31, 28, 29, 30]. *In all cases, the big-bang and big-crunch singularities are resolved in a direct physical sense.* The resulting Planck scale physics has been explored using analytical and numerical solutions to the quantum Einstein equations as well as effective equations which capture the leading quantum corrections. Singularity theorems are avoided *not* because one uses matter violating energy conditions. Indeed, in the models that have been studied in most detail, *all energy conditions hold.* The theorems are inapplicable because quantum geometry effects modify Einstein equations themselves.

Physically, quantum geometry gives rise to a new repulsive force. This force is *utterly* negligible under normal circumstances. It is only when the matter density becomes about 1% of the Planck density or curvature approaches $1/\ell_{\rm Pl}^2$ that the repulsive force —and hence the deviation from classical general relativity— becomes significant. But then the repulsive force rises *very* quickly, overwhelms classical attraction and causes a quantum bounce. Once the density and curvature falls below the scales just mentioned, the force again becomes negligible and classical general relativity again becomes an excellent approximation. Immediately after the bounce there is a robust phase of super-inflation which is not encountered in general relativity [32, 33]. But it is short lived and in absence of a suitable inflaton potential it does not yield a sufficient number of e-foldings. However, in presence of suitable potentials — such as $m^2\phi^2$— super-inflation *funnels the phase space trajectories to initial conditions which virtually guarantee a slow roll inflation with 60 or more e-foldings* [34]. This is in striking contrast to what happens in general relativity where it has been argued [35] that the probability of N e-folding decreases as e^{-N}. The super-inflationary phase is also likely to have other phenomenological consequences —such as production of gravitational waves— that are being analyzed.

The article is organized as follows. In section ,

CONCEPTUAL SETTING

To set the stage, let me use the simplest cosmological models: FLRW space-times with a massless scalar field. These models are instructive because in classical general relativity all their solutions have a big-bang (and/or big-crunch) singularity. Therefore, a quantum resolution of these singularities is non-trivial. It is not difficult to incorporate additional matter fields and anisotropies.

Figure 1 illustrates classical dynamics for k=0 and k=1 models without a cosmological constant. In the k=0 case, there are two classes of trajectories. In one the universe begins with a big-bang and expands continuously, while in the other, it contracts continuously into a big crunch. In the k=1 case, the universe begins with a big bang, expands to a maximum volume and then undergoes a recollapse to a big crunch singularity. Now, in quantum gravity, one does not have a single space-time in the background but rather a probability amplitude for various space-time geometries. Therefore, unlike in the classical theory, one cannot readily use, e.g., the proper time along a family of preferred

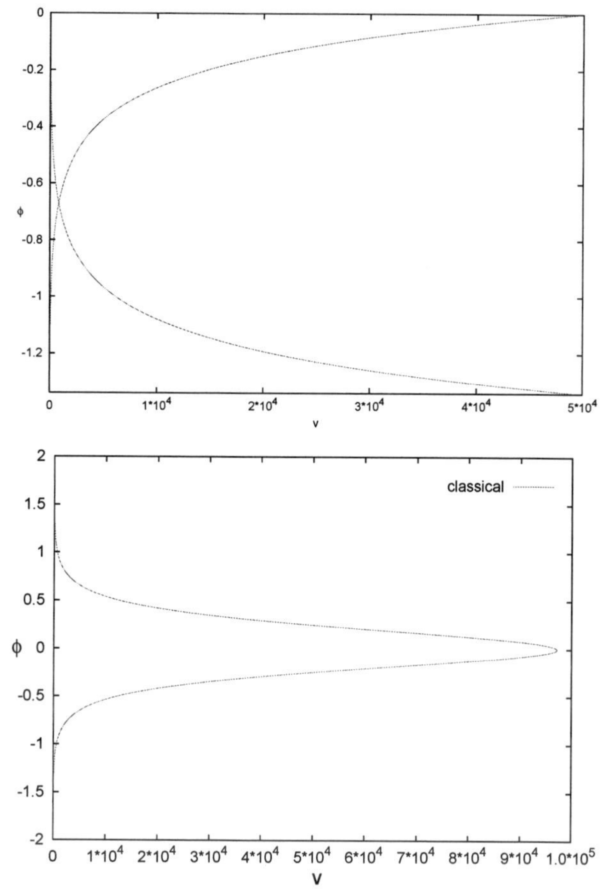

a)

b)

FIGURE 1. Dynamics of FLRW universes with zero cosmological constant and a massless scalar field. Classical trajectories are plotted in the $v-\phi$ plane, where v denotes the volume and ϕ the scalar field. Following the convention in general relativity, the time variable (ϕ) is plotted along the y-axis. *a)* k=0 trajectories. *b)* A k=1 trajectory. In the k=0 case, v is the physical volume of a fiducial cell/box; $v \sim a^3$ where a is the scale factor. The final physical results are of course insensitive to the choice of this cell.

observers as a clock. However, along each dynamical trajectory, the massless scalar field ϕ is monotonic. Therefore, it serves as a good clock or 'emergent time' with respect to which the physical degrees of freedom —the matter density or the volume, anisotropies, curvature scalars and other matter fields, if any— evolve. Note incidentally that, in the k=1 case, because of the classical recollapse, volume is double-valued along any dynamical trajectory. Therefore it cannot serve as a global clock variable, while the scalar field does fulfill this role.

If one has a full quantum theory, one can proceed as follows. Choose a classical solution, i.e. a dynamical trajectory in the v, ϕ plane. The momentum $p_{(\phi)}$ conjugate to ϕ is a constant of motion. Let us suppose its value on our trajectory is $p_{(\phi)} = p^{\star}_{(\phi)}$. Next, choose a point $(v^{\star}, \phi^{\star})$ on any one of these dynamical trajectories where the matter density and space-time curvature is low. This point describes the state of the FLRW universe at a late time where general relativity is expected to be valid. At the 'time' $\phi = \phi^{\star}$ construct a wave packet which is sharply peaked at $v = v^{\star}$ and $p_{(\phi)} = p^{\star}_{(\phi)}$ and evolve it backward and forward in (the scalar field) time. We are then led to two questions:

i) *The infrared issue:* Does the wave packet remain peaked on the classical trajectory in the low curvature regime? Or, do quantum geometry effects accumulate over the cosmological time scales, causing noticeable deviations from classical general relativity? In particular, in the k=1 case, is there a recollapse and if so for large universes does the value $V_{\max}$ of maximum volume agree with that predicted by general relativity [36]?

ii) *The ultraviolet issue:* What is the behavior of the quantum state when the curvature grows and enters the Planck regime? Is the big-bang singularity resolved without any extra input? Or, do we need to supplement dynamics with a new principle, such as the Hartle-Hawking 'no boundary proposal' [2]? What about the big-crunch?

By their very construction, perturbative and effective descriptions have no problem with the first requirement.

However, physically their implications can not be trusted at the Planck scale and mathematically they generally fail to provide a deterministic evolution across the putative singularity. Since the non-perturbative approaches often start from deeper ideas, it is conceivable that they could lead to new structures at the Planck scale which modify the classical dynamics and resolve the big-bang singularity. But once unleashed, do these new quantum effects naturally 'turn-off' sufficiently fast, away from the Planck regime? The universe has had some *14 billion years* to evolve since the putative big bang and even minutest quantum corrections could accumulate over this huge time period leading to observable departures from dynamics predicted by general relativity. Thus, the challenge to quantum gravity theories is to first create huge quantum effects that are capable of overwhelming the extreme gravitational attraction produced by matter densities of some $10^{94}\,\mathrm{gms/cc}$ near the big bang, and then switching them off with extreme rapidity as the matter density falls below this Planck scale. This is a huge burden!

The question then is: How do various approaches fare with respect to these questions? The older quantum cosmology —the Wheeler-DeWitt (WDW) theory— passes the infra-red test with flying colors. But unfortunately the state follows the classical trajectory into the big bang (and in the k=1 case also the big crunch) singularity. The singularity is not resolved because *expectation values of density and curvature continue to diverge* in epochs when their classical counterparts do [18, 19].

For a number of years, the failure of the WDW theory to naturally resolve the big bang singularity was taken to mean that quantum cosmology cannot, by itself, shed significant light on the quantum nature of the big bang. Indeed, for systems with a finite number of degrees of freedom we have the von Neumann uniqueness theorem which guarantees that quantum kinematics is unique. The only freedom we have is in factor ordering and this was deemed insufficient to alter the status-quo provided by the WDW theory.

The situation changed dramatically in LQG. In contrast to the WDW theory, a well established, rigorous kinematical framework *is* available in full LQG [9, 6, 7, 8]. Furthermore, this framework is uniquely singled out by the requirement of diffeomorphism invariance (or background independence) [37, 38]. If one mimics it in symmetry reduced models, one finds that a key assumption of the von-Neumann theorem is violated: *One is led to new quantum mechanics* [16]! This quantum theory is inequivalent to the WDW theory already at the kinematic level. Quantum dynamics built in this new arena agrees with the WDW theory in 'tame' situations but differs dramatically in the Planck regime, leading to a natural resolution of the big bang singularity.

LOOP QUANTUM COSMOLOGY

This section is divided into three parts. In the first I consider the FLRW models without a cosmological constant in some detail, in the second I summarize the more general situation, and in the third I discuss the phenomenon of super inflation and its surprising implications for inflationary scenarios.

Singularity resolution in the FLRW Models

The main LQC results can be summarized as follows [17, 18, 19, 21, 22, 23, 20, 24, 25, 26].

Let first consider the k=0 model without a cosmological constant. Following the strategy outlined in section , let us fix a point at a late time on the trajectory corresponding to an expanding classical universe, construct a quantum state which is sharply peaked at that point, and evolve it using the LQC Hamiltonian constraint. One then finds the following [19].

- The wave packet remains sharply peaked on the classical trajectory so long as the matter density ρ remains below 1% of the Planck density $\rho_{\rm Pl}$. Thus, as in the WDW theory, the LQC evolution meets the infra-red challenge successfully.
- Let us evolve the quantum state back in time, toward the singularity. In the classical solution scalar curvature and the matter energy density keep increasing and eventually diverge at the big bang. The situation is very different in LQC. As mentioned in section , once the density and curvature enter the Planck scale *quantum geometry effects become dominant creating an effective repulsive force which rises very quickly, overwhelms the classical gravitational attraction, and causes a bounce thereby resolving the big bang singularity.* See Fig 2. Numerical simulations show that the density acquires its maximum value $\rho_{\rm max} \approx 0.41\rho_{\rm Pl}$ at the bounce point.
- Although in the Planck regime the peak of the wave function deviates very substantially from the general relativistic trajectory of figure 1, it follows an effective trajectory with very small fluctuations. This effective trajectory

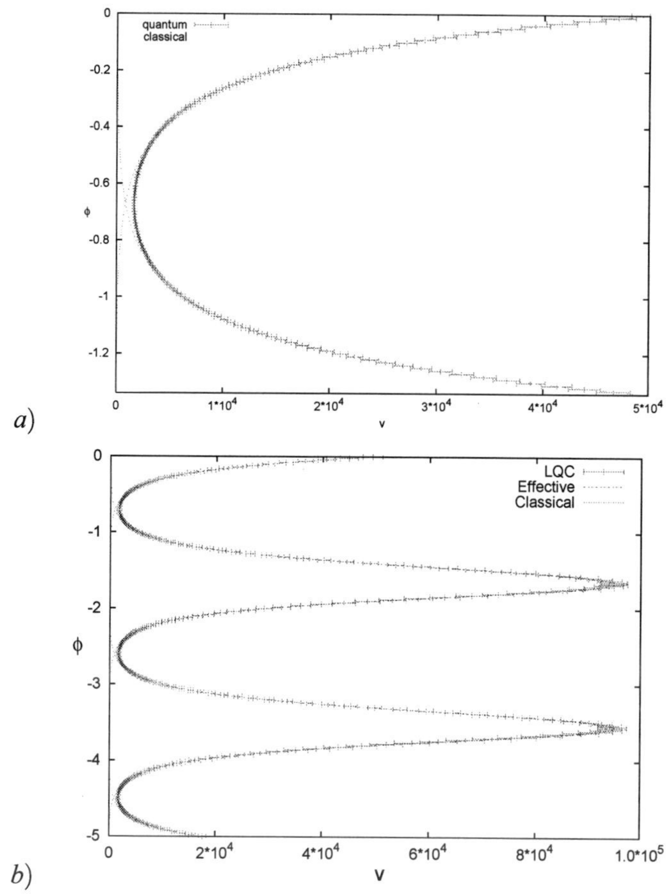

FIGURE 2. In the LQC evolution of models under consideration, the big bang and big crunch singularities are replaced by quantum bounces. Expectation values and dispersion of the volume operator are compared with the classical trajectory and the trajectory from effective Friedmann dynamics. The classical trajectory deviates significantly from the quantum evolution at the Planck scale and evolves into singularities. The effective trajectory provides an excellent approximation to quantum evolution at all scales. *a*) The k=0 case. In the backward evolution, i.e., as ϕ decreases, the wave function follows our post big-bang branch at low densities and curvatures but undergoes a quantum bounce at matter density $\rho \sim 0.41\rho_{\rm Pl}$ and joins on to the classical trajectory that was contracting to the future. *b*) The k=1 case. The quantum bounce occurs again at $\rho \sim 0.41\rho_{\rm Pl}$. Since the big bang and the big crunch are replaced by quantum bounces and the classical re-collapse survives, the evolution undergoes cycles. In these simulations, $p^\star_{(\phi)} = 5\times 10^3$, $\Delta p_{(\phi)}/p^\star_{(\phi)} = 0.018$, and $v^\star = 5\times 10^4$.

was derived [39, 40] using techniques from geometric quantum mechanics. The effective equation incorporates the leading corrections from quantum geometry. They modify the left hand side of Einstein's equations. However, to facilitate comparison with the standard form of Einstein's equations, one moves this correction to the right side through an algebraic manipulation. Then, one finds that the Friedmann equation

$$\left(\frac{\dot a}{a}\right)^2 = \frac{8\pi G}{3}\rho\,. \tag{1}$$

of classical general relativity is replaced by

$$\left(\frac{\dot a}{a}\right)^2 = \frac{8\pi G}{3}\rho\,(1-\frac{\rho}{\rho_{\rm crit}})\,. \tag{2}$$

Here, as usual 'dot' refers to the derivative with respect to *proper time* and $\rho_{\rm crit} = \sqrt{3}/32\pi^2\gamma^3G^2\hbar$, where γ is the Barbero-Immirzi parameter of LQG (whose value $\gamma \sim 0.24$ is determined by the black hole entropy calculations in LQG). By plugging in numbers one finds $\rho_{\rm crit} \approx 0.41\rho_{\rm Pl}$. Thus, $\rho_{\rm crit} \approx \rho_{\rm max}$, found in numerical simulations. Furthermore, one can show analytically [22] that the eigenvalues of the density operator on the *physical* quantum states are bounded above by $\rho_{\rm sup}$, also given by $\rho_{\rm sup} = \sqrt{3}/32\pi^2\gamma^3 G^2\hbar$. Thus, *there is an excellent match between the quantum theory which provides* $\rho_{\rm sup}$ [22], *the effective equations which provide* $\rho_{\rm crit}$ [19] *and numerical simulations which provide* $\rho_{\rm max}$ *[19].*

• In classical general relativity the right side, $8\pi G\rho/3$, of the Friedmann equation is positive, whence $\dot{a}$ cannot vanish; the universe either expands forever from the big bang or contracts forever ending in the big crunch. In the LQC effective equation, on the other hand, $\dot{a}$ vanishes when $\rho = \rho_{\rm crit}$ at which a quantum bounce occurs: To the past of this event, the universe contracts while to the future, it expands. This is possible because the LQC correction $\rho/\rho_{\rm crit}$ *naturally* comes with a negative sign. This is non-trivial. In the standard brane world scenario, for example, Friedmann equation is also receives a $\rho/\rho_{\rm crit}$ correction but it comes with a positive sign (unless one artificially makes the brane tension negative) whence the singularity is not resolved.

• Consider the standard inflationary scenario for the $m^2\phi^2$ potential with phenomenologically determined values of m. Then the standard initial conditions at the onset of inflation are such that the quantum correction $\rho/\rho_{\rm crit}$ is of the order 10^{-11}, and hence completely negligible. Thus, LQC calculations provide *an a priori justification* for using classical general relativity during inflation.

In the closed, k=1 model, the situation is similar but there are two additional noteworthy features [20]. Although they are not important from phenomenological considerations, they reveal surprising properties of the domain of applicability of classical general relativity.

• To start with, classical general relativity is again an excellent approximation to the LQC evolution till matter density ρ becomes about 1% of the Planck density $\rho_{\rm Pl}$ but, as the density increases further, the two evolutions starts diverging rapidly. Again, quantum geometry effects become dominant creating an effective repulsive force which rises very quickly, overwhelms the classical gravitational attraction, and causes a bounce thereby resolving both the big bang and the big crunch singularities. Surprisingly these considerations apply *even to universes whose maximum radius* $a_{\rm max}$ *is only* $23\ell_{\rm Pl}$. For these universes, general relativity is a very good approximation in the range $8\ell_{\rm Pl} < a < 23\ell_{\rm Pl}$! The matter density acquires its minimum value $\rho_{\rm min}$ at the recollapse. The classical prediction $\rho_{\rm min} = 3/8\pi G a_{\rm max}^2$ is correct to one part in 10^5. The fact that general relativity is so accurate in these situations, where one would have expected quantum corrections to dominate, is rather astonishing.

• on the other hand, there are also situations in which one would have at first expected general relativity valid where quantum corrections dominate! Recall that the volume of the closed universe acquires its minimum value $V_{\rm min}$ at the quantum bounce. $V_{\rm min}$ scales linearly with $p_{(\phi)}$. Consequently, $V_{\rm min}$ can be *much* larger than the Planck size. Consider for example a quantum state describing a universe which attains a maximum radius of a megaparsec. Then the quantum bounce occurs when the volume reaches the value $V_{\rm min} \approx 5.7\times 10^{16}\,{\rm cm}^3$, *some* 10^{115} *times the Planck volume.* A more realistic universe would have a much larger maximum radius and hence $p_{(\phi)}$ and then $V_{\rm min}$ would be much lager! Although it is at first surprising that quantum effects can dominate when the universe is so large, a moment's reflection shows that deviations from the classical behavior are triggered when the density or curvature reaches the Planck scale. The size of the volume by itself is not the relevant for quantum gravity effects.

Thus, in the simplest k=0 and k=1 FLRW models, classical singularities are replaced by quantum bounces and LQC provides a rather detailed picture of the physics in the Planck regime. The situation is essentially the same in the open, k=1 model [21]. (Although the detailed analysis in [21] has certain drawbacks, they can be overcome using techniques introduced in [29].) Furthermore, the singularity resolution does not cause infra-red problems: There is close agreement with classical general relativity away from the Planck scale. The ultraviolet-infrared tension is avoided because, although quantum geometry effects are truly enormous in the Planck regime, they die astonishingly quickly.

Generalizations

In this sub-section I will discuss the fate of singularities and Planck scale physics in models with increasing generality.

• *Beyond the big bang and big crunch singularities:* In section I focused on the big bang and big crunch singularities where the scale factor goes to zero and the matter density and curvature diverges at a finite proper time. But for general matter, new types of singularities can arise even with k=0 and $\Lambda = 0$. For example, in the 'big-rip' singularity, the scale factor *diverges* in a finite proper time along with energy density and pressure. Another possibility is the 'sudden death' singularity where the curvature and/or its derivatives diverge at a *finite* values of the scale factor (and proper time). In the spatially homogeneous, isotropic context, if matter has equation of state $p = p(\rho)$, there is a good control on the type of singularities that can arise [41]. From a physical perspective, we can divide them into two broad categories: strong singularities (for example, the big bang and the big crunch) where observers would be crushed and tame, weak ones (such as those resulting from shell-crossing) through which they would be able to propagate. All possible occurrences have been recently analyzed using effective equations of LQC [27]. The result is that all strong singularities are resolved but not necessarily the weak ones. Since physics does not stop at the weak singularities, they are widely regarded as 'harmless'. An interesting situation arises in the context of a phantom model in which general relativity predicts the occurrence of a big-rip singularity. If one uses the effective equations of LQC, the energy density remains bounded but the pressure and the rate of change of the Hubble parameter diverge [42]. At first this might seem as a problem for LQC. However, a closer examination [27] shows that this is a weak singularity beyond which geodesics can be extended. Thus, in this case the quantum geometry effects tame a strong singularity in to a harmless, weak one.

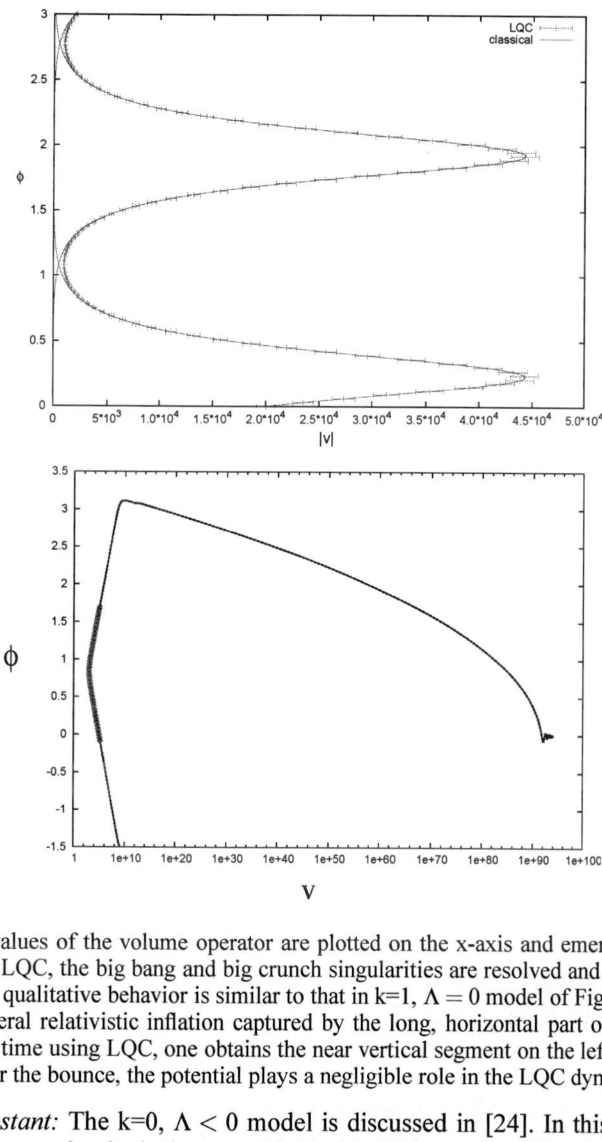

a) *b*)

FIGURE 3. Again, expectation values of the volume operator are plotted on the x-axis and emergent time ϕ on the y-axis. *a*) Negative cosmological constant. In LQC, the big bang and big crunch singularities are resolved and the universe also undergoes a classical re-collapse. Therefore, the qualitative behavior is similar to that in k=1, $\Lambda = 0$ model of Fig. 2.b. *b*) Inflation with $m^2\phi^2$ potential. LQC reproduces the general relativistic inflation captured by the long, horizontal part of the curve that slopes down. However, if we evolve backward in time using LQC, one obtains the near vertical segment on the left showing that the big-bang is replaced by a quantum bounce. Near the bounce, the potential plays a negligible role in the LQC dynamics.

• *Negative cosmological constant:* The k=0, $\Lambda < 0$ model is discussed in [24]. In this case, in classical general relativity the situation is analogous to that in the k=1 model, discussed above. The universe starts out with a big bang, expands to a maximum value of the scale factor, at which the positive energy density matter is exactly balanced by the negative energy density in the cosmological constant. If the universe were to expand further, the total energy density would become negative. But the Friedmann equation of classical general relativity (1) does not allow the total energy density to become negative whence the universe recollapses. In LQC the model has been analyzed in detail both by solving the quantum Einstein's equations numerically and by analyzing the solutions of effective equations. Again, for states which are sharply peaked Gaussians at late times, the effective equations provide an accurate account of full quantum dynamics, including the Planck regime. The quantum corrected Friedmann equation again has the form (2) where ρ now includes also the energy density in the cosmological constant; the big bang and the big crunch are replaced by quantum bounces. The qualitative picture is thus similar to that in the k=1, $\Lambda = 0$ models. The evolution is (approximately) cyclic. The total density is maximum, $\rho = \rho_{\rm crit}$, at the quantum bounce, the universe then expands, and the total density decreases. When it vanishes, there is a classical re-collapse, the universe starts contracting, and the total density increases till it reaches the maximum value, $\rho_{\rm crit}$ when there is again a quantum bounce. The value of $\rho_{\rm crit}$ is the same as in the $\Lambda = 0$ case. (See Fig 3.a)

• *Positive cosmological constant:* In classical general relativity, the situation in the k=0, $\Lambda > 0$ case is similar to that in the k=0, $\Lambda = 0$ case, discussed in section . But the classical trajectory which starts out at the big-bang now expands to an infinite volume at a *finite* value $\phi_{\rm max}$ of the emergent time ϕ. The energy density ρ_ϕ in the scalar field goes to zero at $\phi_{\rm max}$. This opens up an interesting possibility in LQC. Because the evolution in the emergent time ϕ is unitary in LQC, one can continue it beyond the point at which the density goes to zero. States which are semi-classical in the low ρ_ϕ regime again follow effective trajectories which now naturally extend beyond $\phi_{\rm max}$. Since ρ_ϕ remains bounded, it is convenient to draw these trajectories in the ρ_ϕ-ϕ plane. They agree with the classical trajectories in the low ρ_ϕ regime and the extension is just their analytical continuation. It would be very interesting to work out physical ramifications of this unforeseen feature of LQC dynamics. Finally, in this case, LQC predicts that the cosmological constant has a maximum possible value. As one would expect the bound is given by the Planck scale. Therefore it is not of phenomenological interest. However, it has conceptual interest since it is a direct consequence of quantum geometry effects.

• *Inflationary scenarios:* The LQC framework has also been extended to incorporate an inflation. Quantum Einstein's equations have been numerically solved for the $m^2\phi^2$ potential [26]. LQC admits wave functions that are sharply peaked at a standard inflating trajectory. This is just as one would expect because in the standard scenario inflation occurs well away from the Planck regime where general relativity is an excellent approximation to LQC. When such a wave function is evolved *backward* in time, one again finds that, rather than following the classical trajectory into the big bang, the wave function bounces (see Fig. 3.b)). Again, effective equations provide an excellent approximation to the full quantum equations. Numerical simulations as well as the effective equations show that the potential has negligible effect near the big bounce. All the principal features of the LQC evolution reported above are recovered, including the value of $\rho_{\rm crit}$.

• *Anisotropies and gravitational waves:* Inclusion of anisotropies leads to Bianchi models. In these models, the Weyl curvature is not zero because, physically, they admit gravitational waves. Detailed LQC analysis has been carried out in Bianchi I and II models at the analytical level [28, 29, 30] (and the method extends to all class A Bianchi models). It establishes that the big bang singularity is resolved. Effective equations again show that the matter density is bounded above. However, as one might expect, the bound now is lower than $\rho_{\rm crit}$ because there is also some energy in the gravitational waves that is not captured in matter density. In the isotropic models, space-time curvature is completely encoded in the Ricci scalar. In the anisotropic case, we also have non-trivial Weyl curvature. Now, there is not just one bounce. Rather, any time a curvature invariant (more precisely, shear) enters the Planck regime, quantum geometry effects intervene and dilute it, thereby avoiding the singularity that would have occurred had one used general relativity. Thus, the Planck regime is much richer. However, outside this regime, general relativity is again an excellent approximation to LQC. Finally, as in the isotropic case, matter can be chosen to satisfy all energy conditions. The singularity theorems are transcended because Einstein's equations are modified just in the right fashion by LQC.

• *Beyond homogeneity:* In mathematical general relativity, the so called Gowdy cosmological models have drawn much attention because they admit *local, inhomogeneous degrees of freedom* and are yet sufficiently simple to be tractable. These models have been studied in detail using LQC [31]. One can construct a quantum theory of Gowdy models using conventional field theoretical methods which do not take into account quantum geometry of LQC. One then finds that the singularity is not resolved. On the other hand, using LQC to treat just the homogeneous degrees of freedom already suffices to resolve singularities, even if the inhomogeneous modes are treated using more conventional techniques rooted in Fock quantization. The general expectation is that the ultra-violet behavior would further improve if all modes are treated using loop quantum gravity. But already this 'hybrid' scheme introduced by the Madrid group has given some confidence that *the singularity resolution is not tied to the homogeneous models which have only a finite number of degrees of freedom.*

• *General Space-like Singularities:* In general relativity there is a conjecture due to Belinskii, Lifshitz and Khalatnikov (BKL) which says that as one approaches space-like singularities in general relativity, 'spatial derivatives of basic fields become sub-dominant relative to the time derivatives' and dynamics at any spatial point is well approximated by that of homogeneous models [43]. Bianchi I dynamics plays a dominant role and from time to time there are transitions from one Bianchi I solution to another, mediated by a Bianchi II solution. By now there is considerable support for this conjecture both from rigorous mathematical and numerical investigations [44]. This provides support for the hope that the lessons on the quantum nature of singularities we learned from the Bianchi I and Bianchi II models may be valid much more generally. In particular, it suggests that the quantum geometry effects of LQG may well resolve *generic* space-like singularities of classical general relativity. These are precisely the singularities of direct interest to cosmology. There is now a formulation of the BKL conjecture in the framework that underlies LQG [45]. Together with the quantization procedure introduced in the Bianchi II models [29] —which is directly applicable to a wide class of homogeneous models— this form of the BKL conjecture should enable one to obtain concrete results for general space-like singularities.

Super-inflation and inflation

So far, I have focused on the singularity resolution and the new physics in the Planck regime. It turns out that this physics has interesting features with unforeseen implications also to the inflationary scenarios and phenomenology. Because of space limitation, I will outline only one of these and mention a few other avenues that are being pursued both by cosmology and LQC communities. For simplicity and definiteness, in the detailed discussion I use the k=0, $\Lambda=0$ isotropic cosmologies and assume that the kinetic energy of the inflaton is positive.

Let me begin by listing some basic predictions of LQC which are rather surprising. The theory is generally formulated using volume $v \sim a^3$ (of a fiducial cell) —rather than the scale factor a— and its conjugate momentum b [22, 23]. On solutions to Einstein's equations, $b=\gamma H$ where, as before, γ is the Barbero-Immirzi parameter of LQC and $H=\dot{a}/a$ is the Hubble parameter. However, LQC modifies the Einstein dynamics and on solutions to the effective equations we have

$$H = \frac{1}{2\gamma\lambda}\sin 2\lambda b \approx \frac{0.94}{\ell_{\rm Pl}}\sin 2\lambda b \qquad (3)$$

where $\lambda^2 \approx 5.17\ell_{\rm Pl}^2$ is the 'area-gap', the smallest non-zero eigenvalue of the area operator. In the effective theory, b ranges over $(0,\pi/\lambda)$ and general relativity is recovered in the limit $\lambda \to 0$. As I already mentioned, quantum geometry effects modify the left side of Einstein's equations. In particular, the LQC Friedmann equation first emerges as

$$\frac{\sin^2\lambda b}{\gamma^2\lambda^2} = \frac{8\pi G}{3}\rho = \frac{8\pi G}{3}\big(\frac{\dot{\phi}^2}{2}+V(\phi)\big)\,. \qquad (4)$$

To compare with the standard Friedmann equation (1), it is often convenient to rewrite it in the form (2) using (3). By inspection it is clear from Eqs (2), (3) and (4) that, away from the Planck regime—i.e., when $\lambda b \ll 1$ or, $\rho \ll \rho_{\rm Pl}$— we recover classical general relativity. However, the modifications in the Planck regime are drastic. First, Eq.(3) implies that the Hubble parameter $H=\dot{a}/a$ is bounded; $|H| \le 0.94/\ell_{\rm Pl}$. Second, Eq. (4) implies that the density ρ is bounded by $\rho_{\rm crit} \approx 0.41\rho_{\rm Pl}$. Third, if the potential $V(\phi)$ is bounded below $V \ge V_o$, then it follows from (4) that $\dot{\phi}^2$ is bounded by $2\rho_{\rm crit}+2V_o$. Fourth, if the potential grows unboundedly for large $|\phi|$, then $|\phi|$ is also bounded. For example, for $V=m^2\phi^2/2$, we have $\phi_{\rm max} = 1.41\sqrt{\rho_{\rm crit}}/m$. Finally, the derivative with respect to proper time of b and the Hubble parameter are now given by.

$$\dot{b} = -4\pi\gamma G\dot{\phi}^2 \quad \text{and} \quad \dot{H} = -4\pi G(\cos 2\lambda b)\,\dot{\phi}^2 \qquad (5)$$

As a consequence, *b decreases monotonically in every solution*, starting from $b=\pi/\lambda$ and ending with $b=0$ and, when the potential is bounded below, $|\dot{H}|$ is also bounded by $10.3/\ell_{\rm Pl}^2$.

These results imply that if the potential is bounded below, the matter density and curvature are everywhere finite in LQC. One can also show that a solution where our fixed fiducial cell has a finite volume initially cannot evolve to a configuration where the volume becomes zero. Thus, *the solutions are singularity-free irrespective of whether one focuses on matter density, curvature or the scale factor.* Finally, every solution undergoes precisely one bounce at $b=\pi/2\lambda$ where the Hubble parameter vanishes because of Eq. (3) and the density reaches its maximum value $\rho_{\rm crit}$ because of Eq. (2).

Let me now turn to *super-inflation* (see [32] and especially [33]). At the bounce point, we have $b=\pi/2\lambda$ whence it follows from (5) that $\dot{H}$ is positive. Thus, the universe expands to the future of the bounce point and contracts in its past. Eq (5) shows that $\dot{H}$ continues to be positive till b has decreased to $b=\pi/4\lambda$ at which point $\dot{H}$ vanishes and then it remains negative as v increases monotonically to infinity. Thus, *every LQC solution has a super-inflationary phase from* $b = \pi/2\lambda$ to $b = \pi/4\lambda$. Eqs (3) and (5) imply that at the beginning of this phase, the Hubble parameter H vanishes and $\dot{H}$ is very large, while at the end of this phase H assumes its maximum value and $\dot{H}$ vanishes. Note that the occurrence of this phase is universal; it exists even if the inflaton potential is zero! However, detailed considerations show that this phase is very short lived in proper time and therefore does not yield enough e-foldings to be a viable substitute for the standard inflation.

However, it was recently realized that this phase has an unforeseen consequence: it naturally funnels dynamical trajectories to initial conditions that, in presence of usual inflationary potentials, lead to a long slow roll inflation with at least 60 e-foldings [34].

More precisely the situation can be summarized as follows. The task is to calculate the *a priori* probability that there is inflation with at least 60 e-foldings. As in general relativity (see, e.g. [35]), this can be done by introducing the natural Liouville measure on the space of solutions and dividing the Liouville volume of solutions with adequate number of e-foldings with the total Liouville volume of all solutions. In general relativity, it has been argued that the resulting a priori probability for obtaining 60 or more e-foldings is exceedingly small. What is the situation in LQC? Now the equations of motion are

$$\ddot{v} = \frac{24\pi G}{\rho_{\rm crit}} v \left[(\rho - V)^2 + V(\rho_{\rm crit} - V)\right] \qquad \text{and} \qquad \ddot{\phi} + 3H\dot{\phi} + V_{,\phi} = 0 \tag{6}$$

where $3H = 3\dot{a}/a = \dot{v}/v$, and v, ϕ are, in addition, subject to the LQC-modified Friedmann equation (2). Since we know that each solution undergoes exactly one bounce, it is natural to focus on the initial data at the bounce. At the bounce point $\dot{v} = 0$. Furthermore, since Eq.(2) implies that $\dot{\phi}^2 = 2\rho_{\rm crit} - 2V(\phi)$, modulo sign, $\dot{\phi}$ is determined by ϕ. Thus, each solution is characterized by the values $(v_\star, \phi_\star)$ of v and ϕ (and the sign of $\dot{\phi}$) at the bounce point. One can show that in terms of these quantities the natural Liouville measure is simply $d\phi_\star\, dv_\star$ (up to an overall constant which is irrelevant for calculating ratios of two volumes). Furthermore, $|\phi|$ takes values in the bounded interval $(0, \phi_{\rm max} = 0.9\sqrt{\rho_{\rm Pl}}/m)$. The question now is: What is the fraction of the $(v_\star, \phi_\star)$ space that is occupied by initial data which gives rise to a long slow roll inflation.

Recall that the Hubble parameter H takes its maximum value at the end of super-inflation and the friction term in the equation of motion for ϕ is given by H/m. The phenomenologically preferred value of m is of the order $10^{-6} M_{\rm Pl}$. Therefore, at the end of super-inflation, the friction term is generically large and initial conditions are naturally set for a long slow roll. To calculate the probabilities, let us set $m = 10^{-6} M_{\rm Pl}$ and parameterize $\phi_\star$ as $\phi_\star = f\phi_{\rm max}$ where the fraction f is a number between 0 and 1. One can then perform analytical calculations using usual approximations and check their validity via numerical simulations. One finds that 60 or more e-foldings are guaranteed if f is larger than 3×10^{-6} [34]. This in turn implies that the a priori probability of a slow roll with at least 60 e-foldings is greater than 0.99999. Thus the conclusion in LQC is *opposite* of that in general relativity [35]. This comes about because, unlike in general relativity, $\phi, \dot{\phi}$ are bounded in LQC; each solution undergoes a quantum bounce; and the bounce is followed by a period of super-inflation which funnels dynamical trajectories to initial conditions that are well suited for a long slow roll.

There are several other phenomenological implications of super inflation, and more generally, of the LQC quantum corrections to Einstein's equations, particularly on scalar and tensor perturbations (see e.g. [46, 47, 48]). I discussed some of these in my talk. Unfortunately, space limitation does not allow me to discuss these developments here.

OUTLOOK

Singularities of general relativity are perhaps the most promising gates to physics beyond Einstein. They provide a fertile conceptual and technical ground in our search of a new paradigm in cosmology as well as fundamental physics. Consider some of the deepest conceptual questions we face today: the issue of the Beginning and the end End, the arrow of time, and the puzzle of black hole information loss. Their resolutions hinge on the true nature of space-time singularities. In my view, considerable amount of contemporary confusion about such questions arises from our explicit or implicit insistence that singularities of general relativity are true boundaries of space-time; that we can trust causal structure all the way to these singularities; that notions such as event horizons are absolute even though changes in the metric in a Planck scale neighborhood of the singularity can move event horizons dramatically or even make them disappear altogether [49].

LQG is well suited to address the issue of the fate of singularities in quantum gravity because it is fully non-perturbative and does not pre-suppose that we have a smooth geometry all the way up to the big-bang and big crunch

[6, 7, 8]. Therefore LQC has been used to address many long standing questions in remarkable detail [13, 14, 23]. The scalar field serves as emergent time. Strong curvature singularities of classical general relativity are either resolved [19, 20, 21, 24, 25, 26, 22, 23] or converted to harmless weak singularities by the quantum geometry effects [27]. In particular, the big bang and the big crunch are naturally replaced by quantum bounces. On the 'other side' of the bounce there is again a large universe. General relativity is an excellent approximation to quantum dynamics once the matter density falls below one percent of the Planck density. Thus, LQC successfully meets both the 'ultra-violet' and 'infra-red' challenges. Furthermore results obtained in a number of models using distinct methods re-enforce one another. One is therefore led to take at least the qualitative findings seriously: *Big bang is not the Beginning nor the big crunch the End.* Quantum space-times are vastly larger than what general relativity had us believe!

How can the quantum space-times of LQC manage to be significantly larger than those in general relativity when those in the WDW theory are not? Main departures from the WDW theory occur due to *quantum geometry effects* of LQG. There is no fine tuning of initial conditions, nor a boundary condition at the singularity, postulated from outside. Furthermore, matter can satisfy all the standard energy conditions. Why then does the LQC singularity resolution not contradict the standard singularity theorems of Penrose, Hawking and others? These theorems are inapplicable because *the left hand side* of the classical Einstein's equations is modified by the quantum geometry corrections of LQC. What about the more recent singularity theorems that Borde, Guth and Vilenkin [1] proved in the context of inflation? They are not tied to Einstein's equations. But, motivated by the eternal inflationary scenario, they assume that the expansion remains positive if we recede in the past along any geodesic. Because of the pre-big-bang contracting phase, this assumption is violated in the LQC effective theory.

A qualitative picture that emerges is that the non-perturbative quantum geometry corrections are *'repulsive'*. While they are negligible under normal conditions, they dominate when curvature approaches the Planck scale and can halt the collapse that would classically have led to a singularity. In this respect, there is a curious similarity with the situation in the stellar collapse where a new repulsive force comes into play when the core approaches a critical density, halting further collapse and leading to stable white dwarfs and neutron stars. This force, with its origin in the Fermi-Dirac statistics, is *associated with the quantum nature of matter*. However, if the total mass of the star is larger than, say, 5 solar masses, classical gravity overwhelms this force. The new repulsive force of LQC is *associated with the quantum nature of geometry*. It comes into play only near Planck densities but is then so strong that it can counter the classical, gravitational attraction, irrespective of how large the collapsing mass is. It is this force that prevents the formation of singularities.

At first one might think that, since quantum gravity effects concern only a tiny region, whatever they may be, their influence on the global properties of space-time should be negligible whence they would have almost no bearing on the issue of the Beginning and the End. However, as we saw, once the singularity is resolved, vast new regions appear on the 'other side.' New possibilities open up that were totally unforeseen in general relativity [34]. First, matter density, curvature scalars, and the Hubble parameter are all bounded. Second, even in absence of a potential, there is a robust super-inflationary phase immediately after the quantum bounce [32, 33]. As we discussed in section this novel phase has interesting consequences. In particular, it naturally leads to the phenomenon of *Hubble freezing*: the Hubble parameter assumes its largest possible value at the end of super-inflation. As a consequence, if there is a potential such as $m^2\phi^2$, initial conditions are naturally set for a long slow roll inflation [34]. One does not have to appeal to low probability quantum fluctuations to explain how the inflaton managed to get sufficiently high in the potential to seed a slow roll inflation leading to at least 60 e-foldings. Finally, *these conclusions are insensitive to what happened in the pre-big-bang branch.*

Another major direction of current work concerns perturbations that seed structure formation. In the new paradigm provided by LQC, one is led to re-examine the issue of initial conditions for these perturbations. We no longer have to specify them at a singularity. It is most natural to set them in the infinite past where matter density and curvatures tend to zero, whence it makes direct physical sense to speak of vacuum fluctuations. The LQC paradigm leads us to re-asses the old questions from entirely new angles (see, e.g., [46, 47, 48]). This opens up a wealth of interesting challenges and LQC provides novel ideas and tools to meet them.

Acknowledgments: I would like to thank Alex Corichi, Tomasz Pawlowski, Param Singh, David Sloan, Victor Taveras, Kevin Vandersloot and Edward Wilson-Ewing for collaboration and discussions. This work was supported in part by the NSF grant PHY0854743, the Alexander von Humboldt Foundation, The George A. and Margaret M. Downsbrough Endowment and the Eberly research funds of Penn State.

REFERENCES

1. A. Borde, A. Guth and A. Vilenkin, Inflationary spacetimes are not past-complete, Phys. Rev. Lett. **90**, 151301 (2003)
2. J. B. Hartle and S. W. Hawking, Wave Function Of The Universe. Phys. Rev. D **28**, 2960 (1983)
3. M. Gasperini and G. Veneziano, The pre-big bang scenario in string cosmology, Phys. Rept. **373**, 1 (2003)
4. J. Khoury, B. A. Ovrut, P. J. Steinhardt, N. Turok, The Ekpyrotic Universe: Colliding Branes and the Origin of the Hot Big Bang, Phys.Rev. D64 (2001) 123522
5. J. Khoury, B. . Ovrut, N. Seiberg, P. J. Steinhardt, N. Turok, From Big Crunch to Big Bang, Phys.Rev. **D65** 086007 (2002)
6. A. Ashtekar and J. Lewandowski, Background independent quantum gravity: A status report, Class. Quant. Grav. **21**, R53-R152 (2004)
7. C. Rovelli,*Quantum Gravity*. (Cambridge University Press, Cambridge (2004))
8. T. Thiemann, *Introduction to Modern Canonical Quantum General Relativity*. (Cambridge University Press, Cambridge, (2007))
9. A. Ashtekar, J. Lewandowski, D. Marolf, J. Mourão and T. Thiemann, Quantization of diffeomorphism invariant theories of connections with local degrees of freedom. Jour. Math. Phys. **36**, 6456–6493 (1995)
10. R. Rovelli and L. Smolin, Discreteness of area and volume in quantum gravity, Nucl. Phys. **B442**593-622 (1995); Erratum **B456**, 753 (1996)
11. A. Ashtekar and J. Lewandowski, Quantum theory of geometry I: Area operators, Class. Quantum Grav. **14**, A55-81 (1997)
12. A. Ashtekar and J. Lewandowski, Quantum theory of geometry II: Volume operators, Adv. Theo. & Math. Phys. **1**, 388-429 (1997)
13. M. Bojowald, Loop quantum cosmology. Liv. Rev. Rel. **8**, 11 (2005).
14. A. Ashtekar, Loop Quantum Cosmology: An Overview (2007), Gen. Rel. and Grav. (at Press), `arXiv:0812.0177`
15. M. Bojowald, Absence of singularity in loop quantum cosmology, Phys. Rev. Lett. **86**, 5227-5230 (2001), `arXiv:gr-qc/0102069`.
16. A. Ashtekar, M. Bojowald and J. Lewandowski, Mathematical structure of loop quantum cosmology. Adv. Theo. Math. Phys. **7**, 233–268 (2003)
17. A. Ashtekar, T. Pawlowski and P. Singh, Quantum nature of the big bang, Phys. Rev. Lett. **96**, 141301 (2006)
18. A. Ashtekar, T. Pawlowski and P. Singh, Quantum nature of the big bang: An analytical and numerical investigation I, Phys. Rev. **D73**, 124038 (2006)
19. A. Ashtekar, T. Pawlowski and P. Singh, Quantum nature of the big bang: Improved dynamics, Phys. Rev. **D74**, 084003 (2006)
20. A. Ashtekar, T. Pawlowski, P. Singh and K. Vandersloot, Loop quantum cosmology of k=1 FRW models. Phys. Rev. **D75**, 0240035 (2006);
    L. Szulc, W. Kaminski, J. Lewandowski, Closed FRW model in loop quantum cosmology. Class. Quant. Grav. **24**, 2621-2635 (2006)
21. K. Vandersloot, Loop quantum cosmology and the k=−1 RW model, Phys. Rev. **D75**, 023523 (2007)
22. A. Ashtekar, A. Corichi and P. Singh, Robustness of predictions of loop quantum cosmology, Phys. Rev. **D77**, 024046 (2008)
23. A. Corichi and P. Singh, Is loop quantization in cosmology unique? Phys. Rev. **D78**, 024034 (2008)
24. E. Bentivegna and T. Pawlowski, Anti-deSitter universe dynamics in LQC, Phys. Rev. D77, 124025 (2008)
25. A. Ashtekar and T. Pawlowski, Loop quantum cosmology with a positive cosmological constant (in preparation)
26. A. Ashtekar, T. Pawlowski and P. Singh, Loop quantum cosmology in the pre-inflationary epoch (in preparation)
27. P. Singh, Loop quantum cosmos are never singular? Class. Quant. Grav. **26**, 125005 (2009)
28. A. Ashtekar and E. Wilson-Ewing, Loop quantum cosmology of Bianchi type I models, Phys. Rev. **D79**, 083535 (2009)
29. A. Ashtekar and E. Edward-Wilson, Loop quantum cosmology of Binachi II models, `arXiv:0910.1278`
30. A. Corichi and P. Singh, A geometric perspective on singularity resolution and uniqueness in quantum cosmology, Phys. Rev. **D80**, 044024 (2009)
31. M. Martin-Benito, L. J. Garay and G. A. Mena Marugan, Hybrid quantum Gowdy cosmology: combining loop and Fock quantizations, *Phys. Rev.***D78**, 083516 (2008)
    D. Brizuela, G. Mena-Marugan and T. Pawlowski, Big bounce and inhomogeneities, `arXiv:0902.0697`
    G. Mena-Marugan and M. Martin-Benito, Hybrid quantum cosmology: Combining loop and Fock quantization, Intl. J. Mod. Phys. **A24**, 2820-2838 (2009)
32. M. Bojowald, Inflation from Quantum Geometry, Phys. Rev. Lett. **89** 261301 (2002)
33. P. Singh, Loop cosmological dynamics and dualities with Randall-Sundrum braneworlds, Phys. Rev. **D73** 063508 (2006)
34. A. Ashtekar and D. Sloan, Inflation and Loop Quantum Cosmology (in Preparation)
35. G. W. Gibbons and N. Turok, Measure problem in cosmology, Phys. Rev. **D77**, 063516 (2008)
36. D. Green and W. Unruh (2004): Difficulties with recollapsing models in closed isotropic loop quantum cosmology. Phys. Rev. **D70**, 103502
37. J. Lewandowski, A. Okolow, H. Sahlmann and T. Thiemann, Uniqueness of diffeomorphism invariant states on holonomy flux algebras, Comm. Math. Phys. **267**, 703-733 (2006)
38. C. Fleishchack, Representations of the Weyl algebra in quantum geometry, Commun. Math. Phys. **285** 67-140 (2009)
39. J. Willis On the low energy ramifications and a mathematical extension of loop quantum gravity. Ph.D. Dissertation, The Pennsylvaina State University (2004)
40. V. Taveras, LQC corrections to the Friedmann equations for a universe with a free scalar field, *Phys. Rev.* **D78**, 064072 (2008)

41. S. Nojiri, S. Odintsov and S. Tsujikawa, Properties of singularities in the (phantom) dark energy universe, Phys. Rev. **D71**, 063004 (2005)
42. T. Cailleteau, A. Cardoso, K. Vandersloot and D. Wands, Singularities in loop quantum cosmology, Phys. Rev. Lett. **101**, 251302 (2008)
43. V. A. Belinskii, I. M. Khalatnikov, and E. M. Lifshitz, Oscillatory approach to a singular point in relativistic cosmology, Adv. Phys. **19**, 525 (1970);
    A general solution of the Einstein equations with a time singularity, Adv. Phys. **31**, 639 (1982)
44. B. Berger, Numerical Approaches to Spacetime, Liv. Rev. Rel.**5**, 1 (2002)
45. A. Ashtekar, A. Henderson and D. Sloan, Hamiltonian formulation of general relativity and the Belinksii, Khalatnikov, Lifshitz conjecture, Class. Quantum Grav. **26**, 052001 (2009)
46. E. J. Copeland, D. J. Mulryne, N. J. Nunes, M. Shaeri, Super-inflation in Loop Quantum Cosmology, Phys. Rev. **D77**, 023510 (2008) The gravitational wave background from super-inflation in Loop Quantum Cosmology, Phys. Rev. **D79**, 023508 (2009)
47. J. Grain, A. Barrau, Cosmological footprints of loop quantum gravity, Phys. Rev. Lett. **102**, 081301 (2009)
48. J. Grain, Loop Quantum Cosmology corrections on gravity waves produced during primordial inflation, `arXiv:0911.1625`
49. P. Hajicek, Origin of black hole radiation, Phys. Rev. **D36**, 1065 (1987)

Gravitation and Cosmology

SUSY FLAT DIRECTIONS – TO GET A VEV OR NOT? (ABSTRACT)

Anders Basboll

Department of Physics & Astronomy, University of Sussex, Brighton, BN1 9QH, United Kingdom

We investigate the potential of SUSY flat directions (FDs). Large FD vacuum expecta- tion values (VEVs) can delay thermalisation and solve the gravitino problem - if FDs decay per- turbatively. This depends on how many and which directions get the VEVs. Recently the decay of the FDs have been studied with the VEVs as input. Here we look at how the VEVs come about Đ statistically and analytically.

Keywords: Flat Directions,MSSM,Supersymmetry,gaugeinvariance,scalarpotential,Preheating
PACS: 12.60.Jv, 98.80.Cq

NEW UNDERSTANDING OF THE DARK ENERGY IN THE LIGHT OF A NEW SPACE-TIME (ABSTRACT)

Faycal Ben Adda

Applied mathematics department, University of Hail, P.O.Box 2440 Hail, Saudi Arabia

A new mathematical object more general than classical manifold is built, as a union of Hausdorff topological spaces all disjoint or all the same linked together with an internal structure and a diagonal topology. The construction of this mathematical object is made via quantified chain-react of double homeomorphisms over the graphs product of left and right mean functions. From the construction, it turns out that the mathematical object obtained is an expanding manifold where its basic elements are expanding and present new structures at every step of the chain react. A direct calculus of the distance between two distinct basic elements of the fractal manifold from one step to another leads directly to the Hubble's law which advocates to use the fractal manifold model in the description of the dynamic of our universe. A new metric of the space-time is then obtained. The geodesic of the light in this kind of space-time leads to the invisibility of the geometry of the universe, and to a natural distribution of matter into it. The space of our universe appears to be an infinite packed number of expanding basic elements in which the light is reflected only if it interacts with matter. It turns out that at the beginning of the expansion the different basic elements of the space start to expand simultaneously, from zero size, which creates the expansion and the distribution of matter we observe today, and this simultaneous expansion explains the homogeneity observed at large scale. The expansion of the basic elements of the space drives matter to be located on their boundaries, and matter appears then to be maintained everywhere in our universe by an invisible construction. A new interpretation of the dark energy is then obtained.

Keywords: Fractals, Differential manifold
PACS: 98.80.-k, 95.36.+X, 05.45.Df, 02.03Uu

CP1241, *Invisible Universe: Proceedings of the Conference,* edited by J.-M. Alimi and A. Füzfa

BRANE UNIVERSE: GLOBAL GEOMETRY (ABSTRACT)

Victor Berezin

Institute for Nuclear Research, Russian Academy of Sciences, 60th October Anniversary Prospect, 7a, 117312, Moscow, Russia

The global geometries of bulk vacuum space-times in the brane-universe models are investigated and classified in terms of geometrical invariants. The corresponding Carter-Penrose diagrams and embedding diagrams are constructed. It is shown that for a given energy-momentum induced on the brane there can be different types of global geometries depending on the signs of a bulk cosmological term and surface energy density of the brane (the sign of the latter does not influence the internal cosmological evolution). It is shown that in the Randall-Sundrum scenario it is possible to have an asymmetric hierarchy splitting even with a Z^2-symmetric matching of "our" brane to the bulk.

Keywords: brane models

STABILITY CONDITIONS FOR A NONCOMMUTATIVE SCALAR FIELD COUPLED TO GRAVITY VIA THE POSITIVE ENERGY THEOREM (ABSTRACT)

Orfeu Bertolami and Carlos A. D. Zarro

Instituto Superior Técnico, Departamento de Física, Av. Rovisco Pais, 1049-001 Lisboa, Portugal

The stability requirements for a noncommutative scalar field coupled to gravity is investigated through the positive energy theorem. It is shown that for a noncommutative scalar with a polynomial potential, the stability conditions are similar to the ones for the commutative case. This result remains valid even whether the space-time has horizons.

Keywords: Noncommutative geometry, noncommutative scalar field in curved spaces, stability conditions, positive energy theorem.
PACS: 02.40.Gh, 11.10.Nx, 04.62+v

INFLATION IN THE STANDARD MODEL AND νMSM WITH NON-MINIMAL COUPLING TO GRAVITY (ABSTRACT)

F. Bezrukov[1,2]

[1]*Max-Planck-Institut für Kernphysik, PO Box 103980, 69029 Heidelberg, Germany*

[2]*Institute for Nuclear Research of the Russian Academy of Sciences, 60th October Anniversary prospect 7a, Moscow 117312, Russia*

The possibility of inflation in the Standard Model is described. The only addition is the non-minimal coupling of the Higgs boson to gravity. The Higgs boson itself then plays the role if the inflaton field at high field values. The spectral index is predicted by the model, well in agreement with the current experimental data. Amplitude of the tensor perturbations (primordial gravity waves) is expected to be very small. The reheating temperature in the model is naturally high. The quantum corrections predict a window for allowed Higgs boson masses and provide a possible connection between the particle physics experiments and cosmology.

Keywords: inflation, Higgs boson, non-minimal coupling
PACS: 98.80.Cq 14.80.Bn

EXTENSIONS OF BIRKHOFF'S THEOREM IN 6D GAUSS-BONNET GRAVITY (ABSTRACT)

Charalampos Bogdanos

Laboratoire de Physique Théorique Université de Paris-Sud 11 Bâtiment 210, 91405 Orsay CEDEX, France

We present a generalization of Birkhoff's theorem in the context of six-dimensional Gauss-Bonnet theory. Contrary to what we encounter in ordinary General Relativity, the presence of the higher curvature terms leads to novel constraints on the geometry of the admissible black hole horizons. As a result, we can have static solutions which depart from spherical symmetry, but must satisfy specific constraints involving the Weyl tensor of the four-dimensional horizon. After obtaining the general profile of the static solutions in the transverse space, we give explicit examples of such black hole horizons, which are in general of an anisotropic nature.

Keywords: Gauss-Bonnet, higher dimensions, black holes, staticity, exact solutions.
PACS: 04.50.-h, 04.50.Gh

SEARCH FOR QUANTUM GRAVITY INDUCED EFFECTS WITH PKS 2155-304 FLARING PERIOD IN 2006 BY H.E.S.S. (ABSTRACT)

J. Bolmont[1], R. Buhler[2], A. Jacholkowska[1] and S. Wagner[3] (on behalf of the H.E.S.S. Collaboration)

[1]*LPNHE, Université Pierre et Marie Curie Paris 6, Université Denis Diderot Paris 7, CNRS/IN2P3, France*

[2]*Max-Planck-Institut fur Kernphysik, Heidelberg, Germany.*

[3]*Landessternwarte, Universitat Heidelberg, Germany*

Highly energetic, variable and distant sources such as Active Galactic Nuclei provide a good means to evaluate effects due to the emission and the propagation of high energy photons. In this work, studies of possible energy-dependent time-lags with PKS 2155-304 light curve as measured by H.E.S.S. on 28 July 2006 (MJD 52944) are presented. The time-lags could come from the emission processes and also sign a Lorentz Symmetry breaking as predicted in some Quantum Gravity models. After a brief description of the methods in use, the latest results from two different analyses of the flare will be given.

Keywords: H.E.S.S., Quantum Gravity, Active Galactic Nuclei, Lorentz Invariance Violation
PACS: 04.60.Bc, 98.54.Cm

CARDASSIAN-LIKE MODEL FROM RASTALL'S THEORY (ABSTRACT)

Monica Capone[1], Vincenzo Fabrizio Cardone[2] and Matteo Luca Ruggiero[3]

[1] *Dipartimento di Matematica, Universita di Torino, Via Carlo Alberto 10, & INFN, Via Pietro Giuria 1, 10125 - Torino, Italy*

[2] *Dipartimento di Fisica Generale, Universita di Torino, & INFN, Via Pietro Giuria 1, 10125 - Torino, Italy*

[2] *Dipartimento di Fisica, Politecnico di Torino, Corso Duca degli Abruzzi 23, 10129 - Torino, Italy*

In an attempt to look for a viable mechanism leading to a present day accelerated expansion, we investigate the possibility that the observed cosmic speed up may be recovered in the framework of the Rastall's theory, relying on the non-conservativity of the stress-energy tensor, i.e. $T^{\mu}_{\nu;\mu} \neq 0$. We derive the modified Friedmann equations and show that they correspond to Cardassian-like equations. We also show that, under suitable assumptions on the equation of state of the matter term sourcing the gravitational field, it is indeed possible to get an accelerated expansion, in agreement with the Hubble diagram of both Type Ia Supernovae (SNeIa) and Gamma Ray Bursts (GRBs). Unfortunately, to achieve such a result one has to postulate a matter density parameter much larger than the typical $\Omega_M \simeq 0.3$ value inferred from cluster gas mass fraction data. As a further issue, we then discuss the possibility to retrieve the Rastall's theory from a Palatini variational principle approach to $f(R)$ gravity. However, such an attempt turns out to be unsuccessful.

Keywords: cosmology; dark energy; alternative theories of gravity
PACS: 04.50.Kd; 95.35.+d; 98.80.Jk

ANAMORPHOSIS IN HYBRID INFLATION: HOW TO AVOID FINE-TUNING OF INITIAL CONDITIONS? (ABSTRACT)

Sébastien Clesse[1,2]

[1]*Service de Physique Théorique, Université Libre de Bruxelles, CP225, Boulevard du Triomphe, 1050 Brussels, Belgium*

[2]*Center for Particle Physics and Phenomenology, Louvain University, 2 chemin du cyclotron, 1348 Louvain-la-Neuve, Belgium*

In order to generate more than 60 e-folds of accelerated expansion in original hybrid inflation, 2-fields trajectories are usually required to be initially fine-tuned in a very narrow band along the inflationary valley or in some isolated points outside it. From a more precise investigation of the dynamics, these points which can cover a non-negligible proportion of the space of sub-planckian initial field values, depending on the potential parameters, are shown to be organised in connected domains with fractal boundaries. They correspond to trajectories first falling towards the bottom of the potential, then climbing and slow-rolling back along the inflationary valley. The full parameter space, including initial velocities and all the potential parameters, is then explored by using Monte-Carlo-Markov-Chains (MCMC) methods. Results indicate that successful initial conditions (IC) outside the valley are not localized in the parameter space and are the dominant way to realise inflation, independently of initial field velocities. Natural bounds on parameters are deduced. The genericity of our results is confirmed in 5 other hybrid models from various framework.

Keywords: Hybrid inflation, initial conditions, MCMC, fractal dimension
PACS:98.80.Cq

HIGHER-ORDER BRANE GRAVITY MODELS (ABSTRACT)

Mariusz P. Dabrowski and Adam Balcerzak

Institute of Physics, University of Szczecin, Wielkopolska 15, 70-451 Szczecin, Poland

We discuss a very general theory of gravity, of which Lagrangian is an arbitrary function of the curvature invariants, on the brane. In general, the formulation of the junction conditions (except for Euler characteristics such as Gauss-Bonnet term) leads to the powers of the delta function and requires regularization. We suggest the way to avoid such a problem by imposing the metric and its first derivative to be regular at the brane, the second derivative to have a kink, the third derivative of the metric to have a step function discontinuity, and no sooner as the fourth derivative of the metric to give the delta function contribution to the field equations. Alternatively, we discuss the reduction of the fourth-order gravity to the second order theory by introducing extra scalar and tensor fields: the scalaron and the tensoron. In order to obtain junction conditions we apply two methods: the application of the Gauss-Codazzi formalism and the application of the generalized Gibbons-Hawking boundary terms which are appended to the appropriate actions. In the most general case we derive junction conditions without assuming the continuity of the scalaron and the tensoron on the brane. The derived junction conditions can serve studying the cosmological implications of the higher-order brane gravity models.

Keywords: higher-order gravities, braneworld, junction conditions, boundary terms
PACS: 04.50.Kd; 11.25.-w; 98.80.-k; 98.80.Jk

EXOTIC-SINGULARITY-DRIVEN DARK ENERGY (ABSTRACT)

Mariusz P. Dąbrowski and Tomasz Denkiewicz

Institute of Physics, University of Szczecin, Wielkopolska 15, 70-451 Szczecin, Poland

We discuss various types of exotic (non-standard) singularities in the Universe: a Big-Rip (BR or type I), a Sudden Future Singularity (SFS or type II), a Generalized Sudden Future Singularity, a Finite Scale Factor singularity (FSF or type III), a Big-Separation (BS or type IV) and a w-singularity. They are characterized by violation of all or some of the energy conditions which results in a blow-up of all or some of the physical quantities: the scale factor, the energy density, the pressure, and the barotropic index. We relate the emergence of these singularities with physical theories (superstring, brane, higher-order gravity, loop quantum cosmology). We show how the models involving exotic singularities may serve as dark energy by applying the observational data. In particular, we show that some of these exotic singularities (though being of a weak type according to relativistic definitions) may occur in the near future of the universe.

Keywords: Big-Rip, Sudden Future Singularity, Finite Scale Factor singularity, Big- Separation, w-singularity, supernovae, dark energy
PACS:04.20.Dw; 11.25.-w; 98.80.Es; 98.80.Jk

GRAVASTARS OR BLACK HOLES AS CONSEQUENCE OF EINSTEIN'S THEORY OF GRAVITY (ABSTRACT)

R. Chan[1], M.F.A. da Silva[2], P. Rocha[3] and Anzhong Wang[4]

[1]*Coordenação de Astronomia e Astrofísica, Observatório Nacional, Rua General José Cristino 77, São Cristóvão, CEP 20921-400, Rio de Janeiro, RJ, Brazil.*

[2]*Departamento de Física Teórica, Universidade do Estado do Rio de Janeiro, Rua São Francisco Xavier 524, Maracanã, CEP 20550-013, Rio de Janeiro, RJ, Brazil.*

[3]*Instituto de Física, Universidade Federal Fluminense, Av. Litorânea, s/n, Boa Viagem 24210-340, Niterói, RJ, Brazil and Gerência de Tecnologia da Informação, ACERP, TV Brasil, Rádios Nacional e MEC, Rua da Relação 18, Lapa, CEP 20231-110, Rio de Janeiro, RJ, Brazil.*

[4]*GCAP-CASPER, Department of Physics, Baylor University, Waco, TX 76798, USA.*

Dynamical models of prototype gravastars made of phantom energy are constructed, in which an infinitely thin spherical shell of a perfect fluid with the equation of state $p = (1-\gamma)\sigma$ divides the whole spacetime into two regions, the internal region filled with a phantom fluid, and the external Schwarzschild region. It is found that in some cases the models represent the "bounded excursion" stable gravastars, where the thin shell is oscillating between two finite radii, while in anothers cases they collapse until the formation of black holes or normal stars. In the phase space, the region for the "bounded excursion" gravastars is very small in comparison to that of black holes, but not empty, as found in our previous papers. Therefore, although the existence of gravastars can not be completely excluded from current analysis, the opposite is not possible either, that is, even if gravastars exist, they do not exclude the existence of black holes.

Keywords: Dark energy, Gravastars, Black Holes
PACS: 98.80.-k,04.20.Cv,04.70.Dy

PREHEATING THE UNIVERSE FROM THE STANDARD MODEL HIGGS (ABSTRACT)

Daniel G. Figueroa

Instituto de Física Teórica UAM/CSIC *and Departamento de Física Teórica, Facultad de Ciencias, Universidad Autónoma de Madrid, Cantoblanco, Madrid 28049, Spain.*

We discuss Preheating after an inflationary stage driven by the Standard Model (SM) Higgs field non-minimally coupled to gravity. We find that Preheating is driven by a complex process in which perturbative and non-perturbative effects occur simultaneously. The Higgs field, initially an oscillating coherent condensate, produces non-perturbatively W and Z gauge fields. These decay very rapidly into fermions, thus preventing gauge bosons to accumulate and, consequently, blocking the usual parametric resonance. The energy transferred into the fermionic species is, nevertheless, not enough to reheat the Universe, and resonant effects are eventually developed. Soon after resonance becomes effective, also backreaction from the gauge bosons into the Higgs condensate becomes relevant. We have determined the time evolution of the energy distribution among the remnant Higgs condensate and the non-thermal distribution of the SM fermions and gauge fields, until the moment in which backreaction becomes important. Beyond backreaction our approximations break down and numerical simulations and theoretical considerations beyond this work are required, in order to study the evolution of the system until thermalization.

Keywords: Inflation, Preheating, Higgs, Standard Mode
PACS: 11.15.Tk, 98.80.Cq, 98.80.Bp

TOWARDS A NONCOMMUTATIVE VERSION OF GRAVITATION (ABSTRACT)

Nicolas Franco

Groupe d'Applications Mathématiques aux Sciences du Cosmos, University of Namur FUNDP - Département de Mathématiques, Rempart de la Vierge, 8 B-5000 Namur Belgium, email: nicolas.franco@math.fundp.ac.be

Alain Connes' noncommutative theory led to an interesting model including both Standard Model of particle physics and Euclidean Gravity. Nevertheless, an hyperbolic version of the gravitational part would be necessary to make physical predictions, but it is still under research. We shall present the difficulties to generalize the model from Riemannian to Lorentzian Geometry and discuss key ideas and current attempts.

Keywords: Noncommutative geometry
PACS: 02.40.Gh

A NEW TYPE OF SCALAR FIELD INFLATION (ABSTRACT)

Cristiano Germani[1] and Alex Kehagias[2]

[1]*LUTH, Observatoire de Paris, CNRS UMR 8102, Université Paris Diderot, 5 Place Jules Janssen, 92195 Meudon Cedex, France*

[2]*Physics Division, National Technical University of Athens, 15780 Athens, Greece*

In this short note we discuss the possibility of producing an inflationary background by considering a scalar field whose kinetic term is non-minimally coupled to gravity.

Keywords: Inflation; Modified Gravity.

LOOP QUANTUM COSMOLOGY CORRECTIONS ON GRAVITY WAVES PRODUCED DURING PRIMORDIAL INFLATION (ABSTRACT)

J. Grain

Institut d'Astrophysique Spatiale, Université Paris-Sud 11, CNRS, Bâtiments 120-121, 91405 Orsay Cedex, France

Loop Quantum Gravity (L.Q.G.) is one of the two most promising tentative theory for a quantum description of gravity. When applied to the entire universe, the so-called Loop Quantum Cosmology (L.Q.C.) framework offers microscopical models of the very early stages of the cosmological history, potentially solving the initial singularity problem via bouncing solutions or setting the universe in the appropriate initial conditions for inflation to start, via a phase of super-inflation. More interestingly, L.Q.C. could leave a footprint on cosmological observables such as the Cosmic Microwave Background (CMB) anisotropies. Focusing on the modified dispersion relation when holonomy and inverse-volume corrections arising from the L.Q.C. framework are considered, it is shown that primordial gravity waves generated during inflation are affected by quantum corrections. Depending on the type of corrections, the primordial tensor power spectrum is either suppressed or boosted at large len gth scales, and strongly departs from the power-law behavior expected in the standard scenario.

Keywords: Quantum gravity, quantum cosmology
PACS: 04.60.Pp, 04.60.Bc, 98.80.Cq, 98.80.Qc

TYPE II STRING R^4 TERMS FROM THE COSMOLOGICAL PERSPECTIVE (ABSTRACT)

Masao Iihoshi[1,2] and Sergei V. Ketov[1,3]

[1]*Department of Physics, Tokyo Metropolitan University, Hachioji, Tokyo 192-0397, Japan. email: iihosi-masao@ed.tmu.ac.jp*

[2]*University of Liège, Allée du 6 Août 17, 4000 Liège, Belgium*

[3]*email:ketov@phys.metro-u.ac.jp*

We revisit the gravity theory modified by the quartic-order terms of the spacetime curvature, originating from the type II superstring effects of the leading order in the string Regge slope parameter, from the cosmological perspective. We observe that the cosmological considerations severely constrain the well-known ambiguity of the Ricci-dependent terms in the superstring correction to the gravitational action.

Keywords: modified gravity, inflationary cosmology
PACS: 04.50.-h,04.50.Kd,04.60.-m,04.60.Cf,98.80.-k

F(R) SUPERGRAVITY (ABSTRACT)

Sergei V. Ketov

Department of Physics, Tokyo Metropolitan University, Japan

We review the F(R) supergravity recently proposed in Phys. Lett. B674(2009)59 and Class. and Quantum Grav. 26(2009)135006. Our construction supersymmetrizes popular f(R) theories of modified gravity in four space-time dimensions. We use curved superspace of N=1 Poincaré supergravity in its minimal (2nd order) formulation so that our F(R) supergravity action is manifestly invariant under local N=1 supersymmetry. We prove that the F(R) supergravity is classically equivalent to the standard N=1 Poincaré supergravity coupled to a dynamical chiral superfield, via a Legendre-Weyl transform in superspace. A Kähler potential, a superpotential and a scalar potential of the chiral superfield are governed by a single holomorphic function. We find the conditions of vanishing cosmological constant without fine-tuning, which define a no-scale F(R) supergravity.

CURVATONS IN WARPED THROATS (ABSTRACT)

Takeshi Kobayashi[1] and Shinji Mukohyama[2]

[1]*Institute for Cosmic Ray Research, The University of Tokyo, 5-1-5 Kashiwanoha, Kashiwa, Chiba 277-8582, Japan*

[2]*Institute for the Physics and Mathematics of the Universe (IPMU), The University of Tokyo, 5-1-5 Kashiwanoha, Kashiwa, Chiba 277-8582, Japan*

We present a curvaton model from type IIB string theory compactified on a warped throat with approximate isometries. Considering an (anti-)D3-brane sitting at the throat tip as a prototype standard model brane, we show that the brane's position in the isometry directions can play the role of curvatons. The basic picture is that the fluctuations of the (anti-)D3-brane in the angular isometry directions during inflation eventually turns into the primordial curvature perturbations, and subsequently the brane's oscillation excites other open string modes on the brane and reheat the universe. We find in the explicit case of the KS throat that a wide range of parameters allows a consistent curvaton scenario. It is also shown that the oscillations of branes at throat tips are capable of producing large non-Gaussianity, either through curvature or isocurvature perturbations. Since such setups naturally arise in warped (multi-)throat compactifications and are constrained by observational data, the model can provide tests for compactification scenarios. This work gives an explicit example of string theory providing light fields for generating curvature perturbations. Such mechanisms free the inflaton from being responsible for the perturbations, thus open up new possibilities for inflation models. The discussions in this paper are based on Kobayashi 2009.

Keywords: string theory and cosmology, cosmology with extra dimensions, inflation, cosmological applications of theories with extra dimensions
PACS: 11.25.Mj, 98.80.Cq

SFT NON-LOCALITY IN COSMOLOGY: SOLUTIONS, PERTURBATIONS AND OBSERVATIONAL EVIDENCES (ABSTRACT)

Alexey S. Koshelev[1,2]

[1]*Theoretische Natuurkunde, Vrije Universiteit Brussel and The International Solvay Institutes, Pleinlaan 2, B-1050 Brussels, Belgium*

[2]*Postdoctoral researcher of FWO-Vlaanderen, e-mail: alexey.koshelev@vub.ac.be*

In this note cosmological models coming out of the String Field Theory (SFT) in application to the Dark Energy are reviewed. A way of solution construction in case of linear models is outlined, cosmological perturbations and observational evidences of such models are explored.

Keywords: String Field Theory, Rolling tachyon, Cosmological perturbations, Dark Energy
PACS: 11.25.-w,11.25.Sq,98.80.-k,98.80.Jk

ASTROPHYSICAL IMPLICATIONS OF THE SUPERSTRING-INSPIRED E_6 UNIFICATION AND SHADOW THETA-PARTICLES (ABSTRACT)

C.R. Das[1], **L.V. Laperashvili**[2] **and A. Tureanu** [3]

[1]*Centre for Theoretical Particle Physics, Technical University of Lisbon, Avenida Rovisco Pais, 1, 1049-001 Lisbon, Portugal*

[2]*The Institute of Theoretical and Experimental Physics, Bolshaya Cheremushkinskaya, 25, 117218 Moscow, Russia*

[3]*Department of Physics, University of Helsinki and Helsinki Institute of Physics, P.O.Box 64, FIN-00014 Helsinki, Finland*

We have developed a concept of parallel existence of the ordinary (O) and mirror (M), or shadow (Sh) worlds. E_6 unification, inspired by superstring theory, restores the broken mirror parity at the scale $\sim 10^{18}$ GeV. With the aim to explain the tiny cosmological constant, we consider the breakings: $E_6 \to SO(10) \times U(1)_Z$ – in the O-world, and $E_6' \to SU(6)' \times SU(2)'_\theta$ – in the Sh-world. We assume the existence of shadow θ-particles and the low energy symmetry group $SU(3)'_C \times SU(2)'_L \times SU(2)'_\theta \times U(1)'_Y$ in the shadow world, instead of the Standard Model. The additional non-Abelian $SU(2)'_\theta$ group with massless gauge fields, "thetons", has a macroscopic confinement radius $1/\Lambda'_\theta$. The assumption that $\Lambda'_\theta \approx 2.3 \cdot 10^{-3}$ eV explains the tiny cosmological constant given by recent astrophysical measurements. Searching for the Dark Matter (DM), it is possible to observe and study various signals of theta-particles.

Keywords: mirror world, shadow world, shadow axion, theta particle, dark energy, dark matter, unification, e6
PACS: 12.10.-g, 12.10.Kt, 12.60.Jv, 14.80.Mz, 95.35.+d, 95.36.+x, 98.80.Cq

DARK ENERGY AND THE ENTROPY OF THE OBSERVABLE UNIVERSE (ABSTRACT)

Charles H. Lineweaver[1] and Chas A. Egan[2]

[1]*Planetary Scinece Institute, Research School of Astronomy and Astrophysics, and Research School of Earth Sciences Australian National University*

[2]*Research School of Astronomy and Astrophysics, Australian National University*

The initial low entropy of the universe has allowed irreversible processes, such as the reader reading this abstract, to happen in the universe. This initial low entropy is due to a low value for the initial gravitational entropy of the universe. The standard ΛCDM cosmology has a cosmic event horizon and an associated Gibbons-Hawking entropy. We compute the entropy of the universe including the entropy of the current event horizon and the entropy of the matter and photons within the cosmic event horizon. We estimate the entropy of the current cosmic event horizon to be $2.6 \pm 0.3 \times 10^{122}$ k and find it to be 10^{19} times larger than the next most dominant contribution, which is from super massive black holes. We plot an entropy budget as a function of time and find that the cosmic event horizon entropy has dominated other sources of entropy since 10-20 seconds after the big bang. See Egan & Lineweaver (2009) for details and discussion.

Keywords: Entropy, Gravity, Dark Energy
PACS: 996.10, 96.12, 96.55, 98.80

GENERALIZATION OF HAJICEK AND KUCHAŘ'S CANONICAL QUANTIZATION SCHEME TO THE 2+1 GEOMETRIES ADMITTING ONE KILLING VECTOR FIELD (ABSTRACT)

Evangelos Melas

Technological Educational Institution of Lamia , 35100 Lamia , Greece

The Canonical analysis and subsequent Quantization of the 2+1 action of pure gravity plus a Cosmological constant term is considered, under the assumption of the existence of one manifest Killing field (axial symmetry). The proper imposition of the Quantum analogues of the two Linear (momentum) constraints reduces an initial collection of state vectors, consisting of all smooth functionals of the components(and/or their derivatives) of the spatial metric, to particular scalar smooth functionals. The demand that the midi-superspace metric (inferred from the kinetic part of the Quadratic (Hamiltonian) constraint) must define on the space of these states an induced metric whose components are given in terms of the same states, which is made possible through an appropriate re-normalization assumption, severely reduces the possible state vectors to three unique (up to general coordinate transformations) scalar functionals. The Quantum analogue of the Hamiltonian constraint produces a Wheeler-DeWitt equation based on this reduced manifold of states, which is completely integrated.

Keywords: Canonical Quantum Gravity
PACS: 04.60.Ds, 04.60.Kz

COSMIC SUPERSTRINGS: DYNAMICS AND CUSPS (ABSTRACT)

William Nelson[1,2] (this work was done in collaboration with Anne-Christine Davis, Senthooran Rajamanoharan, and Mairi Sakellariadou)

[1] *King's College London, Strand, London, WC2R 2LS, U. K.*

[2] *Institute of Gravitation and the Cosmos, Penn State University, State College, PA 16801, U. S. A.*

Whilst standard field theoretic Cosmic Strings cannot end, Cosmic Superstrings can form three string junctions, at which each string ends. This opens up a new class of possible boundary conditions for such strings and we show that, at least when the junctions are close together, a string ending of two such junctions will generically have cusps. Cusps are of particular interest as they are strong emitters of radiation (both gravitational and particle) and hence are possible observables. The detection of cusps from Cosmic Superstrings between junctions would be a rare observational window into the realm of String theory and Brane inflation models.

Keywords: Cosmic Superstrings, Cusps, Brane Cosmology
PACS: 11.10.Lm, 11.25.Wx, 98.80.Cq

UNIQUENESS THEOREM FOR BLACK OBJECTS (ABSTRACT)

Marek Rogatko

Institute of Physics, Maria Curie-Sklodowska University, 20-031 Lublin, Pl. Marii Curie-Sklodowskiej 1, Poland, email : rogat@kft.umcs.lublin.pl

We shall review the current status of uniqueness theorem for black objects in higher dimensional spacetime. At the beginning we consider static charged asymptotically flat spacelike hypersurface with compact interior with both degenerate and non-degenerate components of the event horizon in n-dimensional spacetime. We gave some remarks concerning partial results in proving uniqueness of stationary axisymmetric multidimensional solutions and winding numbers which can uniquely characterize the topology and symmetry structure of black objects.

Keywords: n-dimensional gravity, uniqueness theorems, black objects
PACS: 04.50.+h, 0470.Bw

HAWKING-LIKE RADIATION IN A FRW UNIVERSE (ABSTRACT)

Tao Zhu[1], Ji-Rong Ren[1] and Douglas Singleton[2,3]

[1] *Institute of Theoretical Physics, Lanzhou University, Lanzhou 730000, P. R. China (use complete addresses)*

[2] *Physics Department, California State University Fresno, Fresno, California 93740, USA*

[3] *Cosmology, Peoples' Friendship University of Russia, Moscow 117198, Russia*

In this article we study Hawking-like radiation from a Friedmann-Robertson-Walker (FRW) universe using the quasi-classical WKB/tunneling method. In calculating the correct temperature of the Hawking-like radiation we emphasize two important points: (i) the canonical invariance of the tunneling amplitude and (ii) existence of a temporal contribution to the tunneling amplitude. We also study back reaction and energy conservation of the radiated particles and find that the tunneling probability and change in entropy are related by $\Gamma \propto exp - (\Delta\Sigma)$ which differs from the standard result of $\Gamma \propto exp + (\Delta\Sigma)$. This discrepancy if accounted for by regarding the whole FRW universe as an isolated adiabatic system whose change in total entropy is zero i.e. $\Delta\Sigma_{total} = \Delta\Sigma_{int} + \Delta\Sigma_{ext} = 0$.

Keywords: Hawking Radiation, Thermodynamics of FRW Universe, Back Reaction, Canonical Invariance
PACS: 11.25.Sq, 11.10.-z, 11.25.Pm.

A WARPED COSMIC STRING (ABSTRACT)

R. J. Slagter

Institute of Physics, University of Amsterdam and ASFYON, Astronomisch Fysisch Onderzoek Nederland, 1405EP Bussum, The Netherlands

We present a cosmic string solution in Einstein-Yang-Mills Gauss-Bonnet theory on a warped 5 dimensional spacetime conform the Randall-Sundrum-2 theory.In a simplipied model, we find an exact solutions with exponential decreasing or periodic warp function. In a more general setting, where the metric- and Yang-Mills components depend on both scales and one of the YM components resides in the bulk, we find a time dependent numerical solution.

PACS:11.25.Wx, 04.50.+h, 98.80.Cq

DYNAMICS AND PROPERTIES OF THE INVISIBLE UNIVERSE

Models of Dark Energy

Edmund J. Copeland

School of Physics and Astronomy, University of Nottingham, University Park, Nottingham, NG7 2RD, England

Abstract. I briefly review models of dark energy. These include models of a cosmological constant arising in string theory, dynamical models where a scalar field may be responsible for the observed late time acceleration through to the possibility that we are not fully in control of the gravity sector and the acceleration may be some manifestation of modified gravity. All cases require some degree of fine tuning for the models to be compatible with observation.

Keywords: Dark energy, Cosmology
PACS: 95.36.+x; 98.80.-k

INTRODUCTION

The universe is expanding, we have known that for many years. The conventional wisdom perhaps through to the end of the 1980's was that the presence of non relativistic matter as a dominant source of energy density led to a Universe which was decelerating in an understandable way. It all made sense, but unfortunately through the 1980's and early 1990's a number of observations of the large scale structure in the universe began to question the accepted paradigm of a cold dark matter (CDM) dominated $\Omega_{\text{matter}} = 1$ universe. For example in a study of the large scale clustering of IRAS galaxies performed in [1] it was realised that it was difficult to reconcile the observed clustering of these galaxies with that predicted in the standard paradigm at the time. In fact it led a number of authors to begin questioning the paradigm and in 1990 Efstathiou et al, [3] actually proposed that a better fit to the data available at that time would be a scenario with a "spatially flat cosmology in which as much as 80 percent of the critical density is provided by a positive cosmological constant." This remarkable conclusion was reached well before the observations of distant Type Ia supernova hit the headlines in 1998. Their reasoning was reinforced in the analysis of the APM Galaxy Survey data in which there was strong evidence for more large-scale structure in the survey than predicted by the standard scale-invariant CDM model [2]. Perhaps we should not be too surprised that there were people back in the 80's and 90's who were turning their attention to the prospect of a cosmological constant. After all, there were many groups of astronomers trying to reconcile the fact that the large scale structure data was looking so inconsistent with a standard CDM scenario. However, I think it is fair to say that not too many people (especially from the particle physics camp) were prepared to face the prospect that we had to have a Λ present today if our cosmology was to make sense. Two exceptions to this were Weinberg [4] and Linde [5], as they argued for a Λ but from an anthropic standpoint. However, as will shortly see the real catalyst which sparked theorists into action was the reported observations by the Supernova groups in 1998 [6, 7] of deviations in the Hubble diagram at large redshifts providing evidence of an accelerating Universe. It seemed to galvanise people into action and has led to an unprecedented amount of attention being lavished on the question of the origin of the dark energy as the source of the accelerating universe. In this brief review I will try and give a flavour of the many attempts that have been made to access the nature of dark energy. Given that there have been over 2000 papers written on the subject since 1998, I apologise from the outset. I can not cover the field properly, nor give proper references. There are a number of reviews on the subject and I urge the interested reader to look there for details. Here are just a few of them [8, 9, 10, 11, 12, 13, 14, 15, 16]. Neither will I go into great details over any particular approach, this is going to be a schematic review, aimed at giving the reader a flavour of the attempts that have been made to understand the nature of the dark energy.

In what follows I will describe attempts to model the dark energy as a pure cosmological constant Λ. This remains the best fit to the data, although from a particle physics perspective it is fraught with problems. I will then discuss a range of dynamical dark energy models in which we assume there is some as yet unknown mechanism which has forced the real cosmological constant to vanish to day, and the dark energy is associated with say the potential energy of some slowly rolling scalar field. This too has problems, not least being the fact such a field generally has to have a very small mass, which means that it is subject to the severe constraints that exist in the form of fifth force experiments.

CP1241, *Invisible Universe: Proceedings of the Conference,* edited by J.-M. Alimi and A. Füzfa

Finally, we will look at attempts to modify the left hand side of Einstein's equations, the geometry part. Perhaps it is not the matter that needs looking at, it is the spacetime to which the matter is coupled. However, we will see that even in this case, it is pretty hard to modify Einstein's theory too much to accommodate the late time accelerating universe. All in all, the dark energy issue remains one of the most interesting and perplexing problems in physics and astrophysics.

A LITTLE BACKGROUND

A brief reminder of the background equations. We have the Friedmann and fluid equations describing the evolution of the scale factor $a(t)$ in an isotropic and homogeneous universe of spatial curvature K containing matter of density ρ and pressure p, related by an equation of state $p = w\rho$.

$$H^2 \equiv \left(\frac{\dot{a}}{a}\right)^2 = \frac{8\pi G\rho}{3} - \frac{K}{a^2}, \tag{1}$$

$$\dot{H} = -4\pi G(p+\rho) + \frac{K}{a^2}, \tag{2}$$

$$\dot{\rho} + 3H(\rho + p) = 0. \tag{3}$$

where H is the Hubble parameter.

Equation (3) can be derived from Eqs. (1) and (2), which means that two of Eqs. (1), (2) and (3) are independent. Eliminating the K/a^2 term from Eqs. (1) and (2), we obtain

$$\frac{\ddot{a}}{a} = -\frac{4\pi G}{3}(\rho + 3p). \tag{4}$$

Hence the accelerated expansion occurs for $\rho + 3p < 0$.

Let us consider the evolution of the universe filled with a barotropic perfect fluid with an equation of state

$$w = p/\rho, \tag{5}$$

where w is assumed to be constant. Then by solving the Einstein equations given in Eqs. (1) and (2) with $K = 0$, we obtain

$$H = \frac{2}{3(1+w)(t-t_0)}, \tag{6}$$

$$a(t) \propto (t-t_0)^{\frac{2}{3(1+w)}}, \tag{7}$$

$$\rho \propto a^{-3(1+w)}, \tag{8}$$

where t_0 is constant. We note that the above solution is valid for $w \neq -1$. The radiation dominated universe corresponds to $w = 1/3$, whereas the dust dominated universe to $w = 0$. In these cases we have

$$\text{Radiation}: \; a(t) \propto (t-t_0)^{1/2}, \quad \rho \propto a^{-4}, \tag{9}$$

$$\text{Dust}: \; a(t) \propto (t-t_0)^{2/3}, \quad \rho \propto a^{-3}. \tag{10}$$

Both cases correspond to a decelerated expansion of the universe.

From Eq. (4) an accelerated expansion ($\ddot{a}(t) > 0$) occurs for the equation of state given by

$$w < -1/3. \tag{11}$$

In order to explain the current acceleration of the universe, we require an exotic energy dubbed "dark energy" with equation of state satisfying Eq. (11). From Eq. (3) the energy density ρ is constant for $w = -1$. This is the famous cosmological constant case and it follows that the Hubble rate is also constant from Eq. (1), giving the evolution of the scale factor:

$$a \propto e^{Ht}, \tag{12}$$

which is the de-Sitter universe.

Now introducing the dimensionless density parameter $\Omega(t) = \rho(t)/\rho_c(t)$ where $\rho_c(t) = 3H^2(t)/8\pi G$ is the critical density (gives a spatially flat universe), we can rewrite the Friedmann equation as an elegant algebraic expression $\Omega_m + \Omega_\Lambda + \Omega_k = 1$ where I have divided the density up into three contributions arising from matter (including baryons, cold dark matter and massive neutrinos), a cosmological constant type term and spatial curvature. Turning briefly to the observational constraints on these three quantities. First Ω_m the contribution from matter. Well a number of surveys of large scale structure, measurements of the Baryon Acoustic Oscillation (BAO) scale, weak lensing tests, cluster baryon abundance measurements using X-ray measurements of intracluster gas, or SZ measurements and numerical simulations of structure growth all appear to indicate that $\Omega_m \sim 0.25$. For example the most recent work measuring the BAO scale by combining 2dFGRS and SDSS data suggests the best fit cosmological parameters are $\Omega_m = 0.25$, $\Omega_b h^2 = 0.0223$ and $h = 0.72$ [17]. The bottom line is that $\Omega_m \ll 1$. Analysing the peak structure of the CMB anisotropies provides evidence for the contribution of Ω_k. In fact to a good approximation, the position of the first peak in ℓ space is given by $\ell_{\text{peak}} \sim \frac{220}{\sqrt{\Omega_m + \Omega_\Lambda}}$.

The most recent beautiful WMAP5 data when combined with with the distance measurements from the Type Ia supernovae and the BAO in the distribution of galaxies probably provides the most powerful constraints to date on the cosmological parameters [18]. For instance assuming the standard ΛCDM paradigm the following values are obtained (including 68% uncertainties): $\Omega_b h^2 = 0.02265 \pm 0.00059$, $\Omega_{\text{cdm}} h^2 = 0.1143 \pm 0.0034$, $\Omega_\Lambda = 0.721 \pm 0.015$, which allows for the following to be derived: $H_0 = 70.1 \pm 1.3$ km s^{-1} Mpc^{-1}, $\Omega_b = 0.0462 \pm 0.0015$, $\Omega_{\text{cdm}} = 0.233 \pm 0.013$ and $\Omega_m h^2 = 0.1369 \pm 0.0037$. Tight constraints (95% CL) follow on the (constant) equation of state for dark energy, $-0.11 < 1 + w < 0.14$ and on the spatial curvature of the universe $-0.0175 < \Omega_k < 0.0085$ implying that the universe is perfectly consistent with being spatially flat $\Omega_k \simeq 0$ and dominated by a true cosmological constant $w = -1$ [18]. However a slight word of caution, always be aware that these numbers come with certain prior assumptions made about some of the other parameters. There could well be some residual dependence on these parameters that has not been properly taken into account. For example when the constraint of a constant dark energy equation of state is relaxed to allow for dynamical dark energy, we find for the present value w_0, $-0.38 < 1 + w_0 < 0.14$ [18], not as tight a bound as before.

Finally, the observations of the Type Ia supernova indicates that the Hubble diagram is no longer well fitted by a CDM dominated universe, and prefers a universe that is dominated with a cosmological constant today – well not actually today. We don't know what the universe is doing right now, we can only know what it was doing around a redshift of $z \sim 0.2 - 1$ as far as the supernova are concerned. For all we know it has actually stopped accelerating today and we just haven't received the information yet!

MODELS OF Λ

Why is the observed value of the cosmological constant a problem? For many it isn't, it is just another parameter that makes up the cosmological parameter set, and its observed contribution should just be taken as an empirical fact. However, for those who believe it has something to do with the world of particle physics then it is a real problem because its energy scale is so small compared to the value we would naively expect it to have. This cosmological constant problem has been with us for a long time. It was assumed the cosmological constant (or Λ) was zero today, but we did not have any real reason to set it to that value, rather we relied on some as yet unknown symmetry to protect it and maintain it at zero. To be a bit more precise, as we have seen today it is of order the critical density inferring an energy density scale of $10^{-47}(GeV)^4$. Now in the world of quantum field theory we expect that the zero point fluctuations of any scalar fields (for example) present will contribute towards an effective cosmological constant. Moreover the scale of that constant will be governed by the cut-off scale associated with the theory. So if I associated that with the Planck scale, I would expect a contribution of vacuum fluctuations to be at a scale of $10^{71}(GeV)^4$ a mere 10^{118} orders of magnitude larger than is observed. If you didn't trust Planck scale physics (and who could blame you!) then how about the electroweak scale at around a TeV. Then we would expect a contribution of around $1(TeV)^4$ which is a mismatch of 10^{61} or so. Thus even invoking supersymmetry (which brings us down to a TeV scale) we still have an incredible mismatch of scales. This is the cosmological constant problem, why has the universe not been completely dominated by these energy sources before now? Why are they effectively decoupling from the cosmological dynamics ? Needless to say there has been a great deal of interest in the possibility that a cosmological constant may pop out of string theory and in the braneworld. This has met with mixed success but has opened up an area of research which has led people into the myriad of possibilities known as the String Landscape. It was known for some time that in conventional supergravity models there were 'No go' theorems which forbid cosmic acceleration in cosmological solutions arising from compactification of pure supergravity models where the internal space is a time-independent, non-singular compact manifold without boundary [19, 20]. Now one way of avoiding the theorem is to relax the conditions under which it applies such as allowing the internal space to be time-dependent, the analogue of a time-dependent scalar fields (radion). However, doing this, whilst leading to some solutions which have an accelerating late time evolution, they generally had mixed fortune in that they were preceded by a kinetic energy dominated evolution rather than a more useful matter dominated era. An alternative approach was developed by a number of authors who were interested in the pressing issue of stabilising the extra dimensions (moduli fields) in string theory [21, 22]. They realised that by using antisymmetric tensor fields to stabilise the extra dimensions (basically by wrapping them in fluxes), the net effect was that the resulting effective cosmological constant was described as a combination of the bare cosmological constant and the energy density of the fluxes (see also earlier work by Linde [5]). Given suitably quantised fluxes they argued that it was possible to have the required precision to account for the observed cosmological constant with of order 100 fluxes. This was easy to formulate in string theory. In fact given that there are believed to be of order 10^{500} possible vaccua in string theory, it opened up the possibility that this new Landscape would inevitably have regions (possibly many of them) where the vacuum energy matched our observed one [23]. This Landscape is good and bad. The good aspect of it, is that should it exist, it seems quite plausible that we can expect there to be many many Λ's and a large number of these will be in the range of values we require. Indeed in a pioneering paper back in the late 1980's Weinberg effectively predicted we would have to find a Λ term today with the kind of scale it appears to have [4]. Anything larger would have meant the universe would expand too rapidly for structures to collapse to form galaxies. The downside is that it is proving almost impossible to calculate anything properly in this framework. We dont yet have exact solutions we can use and so tend to end up adopting Anthropic arguments to justify our claims. The Landscape is clearly an interesting feature of string theory, and the hope for many of us is that one day we will be able to really test the notion that this is the origin of our observed universe. Other string related approaches to Λ include: using the extra dimensions to remove the gravitational effect of the vacuum energy, leaving us with an effectively flat four dimensional theory [24]; of using a mechanism whereby Λ relaxes through the nucleation of branes coupled to a gauge potential [25] and by using the Cyclic universe as a means to give the cosmological constant plenty of time to relax to its present day value [26].

ROLLING SCALAR FIELDS – QUINTESSENCE AND K-ESSENCE

An alternative approach to explain the nature of the dark energy is to say that we don't know why Λ is zero, but accept it is and try to find an alternative explanation for the dark energy in the form of an evolving scalar field – a late time inflaton field if you like. Many of us have had a go at this, far too many to mention, but I will briefly describe some of the neat features associated with it. Many Quintessence scenarios have the attractive feature that there are attractor solutions where the system evolves so as to not dominate the dynamics at early times but come to dominate at late times. This means that the system can evolve from a wide range of initial conditions in the safe knowledge that it will inevitably dominate the evolution at late times. The idea first developed by Ratra and Peebles [27] as well as Wetterich [28] in the mid 1980's was later used for this purpose in [29, 30]. I am not going to go into the details of any of the scalar field models – there are so many of them, but I feel it should be made clear that they all have the same basic fine tuning issues. First of all the parameters in the potential need to be generally adjusted so that today the scalar field potential matches that required for it to be dark energy. The potentials have to be flat enough to have slow roll inflation and this needs to be protected against supergravity corrections. Secondly, it has to come to dominate at the correct time, namely in the recent past. In almost all cases of Quintessence the net affect is that the mass of the Quintessence field has to be very small, $m_Q \simeq 10^{-33}$eV, with an associated energy density of the field being $V_0 \simeq (10^{-3}eV)^4$. Such small mass scale should cause a few alarm bells to go off. A light field implies long range interactions, hence fifth force tests. There are strong limits on the allowed couplings of such fields to matter, and basically any such field needs to be protected from having too strong a coupling to matter otherwise we would have seen it as a fifth force. One possible route therefore is to consider models that have protected masses such as through using Nambu-Goto Goldstone bosons [31, 32, 33]. Following the equation of state for the scalar field during its evolution we see how it tracks that of the background radiation, then matter before coming to dominate like a cosmological constant today. However, it is not a constant, it is evolving. One of the biggest questions we face is how can we test whether dark energy is evolving and moreover can we see that it might be different from a true cosmological constant?

An alternative approach to having a suitably flat potential, is to consider the case of a non-canonical kinetic sector to the theory driving the late time acceleration[34, 35]. Such an approach is called K-essence and has generated a lot of interest because of the fact that non-canonical terms appear often in string models. One of the nice features of the approach is that a long period of perfect tracking is followed by domination of dark energy triggered by the transition to matter domination – an epoch during which structures can form. This solution of the 'why now' problem is therefore addressed because rather than being put in as an arbitrary time scale, the transition occurs around the epoch of radiation-matter equality an epoch we know happened. However, a word of warning. It is not clear that K-essence solves the coincidence problem. The basin of attraction into the regime of tracker solutions is small compared to those where it immediately goes into K-essence domination, and so there may well be a degree of fine tuning required in the scenario that is not obvious at first sight [36].

Before leaving the idea of scalar fields let me just mention that there have been a number of dark energy papers written making use of tachyon fields, phantom fields, chameleon fields, acceleron fields (scalar fields coupled to massive neutrinos) and chaplygin gases amongst others. The idea of the chameleon field is worth describing in a bit more detail [37]. The key idea is that in order to avoid fifth force type constraints on Quintessence models, why not have a situation where the mass of the field depends on the local matter density? It can then be massive in high density regions (i.e. solar system) and light ($m \sim H$) in low density regions (i.e. cosmological scales). In that way it can explain dark energy without violating solar system bounds. There is plenty of work currently going on to investigate the nature and consistency of chameleon fields.

MODIFYING GRAVITY

Up to now we have worked on the assumption that the dark energy is to be explained as a new form of matter, such as the cosmological constant. However, it could well be that it has nothing to do with the matter, but rather the gravity of the universe that we have got wrong. In other words perhaps we need to modify Einstein's gravity. Two recent reviews on the subject are [12, 16]. It is worth mentioning one approach in a bit more detail, but it gives a flavour of the many attempts that have been made over the past few years. In the same spirit of the pioneering work of Starobinsky [38], Carroll et al looked at a modification of the usual Einstein action [39] in which late time acceleration can be realized by terms containing inverse powers of the Ricci scalar added to the Einstein Hilbert action. Unfortunately the original model ($\mathscr{L} \propto R + 1/R$) is not compatible with solar system experiments and this remains the real test of modified gravity models today. A large number of people have generalised the model of Carroll et al to more general $R + f(R)$ type models. However usually $f(R)$ struggles to satisfy both solar system bounds on deviations from General Relativity and late time acceleration. It brings in an extra light degree of freedom which once again implies fifth force constraints come into play. A possible get out clause that has been investigated is to make the scalar degree of freedom massive in the high density solar vicinity and therefore hidden from solar system tests by a chameleon type mechanism. In general it appears to require a form for $f(R)$ where the mass squared of the scalar is large and positive at high curvature. In fact very recently Amendola and Tsujikawa have analysed these models and concluded that the $f(R)$ models satisfying cosmological and local gravity constraints are practically indistinguishable from the Lambda CDM model, at least at the background level [40]. Other modified gravity models include ideas based on Braneworlds, where the extra dimensions can lead to modifications of the four dimensional gravity we are used to. Of particular popularity has been the model of Dvali-Gabadadze-Porrati (DGP) [41] in which the brane is embedded in a *Minkowski* bulk. In a DGP braneworld, gravity remains four dimensional at short distances but can leak into the bulk at large distances leading to infrared modifications to Einstein gravity. There is a cross-over scale around which gravity manifests these higher-dimensional properties. This scenario is a simple one parameter model which can account for the current acceleration of the universe provided the cross-over scale is fine tuned to match observations. However there are problems with the model, perhaps the most significant and potentially lethal one being the fact that in the Self accelerating branch there are ghosts in the spectrum for any value of the brane tension. Such an instability is serious and suggests that all is not well with the model [42].

I will finish on an area which is proving very difficult to rule out – perhaps there is no such thing as dark energy, we don't need to modify General Relativity, rather we just have to abandon the aspect of the cosmological principle that assumes large scale homogeneity. Although this is not the view of the majority of cosmologists who are happy with the ΛCDM paradigm, it has been suggested by a small but increasing number of authors that the apparent acceleration is due to the fact we are in an inhomogeneous universe containing appreciable sized voids. In other words all we need to do is solve for General Relativity in an inhomogeneous background – easy to write, difficult to do of course. As an example Hunt and Sarkar [43] have recently suggested that we can do away with the need for dark energy if we allow for the possibility that we live in an inhomogeneous universe, where inflation proceeded leading to features (bumps) in the primordial spectrum so that it is not scale free. We could be living in a local void where the Hubble flow is 30%

faster than the global rate. They acknowledge a possible problem making their data consistent with the observed baryon oscillations in the large scale structure power spectra but otherwise they fit the data very well. Another inhomogeneous model has also been proposed recently by Wiltshire [44], and alternatives to Dark Matter and Dark Energy have been proposed by Mannheim [45]. Kolb, Matarrese and collaborators [46] have demonstrated the possibility of explaining away dark energy by solving the Einstein equations in a particular class of inhomogeneous Swiss Cheese universes and demonstrating the role averaging and backreaction can play in interpreting the dynamical evolution of these models. I think it is fair to say that we should always be prepared for the unexpected in this field.

CONCLUSIONS

This is a subject in which observations are transforming the field, especially the CMBR and LSS. The data to date suggests that the universe is accelerating and is being driven by a pure cosmological constant. However we don't really know why it is accelerating and what if anything has generated the constant to be the value it is. Moreover we don't even know if it is a cosmological constant and determining that must be one of the most important questions we would like to know the answer to. In particular, is $w = -1$, the cosmological constant and if not, then what value has it? On the other hand is w(z) dynamical? Perhaps we are actually seeing some new Gravitational Physics. There are lots of models of dark energy but it may yet prove too difficult to separate one from another such as cosmological constant. However we need to try though! Rather frustratingly for some of us, perhaps we will only be able to determine Λ from anthropic arguments and not from fundamental theory. And finally, perhaps we are wrong and we do not need a lambda term at all. The next decade is going to be exciting in this field.

ACKNOWLEDGMENTS

I am very grateful to the organisors of the Invisible Universe International Conference and to Professor Jean-Michel Alimi and Dr. Pier Stefano Corasaniti in particular for the invitation to participate in this wonderful meeting. I am also grateful to the Royal Society for financial support through the Wolfson Merit Award.

REFERENCES

1. G. Efstathiou *et al*, *Mon. Not. Roy. Astron. Soc.* **247** 10 (1990).
2. S. J. Maddox *et al*, *Mon. Not. Roy. Astron. Soc.* **283** 1227 (1996).
3. G. Efstathiou *et al*, *Nature* **348** 705 (1990).
4. S. Weinberg, *Rev. Mod. Phys.* **61** 1 (1989).
5. A. D. Linde, *Rept. Prog. Phys.* **47** 925 (1984).
6. S. Perlmutter *et al.*, *Astrophys. J.* **517** 565 (1999).
7. A. G. Riess *et al.*, *Astron. J.* **116** 1009 (1998); *Astron. J.* **117** 707 (1999).
8. V. Sahni and A. A. Starobinsky, *Int. J. Mod. Phys. D* **9** 373 (2000); V. Sahni, *Lect. Notes Phys.* **653** 141 (2004).
9. S. M. Carroll, *Living Rev. Rel.* **4** 1 (2001).
10. T. Padmanabhan, *Phys. Rept.* **380** 235 (2003); *Current Science* **88** 1057 (2005).
11. P. J. E. Peebles and B. Ratra, *Rev. Mod. Phys.* **75** 559 (2003).
12. S. Nojiri and S. D. Odintsov, *eConf* **C0602061** 06 (2006) *Int. J. Geom. Meth. Mod. Phys.* **4** 115 (2007)
13. E. J. Copeland *et al.*, *Int. J. Mod. Phys. D* **15** 1753 (2006).
14. E. V. Linder, *Rept. Prog. Phys.* **71** 056901 (2008)
15. J. Frieman *et al.*, *Ann. Rev. Astron. Astrophys.* **46** 385 (2008)
16. T. P. Sotiriou and V. Faraoni, arXiv:0805.1726 [gr-qc].
17. W.J. Percival *et al.*, *Mon. Not. Roy. Astron. Soc.* **381** 1053 (2007)
18. E. Komatsu *et al.* [WMAP Collaboration], *Astrophys. J. Suppl.* **180** 330 (2009)
19. G.W. Gibbons, *Aspects of Supergravity Theories*, in *Supersymmetry, Supergravity, and Related Topics*, eds. F. del Aguila, J.A. de Azcárraga and L.E. Ibañez (World Scientific 1985) pp. 346-351.
20. D.H. Wesley, *JCAP* **0901** 041 (2009)
21. R. Bousso and J. Polchinski, *JHEP* **0006** 006 (2000)
22. S. Giddings, S. Kachru and J. Polchinski, *Phys. Rev. D* **66** 106006 (2002).
23. L. Susskind, arXiv:hep-th/0302219.
24. S. Kachru *et al.*, *Phys. Rev. D* **62** 045021 (2000); N. Arkani-Hamed *et al.*, *Phys. Lett. B* **480** 193 (2000)
25. J. L. Feng *et al.*, *Nucl. Phys. B* **602** 307 (2001)

26. P. J. Steinhardt and N. Turok, *Science* **312** 1180 (2006)
27. B. Ratra and J. Peebles, *Phys. Rev D* **37** 321 (1988).
28. C. Wetterich, *Nucl. Phys B* **302** 668 (1988).
29. R. R. Caldwell *et al.*, *Phys. Rev. Lett.* **80** 1582 (1998).
30. I. Zlatev *et al.,Phys. Rev. Lett.* **82** 896 (1999); P. J. Steinhardt *et al.*, *Phys. Rev. D* **59** 123504 (1999).
31. J. A. Frieman *et al.*, *Phys. Rev. Lett.* **75** 2077 (1995).
32. K. Choi, *Phys. Rev. D* **62** 043509 (2000).
33. J. E. Kim and H. P. Nilles, *Phys. Lett. B* **553** 1 (2003).
34. C. Armendáriz-Picón *et al.*, *Phys. Rev. Lett* **85** 4438 (2000).
35. C. Armendáriz-Picón *et al.*, *Phys. Rev. D* **63** 103510 (2001).
36. M. Malquarti *et al.*, *Phys. Rev. D* **67** 123503 (2003); M. Malquarti *et al.*, *Phys. Rev. D* **68** 023512 (2003).
37. J. Khoury and A. Weltman, *Phys. Rev. Lett.* **93** 171104 (2004).
38. A. A. Starobinsky, *Phys. Lett. B* **91** 99 (1980).
39. S. M. Carroll *et al.*, *Phys. Rev. D* **70** 043528 (20040.
40. L. Amendola and S. Tsujikawa, *Phys. Lett. B* **660** 125 (2008)
41. G. R. Dvali *et al.*, *Phys. Lett. B* **485** 208 (2000).
42. C. Charmousis *et al.*, *JHEP* **0610** 066 (2006)
43. P. Hunt and S. Sarkar, *Phys. Rev. D* **76** 123504 (2007)
44. D. L. Wiltshire, *New J. Phys* **9** 377 (2007); *Phys. Rev. Lett.* **99** 251101 (2007)
45. P. D. Mannheim, *Prog. Part. Nucl. Phys.* **56** 340 (2006)
46. E. W. Kolb *et al.*, *New J. Phys.* **8** 322 (2006); *Phys. Rev. D* **78** 103002 (2008); arXiv:0901.4566 [astro-ph.CO]; V. Marra *et al.*, *Phys. Rev. D* **76** 123004 (2007)

Cosmic Degeneracies

Luca Amendola

University of Heidelberg, Philosophenweg 16, 69120 Heidelberg, Germany and INAF/Osservatorio Astronomico di Roma

Abstract. The road to unveil the nature of dark energy is plagued by several sources of degeneracies. We discuss in this contribution three forms of degeneracy that currently limit our understanding of dark energy and how possibly they can be overcome in the future

Keywords: Cosmology, dark energy
PACS: 98.80.-k

INTRODUCTION

The enigma of the cosmic acceleration has sparkled explanations that range from the cosmological constant, to new matter components with negative pressure, to modifications of gravity, to large-scale violations of the cosmological principle of homogeneity and isotropy. In a one line summary, we can say that the acceleration has been attributed to vacuum, energy, gravity or geometry.

To decide among these (and possibly other) alternatives we certainly need more observations. In this contribution I would like however to remark that we also need more *observables*: various forms of degeneracies do in fact limit our ability to extract the needed physics from the observations, no matter how precise they are. I will focus in particular on three degeneracies:

1. *Cosmic degeneracy 1*: there are infinite different models that produce a given background expansion. (In a stronger form, not to be discussed here, there are infinite different models that produce the same cosmology at any given finite order).
2. *Cosmic degeneracy 2*: using cosmological experiments alone, we cannot distinguish modifications of matter from modifications of gravity.
3. *Cosmic degeneracy 3*: we cannot distinguish between a real acceleration in a homogeneous and isotropic model from an apparent acceleration in a inhomogenous (but isotropic) universe.

As will be clear from the following, these degeneracies can in some case be broken by the use of new observables.

COSMIC DEGENERACY I

All the distance definitions in cosmology depend on the cosmological parameters through the integral $r(z) = (c/(a_0H_0))\int_0^z d\tilde{z}/E(\tilde{z})$ where

$$E(z) \equiv H(z)/H_0 \,. \tag{1}$$

in flat spaces and on

$$f_K(r) = \frac{1}{\sqrt{-K}}\sinh\left(\sqrt{-K}r\right)\,, \tag{2}$$

in spaces with spatial curvature K. It is therefore clear that all the measures of expansion that we can obtain through measurements of distances, from standard candles to the CMB acoustic peaks, will constrain only the cosmological parameters contained in $E(z)$ and only in the particular combination that appear in $r(z)$ and in $f_K(r)$. If we had distance information only for a given z then all the combinations of cosmological parameters that produce the same $f_K(r)$ would be equally acceptable: the constraints would therefore be fully degenerate along lines (or surfaces) of constant $f_K(r)$.

CP1241, *Invisible Universe: Proceedings of the Conference,* edited by J.-M. Alimi and A. Füzfa

For instance the closed ΛCDM cosmological model defined by the present values $(\Omega_\Lambda^{(0)}, \Omega_m^{(0)}) = (1,1)$ and the open model defined by $(\Omega_\Lambda^{(0)}, \Omega_m^{(0)}) = (0.1, 0.6)$ give practically identical distances at $z = 1$.

If we have information only in a small range of redshifts the degeneracy will be partially broken but still the constraints will appear elongated along the lines of equal distances. But even if the distance measurements are infinitely many and infinitely precise, we can only recover the dimensionless expansion rate $E(z)$. For instance, if we have perfect knowledge of the luminosity distance $d_L(z)$ we can obtain $E(z)$ through the formula

$$E^2(z) = \frac{c^2(1+z)^2\left[(1+z)^2 + \Omega_K^{(0)} H_0^2 d_L(z)^2\right]}{[(1+z)H_0 d_{L,z}(z) - H_0 d_L(z)]^2}, \tag{3}$$

where $\Omega_K^{(0)} = -\frac{K}{(a_0 H_0)^2}$ and $()_{,z}$ means derivation wrt z. We now assume that the universe is dominated by two components, one of which ("matter") is pressureless and uncoupled and therefore scales as $(1+z)^3$. Then we can derive the equation of state of dark energy via the relation (written in flat space and neglecting radiation for simplicity)

$$w_{\rm DE}(z) = \frac{(1+z)(E^2(z))_{,z} - 3E^2(z)}{3\left[E^2(z) - \Omega_m^{(0)}(1+z)^3\right]}. \tag{4}$$

This can actually be taken as a definition of dark energy equation of state in the cases in which the Friedmann equation is not standard due to modification of gravity.

This shows the first problem in the process of dark energy reconstruction: $w_{DE}(z)$ depends on $\Omega_m^{(0)}$ and in general we cannot determine *both* $w_{DE}(z)$ and $\Omega_m^{(0)}$ from background measurements (unless of course we parametrize $w_{DE}(z)$, i.e. we fix its functional form). To estimate the present matter fraction $\Omega_m^{(0)}$ we need to find the fraction of pressureless energy density and the only way to do so is to identify this component with the clustered matter in galaxies and clusters and estimate it through e.g. weak lensing and peculiar velocity fields. However this identification is not guaranteed: we can easily build models in which not all pressureless matter clusters (within a given scale), and models in which also some component with non-zero pressure clusters. In these cases, the background $\Omega_m^{(0)}$ does not coincide with the clustered fraction that we measure.

But even if we reconstruct $w_{DE}(z)$ to a high precision, we still don't know the clustering properties of dark energy and whether it directly couples to matter (beside the minimal gravitational coupling) or whether gravity is modified. To examine this issue we introduce the perturbed metric (we assume from now on a flat space) expressed in $N = \log a$,

$$ds^2 = e^{2N}\left[-(1+2\Psi)\mathscr{H}^{-2}dN^2 + (1+2\Phi)\delta_{ij}dx^i dx^j\right], \tag{5}$$

where $\mathscr{H}$ is the conformal Hubble rate. Then the general form for the perturbation equation of the pressureless matter density contrast is

$$\delta_m'' + \frac{1}{2}(1-3w_{\rm eff})\delta_m' = -\frac{k}{\mathscr{H}^2}^2\Psi - \frac{3}{2}(1-3w_{\rm eff})\Phi' - 3\Phi'' \tag{6}$$

(where primes are derivatives wrt N). This is to be complemented by equations for the background,

$$w_{eff} \equiv \Omega_{DE} w_{DE} = -\frac{2}{3}\left(\frac{\mathscr{H}'}{\mathscr{H}} + \frac{1}{2}\right)$$

(notice that here $\Omega_{DE} = 1 - \Omega_m$ is not the present value but a z-dependent function), by two equations for Φ and Ψ, the Poisson equation in a generic modified gravity model

$$\Phi = \frac{3}{2}\frac{\mathscr{H}^2}{k^2} Q(a,k)(\Omega_m \delta_m + \Omega_{\rm DE}\delta_{\rm DE})$$

and the anisotropic stress relation

$$\eta(a,k) = \frac{\Phi + \Psi}{\Phi}$$

and finally by a conservation equation for δ_{DE} similar to Eq. (6).

In the sub-Hubble limit we obtain from Eq. (6)

$$\delta_m'' + \frac{1}{2}(1-3w_{\rm eff})\delta_m' - \frac{3}{2}\Xi\Omega_m\delta_m = 0 \tag{7}$$

where

$$\Xi = Q(1-\eta)(1+\frac{\Omega_{\rm DE}\delta_{\rm DE}}{\Omega_m\delta_m}). \tag{8}$$

So if we assume unclustered dark energy and standard gravity then $\Xi = 1$ and the observation of δ_m (and its time derivatives) fixes $\Omega_m(z)$ and allows to measure $w_{DE}(z)$ unambiguously. Of course we are talking about obervations that can be done in principle. On the other hand, if we do not assume $\Xi = 1$, we can only derive the combination $\Xi\Omega_m$. Assuming that Ω_m coincides with the observed clustered fraction we can estimate Ξ and compare it to the theoretical models. For instance, scalar-tensor models with Lagrangian $L(\phi,R)$, potential $U(\phi)$ and $F = \partial L/\partial R$ predict [2]

$$Q = \frac{1}{F}\frac{(k^2/a^2)(1-2Q^2)F+M^2}{(k^2/a^2)F+M^2}, \qquad \eta = -\frac{4F(k^2/a^2)Q^2}{(k^2/a^2)(1-2Q^2)F+M^2}, \tag{9}$$

where $M^2 \equiv d^2U/d\phi^2$ and therefore

$$\Xi = \frac{1}{F}(\frac{(k^2/a^2)(1+2Q^2)F+M^2}{(k^2/a^2)F+M^2})(1+\frac{\Omega_{\rm DE}\delta_{\rm DE}}{\Omega_m\delta_m}).$$

If we assume that dark energy does not cluster we can also derive Q,η independently, at least in principle. In fact, we can estimate the so-called lensing potential $\Phi - \Psi = Q(1+\eta/2)\Phi \equiv \Sigma\Phi = \frac{3}{2}\frac{\mathscr{H}^2}{k^2}\Sigma\Omega_m\delta_m$ from the tomographic weak lensing power spectrum

$$P_{ij}(\ell) = \frac{9H_0^3}{4}\int_0^\infty dz \frac{W_i(z)W_j(z)E^3(z)\Omega_m^2(z)}{(1+z)^4}\Sigma^2 P_{\delta_m}\left(\frac{\ell}{\pi r(z)}\right), \tag{10}$$

where the window function is defined as

$$W_i(z) = \int_z^\infty d\tilde{z}\left[1-\frac{r(z)}{r(\tilde{z})}\right]n_i[\tilde{z}], \tag{11}$$

and n_i is the radial selection function for the i- th bin. By estimating both $P_{ij}(\ell)$ and P_{δ_m} we can therefore derive Σ and, together with Ξ, derive Q,η. To estimate P_{δ_m} however requires also a measure of the bias between galaxies and dark matter. Assuming the existence of a linear relation, we define the bias as

$$b^2(k,z) \equiv \frac{P_{gal}}{P_{\delta_m}}. \tag{12}$$

The bias function can be derived by using the redshift distortion induced on a statistically isotropic spectrum by linear peculiar velocities. As is well known [13], the redshift distortion measures the quantity

$$\beta = \frac{\delta_m'}{\delta_m b}. \tag{13}$$

So now we have three observables: $\beta, P_{ij}(\ell), P_{gal}$ and a conservation equation, Eq. (7), i.e. four relations. If δ_{DE} vanishes, we also have four unknowns: Ψ,Φ,δ_m,b (or equivalently Q,η,δ_m,b). This shows that at first order the problem of the first cosmic degeneracy can indeed be solved, at least in principle, and we can reconstruct the full scalar metric, the modified Poisson equation and the anisotropic stress. The peculiar velocity field can also be obtained by the linear Euler equation. This reconstruction requires therefore the combination of weak lensing and of radial and tangential modes of the power spectrum. That is, we need to combine imaging and spectroscopic cosmological probes.

COSMIC DEGENERACY II

The previous discussion exposes immediately another obstacle to understanding the nature of dark energy, namely the fact that we need to assume a vanishing dark energy clustering if we want to fix the properties of gravity (Q,η) at first order. In other words, Ξ can differ from the standard ΛCDM value of unity due to either a modification of gravity or the clustering of dark energy. The same applies to η: the anisotropic stress could be due either to a new force, e.g. scalar interaction, or to an intrinsic anisotropy in the energy-momentum tensor. Unless we are able to observe separately δ_m and δ_{DE} we cannot distinguish between modification of matter and modification of energy. Therefore, as long as we use only cosmological observations based on gravity and electromagnetic radiation, we cannot separate the effects of the dark components δ_m, δ_{DE}.

COSMIC DEGENERACY III

There is finally another fundamental degeneracy that can limit our search for dark energy. This is due to the fact that the most general metric which appears isotropic with respect to an observer is not the FRW but rather the Lemaitre-Tolman-Bondi (LTB) metric, which can be written as (in this section primes and dots refer to partial space and time derivatives, respectively):

$$ds^2 = -dt^2 + \frac{[R'(t,r)]^2}{1+\beta(r)}dr^2 + R^2(t,r)d\Omega^2, \tag{14}$$

where $\beta(r)$ can be loosely thought of as a position-dependent spatial curvature term. The evolution of $R(t,r)$ is governed by Einstein's equation and depends in all generality on two free spatial functions, for instance $\beta(r)$ and the present density $\rho_0(r)$, just like a CDM model depends on two constants, $\Omega_k^{(0)}$ and ρ_0. Once the metric is fixed, the radial and time coordinates $t(z), r(z)$ for a photon received by an observer at the origin are derived by

$$\frac{dt}{dr} = -\frac{R'}{\sqrt{1+\beta}}, \tag{15}$$

and

$$\frac{dz}{dr} = (1+z)\frac{\dot{R}'}{\sqrt{1+\beta}}. \tag{16}$$

In the LTB metric we can define two distinct Hubble parameters corresponding to the radial and perpendicular directions of expansion

$$H_{||} = \dot{R}'/R', \tag{17}$$

$$H_{\perp} = \dot{R}/R. \tag{18}$$

For instance, the luminosity distance and the angular diameter distance are both functions of $H_{\perp}$, while $H_{||}$ can be observed if we have a fixed rod along the line of sight, which is what occurs with baryon acoustic oscillations.

Now, while we can observe both $H_{||}, H_{\perp}$, we can do so only along the line of sight, so that we observe $H_{||}[r(z),t(z)], H_{\perp}[r(z),t(z)]$. The relations $r(z),t(z)$ depend on the particular LTB model: it is clear therefore that by a suitable choice of the LTB model one can obtain any desired form for $H_{||}, H_{\perp}$ and therefore for the observables. We cannot tell then whether the accelerated expansion rate $H(z)$ (or more precisely $H_{\perp}$) we derive from e.g. the luminosity distance is due to a genuine acceleration in a FRW space-time or an apparent acceleration in a LTB structure [12, 20, 5]. This degeneracy can however be broken if we have access to more than one line of sight, as we discuss in the following.

It has been known for a long time [19] that for any expanding cosmology the redshift z of a given source is not a constant in time. In a decelerating universe all redshifts decrease in time. In models predicting a recent (since $z \sim 1$) acceleration, like the ΛCDM model, sources with redshifts $z \leq 2$ actually have positive dz/dt. In effect, the observation of dz/dt gives one a *direct* measurement of the expansion of the universe (rather than an integral of it, as in the luminosity distance) since it measures $H(z)$ (or more precisely $H_{||}$). Another direct probe of $H_{||}$ are longitudinal acoustic oscillations (while yet another method, galaxy ages, measures $H_{\perp}$). The prospect of employing the redshift drift in cosmology was revisited in [15].

If one assumes a FRW metric, the observed redshift of a given source which emitted its light at a time t_s is today (t_0)

$$z_s(t_0) = \frac{a(t_0)}{a(t_s)} - 1, \tag{19}$$

and it becomes, after a time interval Δt_0 (Δt_s for the source),

$$z_s(t_0 + \Delta t_0) = \frac{a(t_0 + \Delta t_0)}{a(t_s + \Delta t_s)} - 1. \tag{20}$$

The observed redshift variation of the source is then

$$\Delta_t z_s = \frac{a(t_0 + \Delta t_0)}{a(t_s + \Delta t_s)} - \frac{a(t_0)}{a(t_s)}, \tag{21}$$

which can be re-expressed, after an expansion at first order in $\Delta t/t$, as:

$$\Delta_t z_s = \Delta t_0 \left(\frac{\dot{a}(t_0) - \dot{a}(t_s)}{a(t_s)} \right) + \mathcal{O}\left(\frac{\Delta t_0}{t_0}\right)^2. \tag{22}$$

We can rewrite the last expression in terms of the Hubble parameter $H(z) = \dot{a}(z)/a(z)$:

$$\Delta_t z_s = H_0 \Delta t_0 \left(1 + z_s - \frac{H(z_s)}{H_0} \right). \tag{23}$$

It is useful to relate the redshift variation to an apparent velocity shift of the source, $\Delta v = c\Delta_t z_s/(1+z_s)$.

This redshift drift, or $\dot{z}$, or "Sandage effect", has been investigated for a variety of dark energy models currently pursued in the literature [7, 3, 8, 14], and it is interesting to note that most of them predict a very similar redshift profile for the effect, all very close to the one generated by the ΛCDM model. In ΛCDM, the redshift drift is positive in the region $0 < z < 2.4$ but becomes negative for higher redshift. On the other hand, a dark-energy mimicking giant void produces a very distinct z dependance of this drift, and in fact one has, as shown below, that $\mathrm{d}z/\mathrm{d}t$ is always negative.

Recently, two high-precision spectrographs were proposed which could in principle be used for measuring the redshift drift: the Cosmic Dynamics Experiment (CODEX) [16, 14, 8] at the European Extremely Large Telescope (E-ELT) [11] and the Echelle Spectrograph for PREcision Super Stable Observations (ESPRESSO) [9, 14, 8] at the Very Large Telescope array (VLT).

The possibility of detecting the redshift drift with CODEX was analyzed in a couple of papers [7, 3, 14, 8]. In particular, it was shown in [3] that using reasonable mission specifications for CODEX, a discrimination amongst many different proposed dark energy models would not be possible in a time-frame of less than 30 years. In [17] it was found however that void models, on the other hand, are much easier to tell apart through the Sandage effect than other dark energy models. Using very similar mission specifications for CODEX, we estimate that a 5σ detection/exclusion is possible with less than 10 years of observation [17].

The achievable accuracy on $\sigma_{\Delta v}$ by the CODEX experiment was estimated (through Monte Carlo simulations) [16] to be

$$\sigma_{\Delta v} = 1.35 \left(\frac{\mathrm{S/N}}{2370}\right)^{-1} \left(\frac{N_{\mathrm{QSO}}}{30}\right)^{-\frac{1}{2}} \left(\frac{1+z_{\mathrm{QSO}}}{5}\right)^{q} \mathrm{cm/s}, \tag{24}$$

with

$$q \equiv -1.7 \text{ for } z \leq 4, \qquad q \equiv -0.9 \text{ for } z > 4, \tag{25}$$

where S/N is the signal-to-noise ratio per pixel, N_{QSO} is the total number of quasar spectra observed and z_{QSO} their redshift. Note also that the error pre-factor 1.35 corresponds to using all available absorption lines, including metal lines; using only Lyα lines enlarges this pre-factor to 2 [14].

In [17] we investigated three possible mission durations: 5, 10 and 15 years. It is important to note that a larger observational time-frame allows not only for a larger redshift drift (which is linear in time) but also for smaller error bars, as more photons are collected and, therefore, a higher S/N (which increases as $\sqrt{\Delta t}$) can be achieved. In other words, the "effective signal" increases with $\Delta t^{3/2}$ if one assumes a proportional telescope time is maintained.

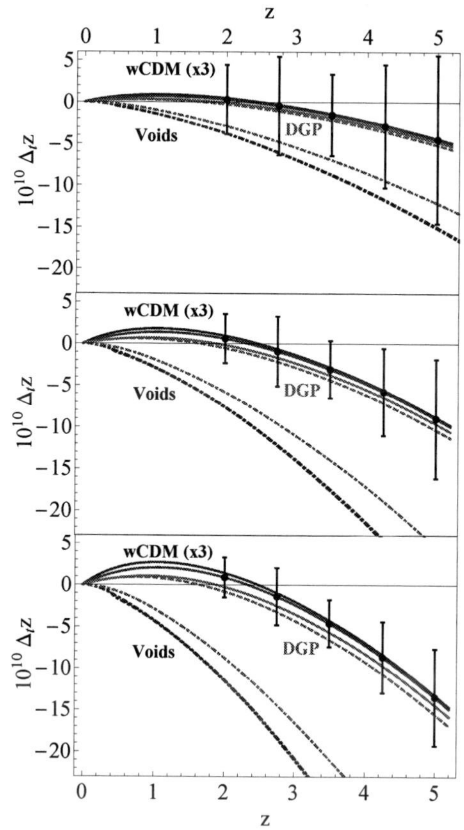

FIGURE 1. Redshift drift for different dark energy models for a total mission duration of 5 (top), 10 (middle) and 15 (bottom plot) years and CODEX forecast error bars (the time-span between each measurement is 2/3 of that – see text). In each plot, the upper 3, solid lines represent wCDM models for $w = -1.25$ (uppermost), $w = -1$ (second) and $w = -0.75$ (third uppermost). The green, dashed line corresponds to a self-accelerating extra-dimensional model (DGP) with $\Omega_{r_c} = 0.13$. The three bottommost, dot-dashed lines stand for the 3 void models considered here: the dark brown (almost indistinguishable) and the red line just above correspond to three different void models considered in [17]. Note that a 4σ separation between voids and ΛCDM can be achieved in a decade. From [17].

Figure 1 depicts the Sandage effect for different dark energy models for the three possible (complete observation) time-spans. Also plotted are the forecasted error bars obtainable by CODEX at the E-ELT. Here we are assuming that the time spent observing each quasar is the same, and this accounts for larger error bars at high redshift due to the lower apparent magnitudes of the corresponding quasars. As can be seen, void models could be detected/ruled out at over 4σ with less than a decade of mission-time.

One very interesting aspect of using the Sandage effect to probe void models is the fact it is model-independent to a good degree. The results in Fig. 1 refer in fact to three different models proposed in literature, which should be good representatives of the whole class of dark-energy mimicking LTB models.

Beside the Sandage effect, which in a sense explores the *outside* of our present light cone, voids can be constrained also by looking at different lines of sights *within* our light cone. This can be achieved by receiving light scattered towards us from source on our light cone. For instance, Ref. [4] proposed to observe deviation from the black-body spectrum of CMB photons scattered toward us by reionized structures living near the void edge: these structures see an anisotropic CMB sky and therefore scatter a superposition of black-body spectra that distorts the spectrum we receive. Analogously, one can constrain voids by looking at the kinematics Sunjayev-Zel'dovich spectrum from distant clusters, which is another way to probe the inner light cone [10].

Finally, it has been suggested that voids can be constrained also by comparing directly $H_{||}, H_{\perp}$, e.g. through radial and transversal modes in galaxy power spectra [6]. However, since this can be done again only along our light cone, one can always design a LTB such that $H_{||}[r(z),t(z)] = H_{\perp}[r(z),t(z)]$ along our present light cone but differ elsewhere (regardless of whether this is likely or not). Therefore this test is unable to distinguish at a fundamental level FRW from LTB.

Centered observers violate the Copernican principle of "non-special" position to the maximal extent. Off-center observers, however, have their own problems. They should see direct evidence of anisotropy, either in the distribution of distant sources or in the Hubble diagram of supernovae or as an additional CMB dipole [1] or even as an apparent parallax of quasars [18] (cosmic parallax). If the observed CMB dipole were entirely due to the LTB anisotropy of an off-center observer, the maximum distance to the center would be severely constrained, $r_c < 10\text{-}20\,\mathrm{Mpc}$, although one cannot exclude the possibility of a chance cancellation with the Sun's own motion.

CONCLUSIONS

In this contribution I have discussed several sources of cosmic degeneracy. Some of them can be lifted, at least in principle, by introducing new cosmological observables (perturbation growth, redshift drift). I believe this shows the need to combine more observations with more observables.

For instance, direct kinematic tests as redshift drift and cosmic parallax are conceptually the simplest probe of expansion and of anisotropy since their interpretation do not rely on calibration of standard candles/rulers nor depend on evolutionary or selection effects (as for galaxy ages and number counts). Also, we have shown that to completely reconstruct the space-time geometry at first order (at least on scales at which dark energy is unclustered) and reveal departures from Einstein's gravity, one needs to combine weak lensing and radial/trasversal galaxy clustering. This in turn requires imaging and spectroscopy.

It's really exciting that great observational facilities like Gaia, Euclid and E-ELT are becoming reality just now when we begin to realize the importance of precise astrometric, imaging and spectroscopic measurements for cosmology.

ACKNOWLEDGMENTS

It's a pleasure to thanks my collaborators on these topics: Amedeo Balbi, Paolo Cabella, Martin Kunz, Miguel Quartin, Claudia Quercellini, Domenico Sapone, Shinji Tsujikawa.

REFERENCES

1. H. Alnes and M. Amarzguioui, Phys. Rev. D 75, 023506 (2007), arXiv:astro-ph/0610331.
2. L. Amendola, M. Kunz, & D. Sapone, JCAP 4 (2008), 13
3. A. Balbi and C. Quercellini, MNRAS **382**:1623-1629 (2007), arXiv:0704.2350v3.
4. R. R. Caldwell and A. Stebbins, Physical Review Letters **100**, 191302 (2008), arXiv:0711.3459.
5. M.-N. Celerier, Astron. Astrophys. **353**:63-71 (2000), arXiv:astro-ph/9907206.
6. C. Clarkson, B. Bassett, T. H.-C. Lu, Phys. Rev. Lett. **101**:011301 (2008), arXiv:0712.3457.
7. P.-S. Corasaniti, D. Huterer, & A. Melchiorri, Phys. Rev. D **75** 062001 (2007).
8. S. Cristiani *et al.* Nuovo Cim. **122B**:1159-1164, Nuovo Cim. **122B**:1165-1170 (2007), arXiv:0712.4152v1
9. V. D'Odorico, for the CODEX/ESPRESSO team, arXiv:0708.1258v2.
10. J. Garcia-Bellido and T. Haugboelle, JCAP **0809**:016 (2008), arXiv:0807.1326.
11. R. Gilmozzi and J. Spyromilio, The Messenger **127**, 11 (2007).
12. N. P. Humphreys, R. Maartens and D. R. Matravers, Astrophys. J. **477**:47 (1997), arXiv:astro-ph/9602033.
13. N. Kaiser, MNRAS 227, (1987), 1
14. J. Liske, A. Grazian, E. Vanzella *et al.*, MNRAS **386**:1192-1218 (2008), arXiv:0802.1532v1.
15. A. Loeb, Astrophysical Journal **499**:L111-L114 (1998), arXiv:astro-ph/9802122v1.
16. L. Pasquini, S. Cristiani, H. Dekker *et al.*, The Messenger **122**, 10 (2005).
17. M. Quartin, & L. Amendola, arXiv (2009), arXiv:0909.4954;
18. C. Quercellini, M. Quartin, & L. Amendola, PhRvL 102, (2009), 151302 ; C. Quercellini, P. Cabella, L. Amendola, M. Quartin, & A. Balbi, PhRvD 80, (2009), 063527.
19. A. Sandage, Astrophysical Journal **136**, p.319 (1962).
20. K. Tomita, Astrophys. J. **529** (2000), 38.

New Physics at Low Accelerations (MOND): an Alternative to Dark Matter

Mordehai Milgrom

The Weizmann Institute Center for Astrophysics, Rehovot 76100, Israel

Abstract. I describe the MOND paradigm, which posits a departure from standard physics below a certain acceleration scale. This acceleration as deduced from the dynamics in galaxies is found mysteriously to agree with the cosmic acceleration scales defined by the present day expansion rate and by the density of 'dark energy'. I put special emphasis on phenomenology and on critical comparison with the competing paradigm based on classical dynamics plus cold dark matter. I also describe briefly nonrelativistic and relativistic MOND theories.

Keywords: MOND, dark matter, galaxies, cosmology
PACS: 95.30.Sf, 95.35.+d, 95.36.+x, 98.62.Dm

"There can be no doubt that the interplanetary and interstellar spaces are not empty but are occupied by a material substance, or body, which is certainly the largest, and probably most uniform, body of which we have any knowledge." (J. C. Maxwell).

"En astronomie, nous voyons les corps dont nous étudions les mouvements, et nous admettons le plus souvent qu'ils ne subissent pas l'action d'autres corps invisibles." (H. Poincaré).

INTRODUCTION

Normal matter in the universe makes itself felt in many ways: It interacts through strong and electromagnet interactions; so it emits and absorbs radiation. It self collides efficiently under astrophysical conditions, and dissipates energy. It produces magnetic fields that can be felt by particle acceleration, or by inducing synchrotron emission, for example. It decays and produces fast particles. It causes mechanical effects as in supernova explosions, or in ram-pressure stripping of gas in galaxies. In contradistinction, the putative, omnipresent dark matter (DM) is not required (and is not known) to do any of these things. The only trait it is required to share with standard matter is gravity: it is invoked only so as to modify and enhance the gravitational field of the visible baryons. It is thus conceivable that DM does not, in fact, exist, and that the extra gravity it is purported to supply is provided by a modification of the standard laws of dynamics.

So, following Poincaré's admonition, and not Maxwell's misguided conviction, we have been pursuing for a quarter of a century now a new paradigm of dynamics, dubbed 'MOND'. Its main raison d'etre remains the explanation of dynamics in the universe without the need for DM, which by and large, it does very well (with some exceptions). In particular, right at its advent it made a number of strong predictions, which have been confirmed over the years, and unearthed a number of unsuspected regularities in the properties of galactic systems. But, an additional motivation for embracing MOND has emerged with the recognition that the constant characterizing the paradigm may have cosmological origin, and that MOND as expressed in local physics has symmetries that may have descended from cosmology.

THE MOND PARADIGM

MOND starts by introducing into physics a new constant, a_0, with the dimensions of acceleration. This constant marks the borderline between the pre-MOND physic–valid approximately for accelerations much larger than a_0–and the MOND regime of much lower accelerations, just as $\hbar$ does in the context of quantum physics, and c in that of relativity. The MOND constant also enters strongly into physics in the MOND regime and is predicted to appear in many apparently unconnected laws and relations in this regime of phenomena. To recover standard physics in the high

CP1241, *Invisible Universe: Proceedings of the Conference,* edited by J.-M. Alimi and A. Füzfa

acceleration regime, we require that when we formally take $a_0 \to 0$ in all the equations of motion, a MOND theory tends to standard physics (like the restoration of classical predictions when $\hbar \to 0$ in quantum predictions for various observables). Finally, to achieve the phenomenological goals of MOND, we require that in the deep-MOND limit of *nonrelativistic* (NR) physics; i.e., formally taking $a_0 \to \infty$, the theory becomes scale invariant [54]; i.e., invariant under $(t,\mathbf{r}) \to \lambda(t,\mathbf{r})$. To see what this means, take, for example, a system of masses m_i with all accelerations much smaller than a_0. Then, if $\mathbf{r}_i(t)$ describes some possible history of the system, $\lambda\mathbf{r}_i(t/\lambda)$ is also a solution of the theory. For example, the length of the planetary year in a deep-MOND planetary system is proportional to the orbital size, not to its $3/2$ power as in the Newtonian Kepler's 3rd law.

Look, indeed, at the quintessential problem of the circular motion of a test particle at a distance R around a point mass M. On dimensional grounds alone, the expression for MOND acceleration of the particle must be of the form

$$a = \frac{MG}{R^2}\nu\left(\frac{MG}{R^2 a_0}\right). \tag{1}$$

The MOND basic tenets require that for $MGR^{-2} \gg a_0$ we have $\nu \to 1$, while for $MGR^{-2} \ll a_0$ we have to have scale invariance, which dictates $\nu(y) \approx y^{-1/2}$ (the normalization is absorbed in the definition of a_0), or

$$a \approx (MGa_0)^{1/2}/R. \tag{2}$$

There are three main points to take home from this simple, introductory case, which are also relevant to more general systems: 1. For a given mass, M, the asymptotic acceleration at large radii goes as R^{-1} compared with the standard R^{-2} (accelerations scale as λ^{-1}). 2. In this region, a is proportional to $M^{1/2}$, not to M. 3. The transition occurs always at the same value of the acceleration (not a fixed radius).

Extensive reviews of various aspects of the MOND paradigm can be found in [71, 74, 6, 53, 78].

MOND THEORIES

The above basic tenets of MOND can be incorporated into various MOND theories. For example, in the NR regime one can modify Newtonian gravity by generalizing the Poisson equation for the gravitational field [8, 57]. Another option, which cannot be described as modified gravity is to modify the kinetic action of particles leading to modification of Newton's second law, or to modified inertia [46].

Nonrelativistic theories

The action governing a Newtonian system made of a gravitating mass distribution $\rho(\mathbf{r})$, which produces a gravitational potential ϕ, can be written as $I = \int(\mathcal{L}_K + \mathcal{L}_P)d^3\mathbf{r}dt$, with $\mathcal{L}_K = \rho\mathbf{v}^2/2$ ($\mathbf{v}$ is the velocity field) and $\mathcal{L}_P = -(8\pi G)^{-1}(\vec{\nabla}\phi)^2 - (1/2)\rho\phi$. Bekenstein and Milgrom [8] generalized this Lagrangian by replacing $(\vec{\nabla}\phi)^2$ with $a_0^2\mathcal{F}[(\vec{\nabla}\phi)^2/a_0^2]$ to obtain a MOND generalization of the Poisson equation

$$\vec{\nabla}\cdot[\mu(|\vec{\nabla}\phi|/a_0)\vec{\nabla}\phi] = 4\pi G\rho, \tag{3}$$

with $\mu(x) = \mathcal{F}'(x^2)$ satisfying $\mu(x \gg 1) \approx 1$, $\mu(x \ll 1) \approx x$. This theory had been, for many years, the only complete MOND theory in stock. It has been applied extensively to many problems, both analytically and numerically. For example, solar system tests [43, 7, 56], forces on massive bodies [47, 50, 25], disc stability and bar formation [18, 83], two-body relaxation [22], dynamical friction [62], escape speed from a galaxy [28, 86], galaxy interactions (Combes and Tiret, this volume, [82, 60]), galaxy collapse [61], triaxial models of galactic systems [85, 87], the external-field effect as applied to dwarf spheroidals and warp induction [19, 20, 1], structure formation (e.g., [37]), and quite a few more.

This formulation of MOND has also been interpreted as resulting for the omnipresence of a gravitationally polarizable medium [9], with a relativistic generalization [10, 11].

It has been realized recently [57] that a larger family of modified-gravity MOND theories exist, which involve two or even more potentials, with only one of them coupling to matter. For example, start from a Palatini-type formulation of

Newtonian gravity, introducing beside ϕ an acceleration field $\tilde{\mathbf{g}}$, and taking instead of the above Poissonian Lagrangian density:

$$\mathscr{L}_P = \frac{1}{8\pi G}(\tilde{\mathbf{g}}^2 + 2\vec{\nabla}\phi \cdot \tilde{\mathbf{g}}) - \frac{1}{2}\rho\phi. \tag{4}$$

It gives, upon variation over $\tilde{\mathbf{g}}$ and ϕ: $\tilde{\mathbf{g}} = -\vec{\nabla}\phi$, and $\vec{\nabla}\cdot\tilde{\mathbf{g}} = -4\pi G\rho$, respectively, which is indeed standard Newtonian gravity. If we generalize this action by replacing $\tilde{\mathbf{g}}^2$ with $a_0^2\mathscr{Q}[\tilde{\mathbf{g}}^2/a_0^2]$ we get a theory that is equivalent to that described by eq.(3). However, if we require a priori that $\tilde{\mathbf{g}}$ is derivable from an auxiliary potential $\tilde{\mathbf{g}} = -\vec{\nabla}\tilde{\phi}$, we get a new theory that is a quasi-linear MOND (QUMOND) theory, which is rather easier to apply because it requires solving only linear differential equations [57]

$$\Delta\tilde{\phi} = 4\pi G\rho, \quad \Delta\phi = \vec{\nabla}\cdot[\nu(|\vec{\nabla}\tilde{\phi}|/a_0)\vec{\nabla}\tilde{\phi}], \tag{5}$$

with $\nu(y) \equiv \mathscr{Q}'(y^2)$. This theory requires solving the Poisson equation twice, with a nonlinear algebraic step in between. It has been used recently, beside eq.(3), to calculate MOND effects in the inner solar system [56].

We can get even a larger class of theories by making $\mathscr{L}_P$ a function of the three scalars $(\vec{\nabla}\phi)^2$, $(\vec{\nabla}\tilde{\phi})^2$, $\vec{\nabla}\phi\cdot\vec{\nabla}\tilde{\phi}$ and also involve more potentials [57]. Look, for example, at actions of the form

$$\mathscr{L} = -\frac{1}{8\pi G}\{\beta(\vec{\nabla}\phi)^2 + \alpha(\vec{\nabla}\hat{\phi})^2 - a_0^2\mathscr{M}[(\vec{\nabla}\phi - \vec{\nabla}\hat{\phi})^2/a_0^2]\} + \rho(\frac{1}{2}\mathbf{v}^2 - \phi), \tag{6}$$

which leads to the field equations

$$\vec{\nabla}\cdot[\mu^*(|\vec{\nabla}\tilde{\phi}|/a_0)\vec{\nabla}\tilde{\phi}] = 4\pi G\rho, \qquad \Delta\phi = \vec{\nabla}\cdot[(1-\alpha^{-1}\mathscr{M}')\vec{\nabla}\tilde{\phi}], \tag{7}$$

with

$$\tilde{\phi} \equiv \phi - \hat{\phi}, \quad \mu^* = \beta - \frac{\alpha+\beta}{\alpha}\mathscr{M}'[(\vec{\nabla}\tilde{\phi}/a_0)^2]. \tag{8}$$

The first equation in (7) is solved for $\tilde{\phi}$, and then the second is a Poisson equation for the MOND potential ϕ, with the known right hand side as source. The parameter range $0 < \beta + \alpha \leq 1$ is excluded. The limiting case $\beta + \alpha = 0$ (we can then take $\beta = 1$ without loss of generality) is particularly simple. The theory then reduces to the QUMOND theory (5) with $\nu(y) = 1 + \mathscr{M}'(y^2)$. These theories have lead to a class of relativistic, bimetric MOND (BIMOND) theories (see below).

Modified inertia

An altogether different route to constructing MOND theories(e.g., [46, 53]) is to replace the kinetic action $\sum_i m_i\mathbf{v}_i^2/2$ (written now for a collection of point masses) by a more general functional of the trajectories. Such theories have, generically, to be time nonlocal if they are to obey Galilei invariance and the proper Newtonian and MOND limits [46].

Interestingly, such theories predict a universal equation that determines the orbital motion on circular trajectories in axisymmetric potentials, relating the orbital speed, V, and radius, R, by

$$\mu(V^2/Ra_0)V^2/R = -\frac{\partial\phi_N}{\partial R}, \tag{9}$$

where ϕ_N is the *Newtonian* potential, and $\mu(x)$ is a function that derives from the action of the theory as specialized to circular, constant-speed orbits. This relation is unique for a given theory (i.e., system independent). All MOND rotation-curve analyses to date employ this relation [not the relation derived from theories such as eq.(3) or eq.(5), which would require a new numerical calculation for each system analyzed].

Such 'modified-inertia' theories can differ greatly from modified-gravity formulations of MOND concerning some phenomena, even in the NR regime. Whereas in the latter, the anomalous MOND acceleration of test particles in the field of a given mass depends only on the position in the (modified) field, the time nonlocality of the former theory can produce retardation and hysteresis effects, and makes the anomalous acceleration at any location depend on properties of the whole orbit. For example, if the accelerations are small on some segments of a trajectory, MOND effects can be felt also on segments where the accelerations are high. This can give rise, for example, to different MOND effects on bound and unbound orbits, or on circular and highly elliptic orbits. In the solar system this can differentiate between planets on one hand, and long period comets and unbound spacecraft, such as the Pioneers, on the other (Milgrom, in preparation).

Also, in such theories the predicted behavior of different MOND effects, such as the external-field effect (EFE), can be rather different from that in modified-gravity theories (Milgrom, in preparation).

Relativistic theories

One naturally wants to incorporate the MOND principles in a relativistic extension. The most studied theory following this effort is Bekenstein's Tensor-Vector-Scalar (TeVeS) theory [5], which is an outgrowth of Sanders's stratified gravity approach [67]. There is a large body of work elaborating on, extending, reinterpreting, and criticizing TeVeS. See, e.g., [6, 88, 89, 21, 23, 65, 64], and, in particular, see the extensive reviews [6, 78].

There is also preliminary investigation of relativistic bimetric MOND (BIMOND) theories, which I describe succinctly below [58].

All such theories, even if they turn out to work well phenomenologically, must be only effective, approximate theories as evinced by the appearance of a free function in them, which is to reproduce the interpolating function in equations such as eq.(1). The exact form of this function will hopefully some day emerge from a deeper theory underlying the effective ones.

TeVeS

Here are the main features of this much discussed theory.

- Gravity in TeVeS [5] is described by a metric $g_{\alpha\beta}$, as in General Relativity (GR), plus a vector field, $\mathscr{U}_\alpha$, and a scalar field ϕ. (In other formulations the scalar is eliminated [88].
- Matter is coupled to one combination of the fields: the physical metric $\tilde{g}_{\alpha\beta} \equiv e^{-2\phi}(g_{\alpha\beta} + \mathscr{U}_\alpha \mathscr{U}_\beta) - e^{2\phi}\mathscr{U}_\alpha \mathscr{U}_\beta$.
- $g_{\alpha\beta}$ is governed by the usual Hilbert-Einstein action, the vector field (constrained to have a unit length) by a Maxwell-like action,[1] and the scalar action can be written as $S_s = -\frac{1}{2Gk\hat{k}}\int Q\big[\hat{k}(g^{\alpha\beta} - \mathscr{U}^\alpha\mathscr{U}^\beta)\phi_{,\alpha}\phi_{,\beta}\big](-g)^{1/2}d^4x$.
- There are three constants: k, $\hat{k}$, and a parameter K appearing in the vector action, and one free function $Q(x)$, which engenders the interpolating function of MOND in the NR limit.
- For NR systems TeVeS reproduces the NR MOND phenomenology with the MOND potential being the sum of two potentials, one satisfying the (linear) Poisson equation, the other satisfying eq.(3) with $a_0 \propto k\hat{k}^{-1/2}$.
- For weak fields ($\phi \ll c^2$) TeVeS gives lensing according to the standard GR formula but with the MOND potential.
- Structure formation and CMB: According to preliminary work, TeVeS has the potential to mimic aspects of cosmological DM [69, 26, 76, 79, 90].

Bimetric MOND theories

Inspired by bi-potential theories governed by the Lagrangian density (6), the BIMOND theories [58] involve two metrics as independent degrees of freedom. One, $g_{\mu\nu}$, is the MOND metric, which alone appears in the matter action, and which couples in the standard way to matter. The other metric, $\hat{g}_{\mu\nu}$, is an auxiliary one. In constructing an action for the theory we now have at our disposal the usual scalars made of the curvature tensors of the two metrics such as the Ricci scalars R and $\hat{R}$. But, in addition we can construct scalars using the tensor difference between the Levi-Civita connections of the two metrics:

$$C^\alpha_{\beta\gamma} = \Gamma^\alpha_{\beta\gamma} - \hat{\Gamma}^\alpha_{\beta\gamma}. \tag{10}$$

This is particularly germane in the context of MOND: Connections, and hence the $C^\alpha_{\beta\gamma}$, play the role of gravitational accelerations. So, without introducing new constants in the relativistic context we can use the MOND constant a_0 to

[1] This was later modified to obviate some inconsistencies in the original version [77].

form dimensionless tensors, $a_0^{-1}C^{\alpha}_{\beta\gamma}$, and from these construct scalars.[2] Functions of these scalar can then be used as Lagrangian densities.

Examples of scalars that are quadratic in $C^{\alpha}_{\beta\gamma}$ are

$$\Upsilon_i^{(2)} = g^{\mu\nu}C^{\gamma}_{\mu\lambda}C^{\lambda}_{\nu\gamma}, \quad \bar{C}^{\gamma}C_{\gamma}, \quad g_{\mu\nu}\bar{C}^{\mu}\bar{C}^{\nu}, \quad g^{\mu\nu}C_{\mu}C_{\nu}, \quad g_{\alpha\lambda}g^{\beta\mu}g^{\gamma\nu}C^{\alpha}_{\beta\gamma}C^{\lambda}_{\mu\nu}, \tag{11}$$

where $\bar{C}^{\gamma} \equiv g^{\mu\nu}C^{\gamma}_{\mu\nu}$, $C_{\gamma} \equiv C^{\nu}_{\gamma\nu}$, as well as others, e.g., contractions with $\hat{g}_{\mu\nu}$, in addition to higher, m powers $\Upsilon_i^{(m)}$. We thus seek to construct a covariant Lagrangian density that is a function of R, $\hat{R}$, and scalars Υ_i. I take the Lagrangian to be linear in the curvature scalars, so as not to end up with a higher derivative theory. The tensor $C^{\alpha}_{\beta\gamma}$ and the scalars Υ_i contain only first derivatives of the metrics, so we can take functions of them that interpolate between the MOND $(\Upsilon_i^{(2)}/a_0^2 \ll 1)$ and standard $(\Upsilon_i^{(2)}/a_0^2 \gg 1)$ regimes.

I have thus considered [58] actions of the form

$$I = -\frac{1}{16\pi G}\int[\beta g^{1/2}R + \alpha\hat{g}^{1/2}\hat{R} - 2g^{1/4+q}\hat{g}^{1/4-q}a_0^2\mathcal{M}(\Upsilon/a_0^2)]d^4x + I_M(g_{\mu\nu},\psi_i) + \hat{I}_M(\hat{g}_{\mu\nu},\chi_i), \tag{12}$$

where for brevity's sake I write $\mathcal{M}$ as a function of only one quadratic scalar. Its derivative $\mathcal{M}'$ plays the role of an interpolating function between the MOND and conventional regime. Also, I_M is the matter action, with matter degrees of freedom represented by ψ_i, coupling only to $g_{\mu\nu}$ (g and $\hat{g}$ are minus the determinants of the two metrics). I also permit twin matter described by degrees of freedom χ_i, which couples only to $\hat{g}_{\mu\nu}$.

We see that the modification of GR entailed by MOND does not enter here by modifying the 'elasticity' of space-time (except perhaps its strength), as is done in $f(R)$ theories and the like. The modification is introduced through the interaction between the space-time on which 'our' ψ matter lives and the auxiliary one. In a membrane description of gravity, we can say that matter lives on one membrane of a pair, with the two membranes, each with its own standard elasticity, coupled together. The way the shape of the home membrane is affected by matter then depends on the combined elasticity properties of the double membrane. However, matter response depends only on the shape of its home membrane. Such heuristics may, in fact, lead to a fundamental understanding of the origins of the MOND paradigm, and the meaning of the length $\ell = c^2/a_0$, which enters NR physics as a_0.

In particular, I found the choice of scalar argument of $\mathcal{M}$: $\Upsilon \propto g^{\mu\nu}(C^{\gamma}_{\mu\lambda}C^{\lambda}_{\nu\gamma} - C^{\gamma}_{\mu\nu}C^{\lambda}_{\lambda\gamma})$ to lead to very simple theories. This scalar is constructed from $C^{\alpha}_{\beta\gamma}$ in the same way as the first-derivative part in the Ricci scalar is constructed from the connections. Also, the subclass of theories with $\beta + \alpha = 0$, and arbitrary scalar argument, is interesting and simple, and has been studied in more detail.

Such theories have the following properties:

- Beside Newton's G (and c) they involves only a_0 as a new constant.
- The equations of motion involve no higher than second derivatives in the metrics.
- The nonrelativistic limit: On a locally double-Minkowski background, the first order NR metric is given by $g_{\mu\nu} = \eta_{\mu\nu} - 2\phi\delta_{\mu\nu}$, where ϕ is the MOND potential that is determined by the theory given by eq.(7), and $\hat{g}_{\mu\nu} = \eta_{\mu\nu} - 2\hat{\phi}\delta_{\mu\nu}$, with $\phi^* = \phi - \hat{\phi}$ being the Newtonian potential. QUMOND is gotten for $\beta + \alpha = 0$. I also considered backgrounds other than double Minkowski.
- Gravitational lensing by slowly moving masses: This has always been a holy grail for relativistic MOND theories. It is thus reassuring that the BIMOND theories predict enhanced, MOND-like lensing effortlessly. Since the relation between the full first order MOND metric and the MOND potential is the same as the relation between the GR metric and the Newtonian potential, the MOND potential controls lensing exactly as the Newtonian potential does in GR. In other words, lensing and massive-particle analysis (assuming GR) of the gravitational field of a NR mass (e.g., a galaxy) would give the same MOND potential. With choices of the scalar argument Υ different from the above this result can change somewhat; but, lensing is still enhanced and MOND-like, i.e., underpinned by an asymptotic logarithmic potential $\propto (MGa_0)^{1/2}ln(r)$.
- The GR limit: We can choose the form of $\mathcal{M}$ so that in the formal limit $a_0 \to 0$ we obtain GR exactly, possibly with a 'dark energy' term of order $\mathcal{M}(\infty)a_0^2$. For instance, in the case $\beta = 1$ (and G the Newton constant) the requirement is that $\mathcal{M}'(z) \to 0$ for $z \to \infty$. This limit can be approached as fast as required; so departures from GR in the inner solar system and short period binary pulsars can be made as small as desired.

[2] I work in units where $c = 1$, otherwise we use the MOND scale-length $\ell = c^2/a_0$ in the dimensionless tensors $\ell C^{\alpha}_{\beta\gamma}$.

- These theories give naturally 'dark energy' effects with $\Lambda \sim a_0^2$. For example, the theory described by the simple action (12) above has a vacuum solution with $g_{\mu\nu}$ describing a de Sitter or an Anti de Sitter space-time, and $\hat{g}_{\mu\nu} = \Omega^{-1} g_{\mu\nu}$, with $\Omega = \alpha(1+4q)/\beta(1-4q)$. The cosmological constant $\Lambda = -(2q+1/2)\beta^{-1}\Omega^{4q-1}a_0^2\mathcal{M}(0)$ is then of order a_0^2 if $\mathcal{M}(0)$ is of order unity. More generally, for cosmology with matter, there appears a weakly variable 'dark energy' term. Thus even without any connection with MOND such theories may provide alternatives to 'dark energy', to be investigated alongside existing schemes, such as $f(R)$ theories.
- The theory has the (yet unproven) potential to account for all aspects of the 'dark sector' (galactic 'DM', cosmological 'DM', and 'dark energy') in one fell swoop, with all controlled by a_0.

Another avenue for exploration brought to mind by these theories is the interesting possibility that the twin χ matter exists, and interacts 'gravitationaly' with our ψ matter. This effective interaction turns out to be different from standard gravity.

Be all this promising as it may, it represents only preliminary work as there are still important aspects of the theory to be checked. These include questions of causality, existence of ghosts, etc.. Bimetric theories have a long history and there is a considerable body of related literature discussing, among other things, the above questions of principle for various classes of bimetric theories. (See, for example, the recent treatment in [12, 13], where the authors consider bimetric coupling involving the metrics themselves, not their derivatives, as in BIMOND theories.) In particular, it was shown [16] that bimetric theories with derivative coupling, satisfying some general assumptions, suffer generically from the existence of ghosts. It remains to be seen, however, how relevant and/or deleterious such results are to BIMOND, especially since a major assumption underlying them does not apply in BIMOND.

MOND PHENOMENOLOGY

A MOND theory should predict the motions of test and non-test bodies in the field of an arbitrary mass distribution, such as a galaxy. For example, rotation curve analysis of disc galaxies partly tests such predictions, as it probes the accelerations only in the symmetry plane of the disc and only for nearly circular orbits. However, beyond such tests of MOND predictions for specific, individual systems, MOND predicts a considerable number of relations that should hold between galaxy properties, and which are the analogous, in the realm of the galaxies, to Kepler's laws of planetary motions. In such MOND laws, a_0 appears in several roles, similar to the appearance of $\hbar$ in many quantum phenomena, such as the black-body spectrum, the photoelectric effect, atomic spectra, superconductivity, etc., or the appearance of the speed of light in relativistic phenomena, such as black-hole physics, time dilation, or the relation between velocity and momentum. Without the unifying force of the underlying theory the appearance of the same constant in disparate, and apparently unrelated, phenomena does not make sense.

MOND laws of galactic motion

I now list briefly some of these predicted MOND laws of galactic motions. These, along with several others, are discussed, explained, and referenced in more detail in other publications on MOND (e.g. [53]).

- The rotational speed around an isolated mass becomes constant at large radii (asymptotic flatness of rotation curves): $V(r) \to V_\infty$.
- V_∞ depends only on the total ('baryonic') mass of the body via $V_\infty^4 = MGa_0$. This predicted the so called 'baryonic Tully-Fisher relation' [40, 81, 84] (see Fig.1).
- For quasi-isothermal systems, such as elliptical galaxies, the velocity dispersion depends only on the total mass (in contradistinction with the Newtonian virial relation) via $\sigma^4 \sim MGa_0$. This underlies the Faber-Jackson relation for elliptical galaxies.
- The mass discrepancy (transition from baryon dominance to 'DM' dominance) appears always around $V^2/R = a_0$.
- Isothermal spheres (e.g., as models of ellipticals) have mean surface densities $\bar{\Sigma} \le a_0/G$, as is well supported by the data (see, e.g., Figs. 1, 2 in [33]).
- The central surface density of phantom 'dark halos' is $\approx a_0/2\pi G$ [55, 32], in accordance with the findings in [27].
- Discs with $\bar{\Sigma} \le a_0/G$ have added stability.

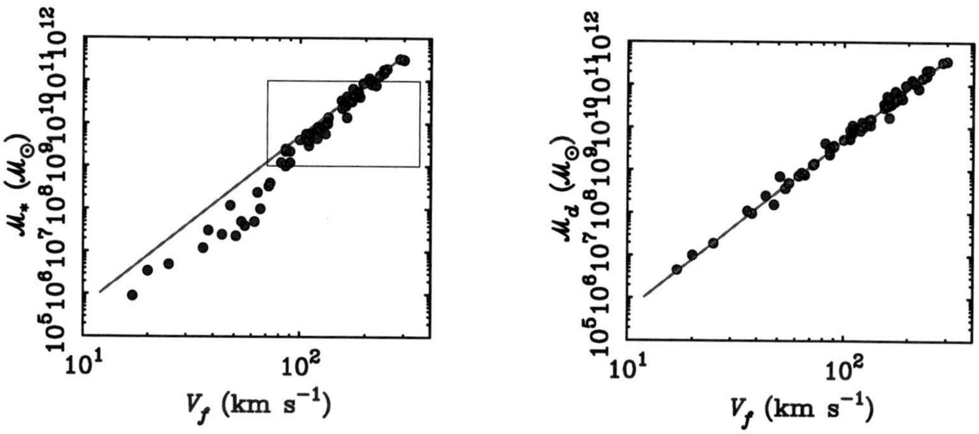

FIGURE 1. Galaxy mass plotted against the rotation curve plateau velocity. Left: analog of the traditional Tully-Fisher plot with mass in stars only. Right: The total mass including that of gas. The solid line has the log-log slope of 4, predicted by MOND, and is not a fit [39] (the small rectangle shows where past analysis had concentrated).

- Disc galaxies should exhibit a distinct disc component of 'DM', with predicted properties, in addition to the extended, spheroidal 'DM' component [49], as deduced by [34].
- Interpreting MOND with 'DM' will result in negative DM densities in some, well specified, locations.

Several important facts about these predictions have to be noted: (i) They follow as inevitable consequences of the basic MOND tenets, and are oblivious to the exact way(s) in which galaxies or other galactic systems formed. (ii) They are independent predictions in the sense that without MOND none of them follows from the others.[3] (iii) Inasmuch as they have been tested they are well consistent with the data. (iv) In the framework of DM such laws must follow from very strict connection between the amount and distribution of DM and those of the baryons, because they relate baryonic properties–such as (baryonic) masses–to properties determined mainly by the DM–such as speeds. (v) Even without further theoretical development, or interpretation, MOND has already directed the eye to many regularities not suspected before, including the appearance of an acceleration constant, a_0, in many a priori unrelated facets of galaxy dynamics.

I now discuss briefly some of these laws: The independence of speed on orbital radius in the limit of large radii follows directly from the scale invariance of the deep-MOND limit: under space-time scaling sizes change, but velocities do not.

The MOND mass-asymptotic-speed relation, underlies the Tully-Fisher relation (TFR), whose existence was known before the advent of MOND. However, the traditional TFR correlates some luminosity measure of galaxies with some velocity measure, with different choices giving different results. To my knowledge the TFR has no creditable theoretical basis in the DM doctrine. MOND has specified exactly what is to be correlated with what: the total baryonic mass with the asymptotic, constant rotational speed. This has lead to the so-called baryonic TF correlation, which indeed is very tight, and which conforms exactly with the predicted MOND relation: the power of 4 and the proportionality factor being Ga_0, well consistent with other determinations of a_0 [40, 39, 81]. Figure 1 shows the results of one such analysis.

The MOND constant, a_0, defines a scale of mass surface density a_0/G. This is why several of the MOND laws predict a special value of the surface density in different contexts. For example, one of the above laws, which follows from detailed analysis, states that isothermal spheres–which in MOND have a finite mass, unlike their Newtonian analogs–cannot have mean surface densities much exceeding a_0/G. This simply reflects the fact the IS disobeying this inequality have mean accelerations larger than a_0, and so are Newtonian, and so cannot exist as finite mass objects.

This last MOND prediction assigns a special surface density to the normal matter component of a galactic object. But there is also an interesting MOND prediction that has come to light recently, which pertains to a property of the pure,

[3] It is possible to build families of galaxies of baryons plus DM that will satisfy any subset of these law, but not the rest. So in the framework of DM they'll each require a separate explanation.

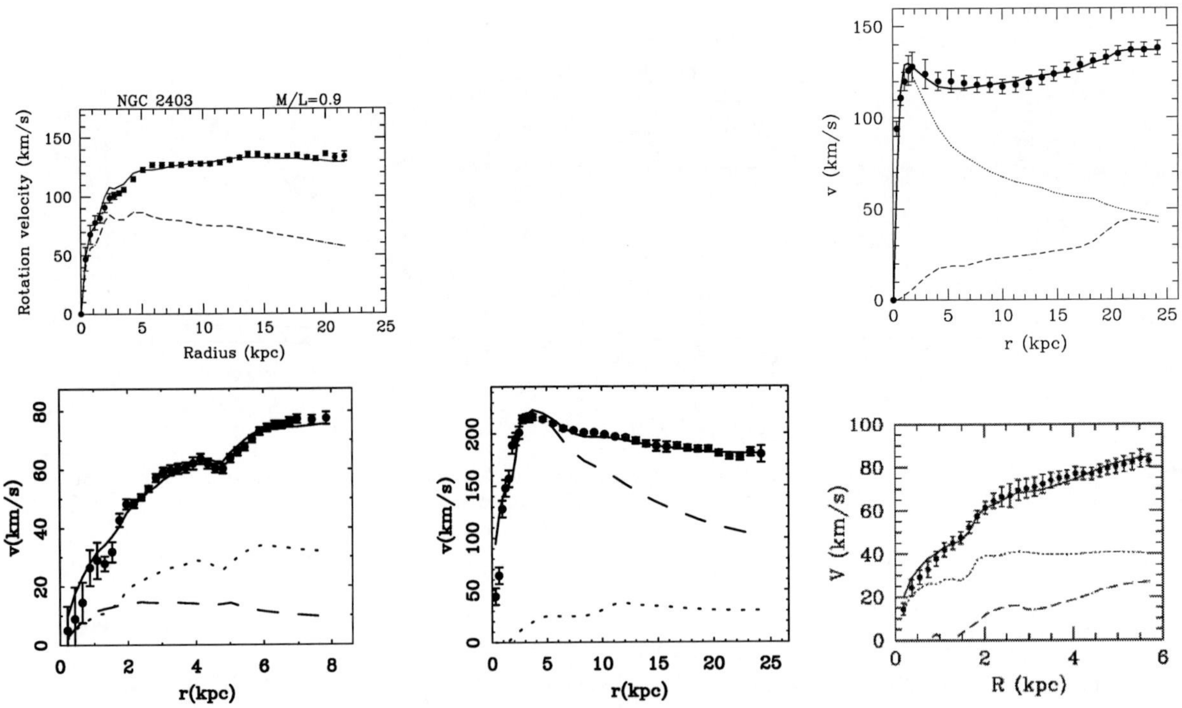

FIGURE 2. The rotation curves of a variety of galaxies from low-mass-low-speed galaxies that are everywhere of low-acceleration (such as the galaxies shown at the lower right and left) to high-mass-high-speed galaxies that have high accelerations in the inner parts transiting to low accelerations at a finite radius (such as the middle plot). The data is shown as points with error bars. The line through the data is the MOND curve. Upper left (curtesy of Bob Sanders): dashed line shows the predicted Newtonian curve fitted to match the inner points. Lower left (from [71]): dotted curve is the Newtonian curve calculated for the gas alone, and the dashed line the same for the stars (they add in quadrature). Middle (from [71]): dashed line the Newtonian curve for the stars, and the dotted curve for the gas. Upper right (from [70]) and Lower right (curtesy of Bob Sanders): the dotted line is the Newtonian curve for the stars, the dashed curve for the gas.

fictitious 'DM halo' needed to explain MOND results with DM. This prediction identifies $a_0/2\pi G$ as a special central surface density of such phantom halos that is of universal significance [55, 32]. The discovery of this prediction was incited by a recent finding to this effect [27]. The two predictions follow from completely different aspects of MOND, but both come up with a special surface density determined by a_0/G, which is the only scale available in MOND. The fact that the same value of a_0 determined from other predictions gives a very good agreement with the findings of [27], can count as another success of MOND.

Rotation Curves of Disc Galaxies

The quintessential MOND achievement, however, is the prediction of detailed rotation curves (RCs) of many individual galaxies, from their baryon distribution alone. Interestingly, the first systematic analysis of such predictions appeared only some four years after the advent of MOND, for it had had to await the appearance of extended HI rotation curves. The first systematic study [35] was followed by amending analysis in [45]; then, many such analyses followed; e.g., [3, 66, 73, 38, 14, 15, 4, 29, 31, 24, 2, 72, 59].

I show in Fig.2 examples of such analysis for five disc galaxies that span the gamut of galaxy properties: from low-mass, low-speed, gas-rich galaxies with a still rising RC, to high-mass, high-speed, gas-poor galaxies, with a still declining RC.

The MOND curves shown involve only one parameter, the mass-to-light ratio, M/L, which is the conversion factor from stellar light to mass (the gas mass is measured directly with no need for conversion). In cases where the gas strongly dominates the mass budget, such as for the galaxy shown on the lower left, the MOND prediction hardly

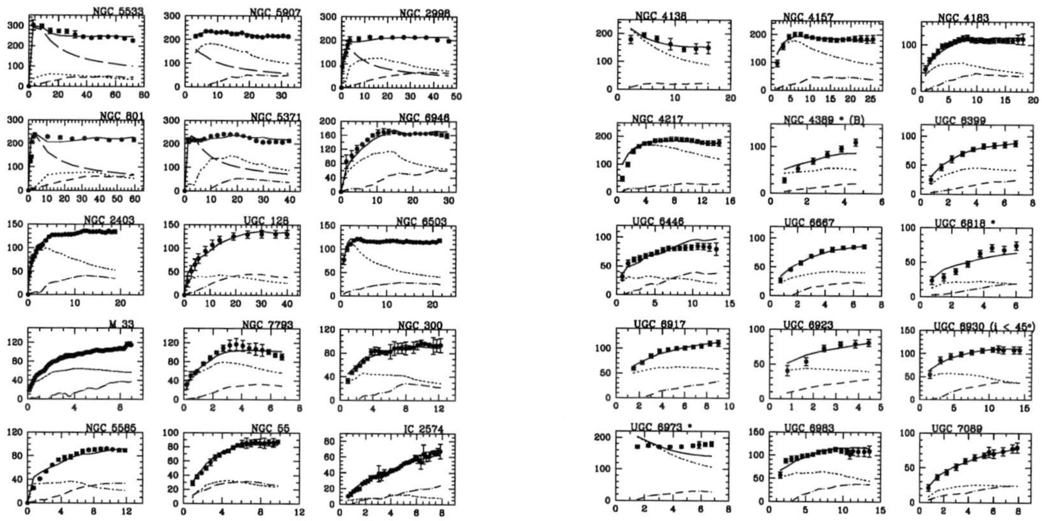

FIGURE 3. Additional MOND rotation curves from [66] and [14] (left) and from [73] (right). (MOND curves in solid; stellar disc Newtonian curves in dotted; gas in dot-dash; and stellar bulge in long dashed.)

depends on the assumed M/L value, and the MOND curve is practically an absolute prediction. When the stars dominate, M/L may be viewed as a parameter chosen to fit the first few points on the rotation curve; the rest of the curve is then predicted uniquely. Note also, that M/L is not a completely free parameter: it can be deduced for a galaxy of a given spectrum from theoretical calculations of stellar population synthesis, within some margins. The values required by the MOND fits were found to agree with those theoretical values (e.g., [73]). In DM fits one usually has full freedom to adjust the radial scale and the normalization of the a priori unknown contribution of the DM halo, in addition to the stellar M/L value.

We see that some measured RCs exhibit features–such as a dip, or a sudden rise–which are well reproduced by the predicted MOND curves. These features are of prime importance in assessing the performance of competing paradigms. They result from relatively sharp features in the baryonic mass distribution of the disc. As this is the only source for the MOND curve, the feature appears in this curve in full strength (as it does in the Newtonian curve without DM). However, adding a Dominant DM component, which does not have thae feature (indeed, which cannot have the feature if it is spherical) washes the feature out in the best-fit DM curve.

Figures 3, 4 show more results of MOND rotation-curve analyses.

round systems

Dwarf spheroidal satellites of the Milky Way are very low acceleration systems. They indeed show large mass discrepancies, as was predicted from MOND[42] before these discrepancies were measured. The state of the art analysis of these systems and references to earlier work can be found in [1, 75], where it is found that, with one or two exceptions, perhaps, MOND indeed accounts for the large mass discrepancies deduced with Newtonian dynamics.

As has been realized through many analyses starting in the early 1990s, the only system type where MOND has failed systematically to account for the full mass discrepancy comprise galaxy clusters. The situation has been reviewed recently in [52] (and many references therein). Briefly: the typical global mass discrepancy in clusters (within say 2 Mpc) is about a factor 7-10). MOND corrects this by reducing the required mass by a factor 3-5, but leaves still a factor of two discrepancy. So roughly as much 'dark mass' is still required as there is presently visible baryonic mass. MOND adherents have attributed this dark mass to possibly neutrinos or to some form of yet undetected cluster baryonic matter (CBDM). The distribution of the CBDM in the cluster is rather like that of the galaxies, being rather

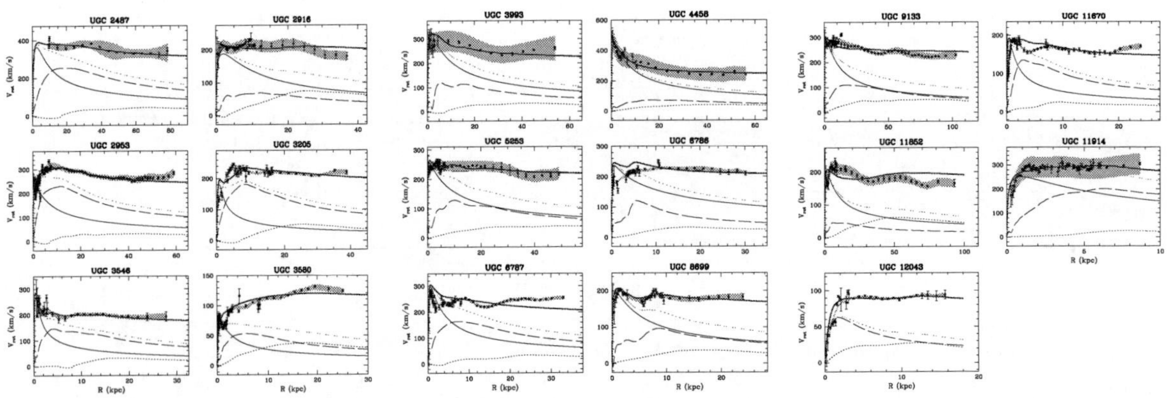

FIGURE 4. Additional MOND rotation curves from [72]. The grey shaded bands give the allowed range due to inclination uncertainties. Thin black solid, dashed and dotted lines give the contributions from stellar bulge, disc and gas respectively. The thin green (grey) line gives the Newtonian sum of the individual components and the bold blue (grey) lines gives the total MOND rotation curve.

more centrally concentrated than the hot gas, which is makes the lion's share of the visible component. This makes the mass discrepancy that remains after the MOND correction rather stronger in the cluster core (within a few hundred kpc). The appearance of the post-collision 'bullet cluster', which has been bruited as 'direct proof of the existence of DM', conforms exactly to what is expected from this picture, long known from studies of individual clusters. It only reiterates the fact that a still undetected mass component exists in cluster cores, but does not pinpoint it as the putative particle DM.

The solar system

We know of two concrete solar system phenomena that MOND could impact. The first is the so called Pioneer anomaly: a yet unexplained constant acceleration of the two Pioneer spacecraft towards the sun, of a magnitude $\approx \bar{a}_0 \equiv 2\pi a_0$. While this anomaly may yet turn out to result from a mundane, underestimated systematics, and not from new physics, MOND does have the potential to explain such an anomaly. Because a similar anomaly in the motions of the planets can be ruled out by a large margin, a new-physics explanation must hinge on the differences between the planetary orbits (bound, nearly circular) and those of the spacecraft (unbound, highly hyperbolic) (see [53] and references therein for details).

I have recently pointed out [56] that certain MOND theories predict an unavoidable MOND effect in the inner solar system (planetary zone) having to do with the galactic acceleration field in which the solar system is falling (the external-field effect of MOND). Near the sun–i.e. at distances much smaller than the MOND radius of the sun, $R_M \equiv (M_\odot G/a_0)^{1/2} \approx 8\times 10^3$a.u.–this appears as an anomalous quadrupole field with anomalous potential $\phi_{an} = (q/4)a_0 R_M^{-1}(2z^2 - x^2 - y^2)$. Here, the z-axis is in the direction of the galactic center, and $q \sim -0.1$ depends on the strength of the galactic acceleration field at the position of the sun, and on the form of the MOND extrapolating function of the theory. This anomaly causes very distinctive effects on planetary motions, such as an anomalous perihelion precession with characteristic dependence on the length of the semi-major axis and on its orientation relative to the galactic center's direction. This effect seems to be below present day detection capabilities, but not inaccessibly below.

MOND VS. COLD DARK MATTER

MOND and DM were initially advanced to explain similar phenomena: the mass discrepancies in galactic systems. This may create the impression that they are twin paradigms, that the deductions we make from both are of similar force, and that the only way to decide between them is to compare the performance of their predictions against the data in cases where the predictions differ. This is far from being the case.

I concentrate henceforth on cold dark matter (CDM), the eminently favored version of the DM paradigm today. Deductions made in CDM and in MOND, pertaining even to the same phenomena and measurements, such as galaxy rotation curves, the existence and nature of a TF relation, the distribution of the mass discrepancy in a given system, etc., are of a totally different nature. In MOND, these are strict and inevitable predictions of all aspects of the mass discrepancy in any given object, based only on the distribution of of normal (baryonic) matter as it is now observed. Such predictions can be made for any given, individual system. In the CDM paradigm, all such deductions for a given object would depend strongly on its detailed history: initial collapse, merger history, cannibalism of satellites, gas accretion, star formation, baryon ejection by supernova explosions and ram pressure effects, dissipation, angular momentum exchange, cooling and heating, etc., etc.–details that we do not know, and cannot know for any individual object, with very few, important exceptions (see below). To make matters worse, the two major components, baryons and the putative, weakly interacting CDM, partake very differently in all the above processes. Despite many works in the literature that attempt to account for these via so called 'semi analytic' calculations, the bare fact is that CDM is incapable of truly predicting the mass discrepancies in galactic systems (again with that important exception to be discussed below). CDM is utterly incapable of predicting even the most basic number: the baryon fraction in a given galaxy.[4] But such numbers are the preliminaries for arriving at a deduced TF relation for example, to say nothing of predicting full rotation curves. What we then see often as CDM analysis of rotation curves are merely best fits, which assume, in addition to the baryons, a certain DM halo mass distribution with the total mass and radial scale of the distribution left as free parameters. The MOND rotation curves have non of this. At worst they involve one free parameter for converting stellar light into baryonic mass (which CDM fits also have in addition to the other parameters). But even this is not needed in many instances, as explained above.

The other important point to note is that even if CDM is one day found capable of deducing from first principles some statistical, population properties pertaining to the relations between baryons and DM (such as the TF relation), it it highly unlikely to ever make predictions for individual systems, as MOND does.

To recapitulate, the fact that galaxies follow closely the MOND predictions argue against CDM in two ways. In the first place because they show that there is a paradigm that can predict all those observed regularities that CDM cannot predict. But, in addition, the fact that such strict regularities exist at all–even irrespective of the existence of MOND–argues against CDM, because the relations between the baryonic and DM components is expected to be highly haphazard in the CDM paradigm.

I mentioned above that there is one important exception to CDM's incapability of predicting baryon-to-DM relations. This concern so called tidal dwarfs: the small 'phoenix' galaxies that are born from the gaseous tidal tails spewed in the aftermath of high speed collisions between ready made galaxies. Such tidal dwarfs are an exception to the above rule because their history is brief and involves only a few, rather well understood processes. Simulations show that hardly any of the CDM purportedly surrounding the two colliding, parent galaxies makes its way into the the tidal tails and the tidal dwarfs that form in them. CDM then predicts no mass discrepancies to speak of in tidal dwarfs. In MOND, the criterion for the appearance of a mass discrepancy is only the accelerations involved, not, in any way, the process of its formation.

So, what is known about mass discrepancies in tidal dwarfs? There are now two possible relevant instances. In the first, Bournaud et al. [17] have identified and analyzed three tidal dwarfs in the tidal debris around NGC 5291. They find, in contrast with the expectations from the CDM paradigm, that the three dwarfs do exhibit mass discrepancies of about a factor three. Subsequent analysis [51, 30] showed that these agree well with the predictions of MOND for these systems. In another possible instance, it has been argued recently [41] that most of the dwarf spheroidal satellites of the Milky Way are remnant tidal dwarfs in a single historical tidal tail. This deduction is based on the fact that both the position of these dwarfs and their motions indicate that they live and move in a disc around the Milky Way. This, and other aspects of the population, are difficult to understand unless, it is argued, they formed as tidal dwarfs. If so, the CDM paradigm implies they should not exhibit the large mass discrepancies they are known to show. In MOND, as said above, we do expect them to show large discrepancies because they are very law acceleration systems.

We saw that even the most common tests, such as RC analysis, hold a strong potential for deciding between MOND and CDM, for those who are willing to grasp better the differences in the nature of these deductions in MOND and CDM. However, in addition there are quite a few phenomena on which MOND and CDM make disparate predictions. I already discussed above the case in point concerning tidal dwarfs. A few other examples are: 1. MOND, but not CDM,

[4] We know that this fraction in a galaxy today is typically much smaller than the cosmic fraction with which proto-galaxies must have started their life. So most of the baryons must have been lost on the way, somehow (in the CDM paradigm), exactly how, and what small fraction of it is left, are the grist for 'semi analytic' models.

predicts that in disc galaxies there will be a strong disc component of 'phantom' DM in addition to the extended spheroidal one. The predicted disc component appears only where the baryonic disc is, and only where the orbital accelerations in the disc are small in the MOND sense [49]. 2. Insistence on interpreting MOND results with DM will lead to negative DM densities in some locations [44]. 3. We saw already that some MOND theories predict an anomalous quadrupole field in the inner solar system, which would be quite unexpected from CDM.

To summarize this section: The success of MOND predictions implies that baryons alone strictly determine the acceleration fields of galactic objects. This conflicts with the expectations in the CDM paradigm (where these fields are, by and large, dominated by the contribution of the DM) because of the haphazard formation and evolution of galactic objects, and because baryons and DM are given to very different behaviors during the evolution. This is evinced by the very different characteristics of the baryonic and the putative DM components in galaxies today (e.g., the very small baryon-to-DM fraction in galaxies compared with the cosmic value, the highly concentrated, and oftentimes disc-like, baryonic component compared with the much more extended, spheroidal DM component). It is thus highly unlikely that DM will someday reproduce MOND.

THE COSMOLOGICAL CONNECTION

So far I discussed MOND only from the phenomenological point of view: It as been advanced to obviate the need for DM, and it has, indeed, managed to predict, explain, and organize large amounts of observations in individual systems, and of population laws and regularities, without invoking DM. However, there is an additional aspect revealed by MOND that may hint at more profound implications: The value of the MOND constant a_0, as determined from the various laws listed above, is $a_0 \approx 1.2 \times 10^{-8}$cm s^{-2}. This value is tantalizingly close to acceleration parameters that characterize cosmology [42, 48]. Define, more conveniently, $\bar{a}_0 \equiv 2\pi a_0$, then one finds

$$\bar{a}_0 \approx a_H = cH_0, \qquad \bar{a}_0 \approx a_\Lambda = c(\Lambda/3)^{1/2}. \tag{13}$$

The first acceleration parameter, a_H, is associated with the present day expansion rate H_0, while the second is defined by the present density of 'dark energy' as determined by Λ. The mysterious fact that today the two cosmic acceleration parameters are nearly the same defines the 'cosmic coincidence' puzzle (another coincidence puzzle being the proximity of the baryon and DM densities, when the two components are thought to have formed by totally different and unrelated processes). The fact that $\bar{a}_0$–an acceleration that emerges very clearly and forcibly from the phenomenology of galactic systems–also nearly equals these two cosmic accelerations should be viewed, in my opinion, as an added cosmic coincidence.

Much as been said about the significance of this last 'coincidence' and its possible origins and implications (e.g., in [42, 67, 48, 10, 11, 54, 58]). Here I only touch briefly on a recent observation that might connect MOND with cosmology even more firmly [54]. If the 'dark energy' is a cosmological constant (CC), then our universe is nearly of a de Sitter (dS) geometry in the vicinity of the present time. In the future, as the CC becomes increasingly dominant over matter, the geometry will tend towards exact dS geometry. Now, the second coincidence in eq.(13) is tantamount to the MOND acceleration being related to the radius, ℓ, of this asymptotic dS cosmos by $\bar{a}_0 \approx c^2/\ell$. The possible tighter ties of MOND with cosmology may then be reflected in the symmetries that characterize these two structures. The symmetry (isometry) group of a 4-D dS space-time is the group of rotations in Minkowski 5-D space time; i.e. $SO(4,1)$. If MOND, and in particular the deep MOND limit, is in some way a reflection of the near dS nature of our cosmos, the symmetries characterizing MOND might reflect the symmetries of the dS space-time, or perhaps both symmetries are a result of the symmetries of the theory that underlies both. Our actual universe is not quite of dS geometry; so we do not expect MOND at large to share symmetries with that geometry. Indeed MOND with its full coverage is not known to have symmetries beyond the standard rotation, and space- and time-translation invariance. But the deep MOND limit itself does have added symmetries in certain formulations. For example, the deep MOND limit of the nonlinear Poisson formulation eq.(3) has been found to have the full conformal symmetry in three space dimensions. Together with the space rotations and translations these conformal symmetries form a larger group which is equivalent to the SO(4,1) geometrical symmetry group of exact dS space-time. This has been proposed [54] as a possible indication of a connection similar to the celebrated duality of conformal field theories on the boundary of an Anti-dS (or dS) space-time and gravity in that space-time. Here as well, the space on which the extended deep-MOND theories act is the Euclidean three-dimensional space (or rather its compactification to a Euclidean 3-D sphere), and this is indeed the past (or future) boundary of a dS space-time.

OPEN QUESTIONS

Beside the yet unexplained remaining factor-of-two discrepancy in galaxy clusters, the major challenge to MOND is to account fully for the need for cosmological DM. In other words we need a MOND inspired mechanism that will account for the effects in galaxy formation, CMB, etc., of what in standard dynamics is attributed to cosmological DM with a contribution of about 0.2 to Ω. One effect of cosmological DM is to enhance structure formation because, being neutral, it would have started to collapse before recombination. MOND clearly has the ability to enhance gravitational collapse over standard dynamics without DM. Several studies of structure formation in MOND inspired schemes, some using somewhat crude NR models, some using relativistic theories such as TeVeS, have been conducted, and have shown such enhancing effects: there are indeed aspects of MOND that could replace cosmological DM effects [68, 63, 80, 36, 79, 26, 90, 37]. It has to be realized, however, that the theory of structure formation, interaction of matter with the CMB, etc. in MOND is still much less developed and focussed then in standard dynamics. In my opinion a satisfactory understanding of all these will come only after we better understand the nature of the MOND-cosmology connection discussed above, and when we then have a full fledged relativistic version of MOND, which, arguably, we are still lacking.

In any event, one should certainly not be dismayed by the tasks still left to be accomplished. The very simple MOND idea proposed a quarter of a century ago, has already achieved much more then could be expected of it at the time of its inception. It can in fact be argued that the richness and variety, and sheer quantity of successful predictions in the context of galaxy dynamics is many fold what is still required in cosmology. Understanding a single galaxy is not as important as understanding cosmology. But, epistemologically, the detailed prediction of a field of a single galaxy–accounting exactly for what otherwise would result from a detailed amount and distribution of DM–is, in my opinion, a feat on a par with predicting the few statistical properties attributed conventionally to cosmological DM.

All in all, with the encouragement from past successes and the attraction of remaining challenges, times are interesting for those who choose to work on MOND.

ACKNOWLEDGMENTS

This research was supported by a center of excellence grant from the Israel Science Foundation.

REFERENCES

1. G. W. Angus, *MNRAS* **387**, 1481 (2008)
2. E. I. Barnes, A. Kosowsky, and J. A. Sellwood, *AJ* **133**, 1698 (2007).
3. K. G. Begeman, A. H. Broeils, and R. H. Sanders, *MNRAS* **249**, 523 (1991).
4. A. Begum, and J. Chengalur, *AA* **413**, 525 (2004).
5. J. Bekenstein, *Phys. Rev. D* **70**, 083509 (2004).
6. J. Bekenstein, *Contemp. Phys.* **47**, 387 (2006).
7. J. Bekenstein, and J. Magueijo, *Phys. Rev. D* **73**, 103513 (2006).
8. J. Bekenstein, and M. Milgrom, *ApJ* **286**, 7 (1984).
9. L. Blanchet, *Class. Quant.Grav.* **24**, 3541 (2007).
10. L. Blanchet, and A. Le Tiec, *Phys. Rev. D* **78**, 024031 (2008).
11. L. Blanchet, and A. Le Tiec, *Phys. Rev. D* **80**, 023524 (2009).
12. D. Blas, C. Deffayet, and J. Garriga, *Class. Quant. Grav.* **23**, 1697 (2006).
13. D. Blas, C. Deffayet, and J. Garriga, *Phys. Rev. D* **76**, 104036 (2007).
14. W. J. G. de Blok, and S. S. McGaugh, *ApJ* **508**, 132 (1998).
15. R. Bottema, J. L. G. Pestaña, B. Rothberg, and R. H. Sanders, *AA* **393**, 453 (2002).
16. N. Boulanger, T. Damour, L. Gualtieri, and M. Henneaux *Nucl.Phys. B* **597**, 127 (2001)
17. F. Bournaud, P.-A. Duc, E. Brinks, M. Boquien, M. Amram, U. Lisenfeld, B. S. Koribalski, F. Walter, and V. Charmandaris, *Science* **316**, 1093 (2007).
18. R. Brada, and M. Milgrom, *ApJ* **519**, 590 (1999).
19. R. Brada, and M. Milgrom, *ApJ* **541**, 556 (2000).
20. R. Brada, and M. Milgrom, *ApJ* **531**, L21 (2000).
21. J.-P. Bruneton, and G. Esposito-Farese, *Phys. Rev. D* **76**, 124012 (2007).
22. L. Ciotti, and J. Binney, *MNRAS* **351**, 285 (2004).
23. C. R. Contaldi, T. Wiseman, and B. Withers, *Phys. Rev. D* **78**, 044034 (2008).
24. E. Corbelli, P. Salucci, *MNRAS*, **374**, 1051 (2007).

25. D. C. Dai, R. Matsuo, and G. Starkman, *arXiv:0811.1565*
26. S. Dodelson, and M. Liguori, *Phys. Rev. Lett.*, **97**, 231301 (2006).
27. F. Donato, G. Gentile, P. Salucci, C. Frigerio Martins, M. I. Wilkinson, G. Gilmore, E.K. Grebel, A. Koch, and R. Wyse, *MNRAS* **397**, 1169 (2009).
28. B. Famaey, J.-P. Bruneton, and H. S. Zhao, *MNRAS* **377**, L79 (2007).
29. G. Gentile, P. Salucci, U. Klein, D. Vergani, and P. Kalberla, *MNRAS* **351**, 903 (2004).
30. G. Gentile, B. Famaey, F. Combes, P. Kroupa, H. S. Zhao, and O. Tiret, *AA* **472**, L25 (2007).
31. G. Gentile, P. Salucci, U. Klein, and G. L. Granato, *MNRAS* **375**, 199 (2007).
32. G. Gentile, B. Famaey, H. S. Zhao, and P. Salucci, *Nature* **461**, 627, (2009).
33. A. W. Graham, D. Merritt, B. Moore, J. Diemand, and B. Terzić, *AJ* **132**, 2711 (2006).
34. P.M.W. Kalberla, L. Dedes, J. Kerp, and U. Haud *AA* **469**, 511 (2007).
35. S. M. Kent, *AJ* **93**, 816 (1987).
36. A. Knebe, and B. K. Gibson, *MNRAS* **347**, 1055 (2004).
37. C. Llinares, A. Knebe, and H. S. Zhao, *MNRAS* **391**, 1778 (2008).
38. S. S. McGaugh, and W. J. G. de Blok, *ApJ* **499**, 66 (1998).
39. McGaugh, S.S., 2005b, Proceedings of the XXIst IAP Colloquium "Mass Profiles and Shapes of Cosmological Structures", Eds. G. Mamon, F. Combes, C. Deffayet, B. Fort, EDP Sciences; arXiv:astro-ph/0510620.
40. S. S. McGaugh, J. M. Schombert, G. D. Bothun, and W. J. G. de Blok, *ApJ* **523**, L99 (2000).
41. M. Metz, P. Kroupa, and H. Jerjen, *MNRAS* **394**, 2223 (2009).
42. M. Milgrom, *ApJ* **270**, 365 (1983).
43. M. Milgrom, *ApJ* **302**, 617 (1986).
44. M. Milgrom, *ApJ* **306**, 9 (1986).
45. M. Milgrom, *ApJ* **333**, 689 (1988).
46. M. Milgrom, *Ann. Phys.* **229**, 384 (1994).
47. M. Milgrom, *Phys. Rev. E* **56**, 1148 (1997).
48. M. Milgrom, *Phys. Lett. A* **253**, 273 (1999).
49. M. Milgrom, *MNRAS* **326**, 1261 (2001).
50. M. Milgrom, *J. Phys. A* **35**, 1437 (2002).
51. M. Milgrom, *ApJL* **667**, L45 (2007).
52. M. Milgrom, *New Astron.Rev.* **51**, 906 (2008).
53. M. Milgrom, In Proceedings XIX Rencontres de Blois; arXiv:0801.3133 (2008).
54. M. Milgrom, *ApJ* **698**, 1630 (2009).
55. M. Milgrom, *MNRAS* **398**, 1023 (2009).
56. M. Milgrom, *MNRAS* **399**, 474 (2009).
57. M. Milgrom, Submitted to MNRAS (2009).
58. M. Milgrom, preprint (2009).
59. M. Milgrom, and R.H. Sanders, *ApJ* **658**, L17 (2007).
60. C. Nipoti, P. Londrillo, and L. Ciotti, *MNRAS* **381**, L104 (2007).
61. C. Nipoti, P. Londrillo, and L. Ciotti, *ApJ* **660**, 256 (2007).
62. C. Nipoti, L. Ciotti, J. Binney, and P. Londrillo, *MNRAS* **386**, 2194 (2008).
63. A. Nusser, *MNRAS* **331**, 909, (2002).
64. E. Sagi, *Phys. Rev. D* **80**, 044032 (2009).
65. E. Sagi, and J. D. Bekenstein, *Phys. Rev. D* **77**, 024010 (2008).
66. R. H. Sanders, *ApJ* **473**, 117 (1996).
67. R. H. Sanders, *ApJ* **480**, 492 (1997).
68. R. H. Sanders, *ApJ* **560**, 1 (2001).
69. R. H. Sanders, *MNRAS* **363**, 459 (2005).
70. R. H. Sanders , Third Aegean Summer School, *The Invisible Universe: Dark Matter and Dark Energy*, arXiv:astro-ph/0601431 (2006).
71. R. H. Sanders, and S. S. McGaugh, *ARAA* **40**, 263 (2002).
72. R. H. Sanders, and E. Noordermeer, *MNRAS* **379**, 702 (2007).
73. R. H. Sanders, and M. A. W. Verheijen, *ApJ* **503**, 97 (1998).
74. R. Scarpa, "Modified Newtonian Dynamics, an Introductory Review" in *First crisis in cosmology*, edited by E.J. Lerner, J.B. Almeida, AIP Conference 822, American Institute of Physics, New York, 2006, pp. 253
75. A. L. Serra, G. W. Angus, and A. Diaferio, arXiv:0907.3691
76. C. Skordis, *Phys. Rev. D* **74**, 103513 (2006).
77. C. Skordis, *Phys. Rev. D* **77**, 123502 (2008).
78. C. Skordis, *Class. Quant. Grav.* **26 (14)**, 143001 (2009).
79. C. Skordis, D. F. Mota, P. G. Ferreira, and C. Boehm, *Phys. Rev. Lett.* **96**, 011301 (2006).
80. S. Stachniewicz, and M. Kutschera, *Acta physica Polonica* **B32**, 362, (2002).
81. D. V. Stark, S. S. McGaugh, and R.A. Swaters, *AJ* **138**, 392 (2009).
82. O. Tiret, and F. Combes, in *Formation and Evolution of Galaxy Disks*, edited by J. G. Funes, and E. M. Corsini, arXiv:0712.1459 (2007).

83. O. Tiret, and F. Combes, *AA* **483**, 719 (2008).
84. C. Trachternach, W. J. G. de Blok, S. S. McGaugh, J. M. van der Hulst, and R. -J. Dettmar, arXiv:0907.5533 (2009).
85. Y. Wang, X. Wu, and H. S. Zhao, *ApJ* **677**, 1033 (2008).
86. X. Wu, H. S. Zhao, B. Famaey, G. Gentile, O. Tiret, F. Combes, G. W. Angus, and A. C. Robin, *ApJ* **665**, L101 (2007).
87. X. Wu, H. S. Zhao, Y. Wang, C. Llinares, and A. Knebe, *MNRAS*, in press (2009).
88. T. G. Zlosnik, P. G. Ferreira, and G. D. Starkman, *Phys. Rev. D* **74**, 044037 (2006).
89. T. G. Zlosnik, P. G. Ferreira, and G. D. Starkman, *Phys. Rev. D* **75**, 044017 (2007).
90. T. G. Zlosnik, P. G. Ferreira, and G. D. Starkman, *Phys. Rev. D* **77**, 084010 (2008).

MOND and the Galaxies

F. Combes[a] and O. Tiret[b]

[a]*Observatoire de Paris, LERMA, 61 Av de l'Observatoire, F-75014 Paris, France*
[b]*SISSA, via Beirut 4, I-34014 Trieste, Italy*

Abstract. We review galaxy formation and dynamics under the MOND hypothesis of modified gravity, and compare to similar galaxies in Newtonian dynamics with dark matter. The aim is to find peculiar predictions both to discriminate between various hypotheses, and to make the theory progress through different constraints, touching the interpolation function, or the fundamental acceleration scale. Galaxy instabilities, forming bars and bulges at longer term, evolve differently in the various theories, and help to bring constraints, together with the observations of bar frequency. Dynamical friction and the predicted merger rate could be a sensitive test of theories. The different scenarios of galaxy formation are compared within the various theories and observations.

Keywords: Galaxies, kinematics and dynamics, dark matter, spiral, bars, interactions, ellipticals
PACS: 98.35.Df, 98.35.Hj, 98.52.Fh, 98.52.Nr, 98.52.Wz, 98.62.Dm, 98.62.Hr

I- INTRODUCTION

Milgrom (1983) has proposed an empirical modification of the law of gravity, as an alternative to dark matter in galaxies. The basic idea of this MOdified Newtonian Dynamics (MOND) derives from the observation that the missing mass problem occurs only in the weak field regime, at low acceleration (see Figure 1). In addition, galaxies are following the baryonic Tully-Fisher relation (McGaugh et al 2000), which tells us that the maximum rotational velocity of a galaxy runs as the 1/4th power of its mass, and does not depend on the radius, suggesting an asymptotic acceleration in 1/r. The proposition is thus that at acceleration below a0= 10^{-10}m/s^2, the gravitational attraction will tend to the formulation a = (a0 aN)$^{1/2}$, where aN is the Newtonian value. This effectively produces an acceleration in 1/r, implying a flat rotation curve in the limiting regime, and automatically the Tully-Fisher relation. The transition between the Newtonian and MOND regime is controlled by an interpolation function $\mu(x)$, of x=a/a0, which standard form is $\mu(x) = x/(1+x^2)^{1/2}$, that essentially tends to x in the MOND regime, when x is smaller than 1, and to unity in the Newtonian regime. This phenomenology has a large success explaining rotation curves and kinematics of galaxies, from irregular dwarf dominated by dark matter (and therefore in the MOND regime), to the giant and ellipticals, dominated by baryons (e.g. the review Sanders & Mc Gaugh 2002).

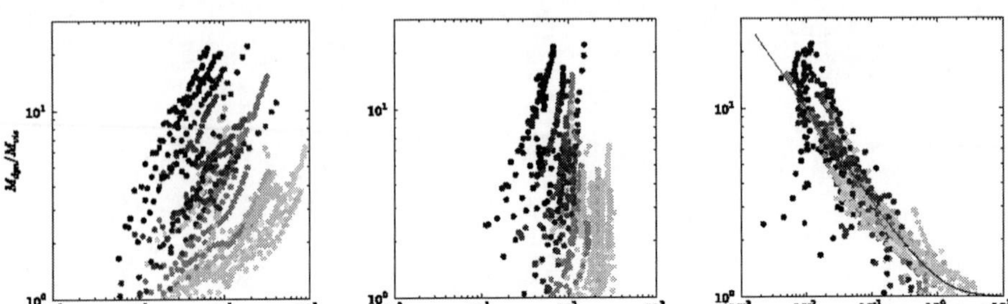

FIGURE 1. The missing mass problem. In the three panels, the dynamical mass to visible mass ratio is plotted versus radius (left), velocity (middle), and acceleration (right), for a series of nearby galaxies, with clean HI rotation curves. Only the acceleration reduces the scatter (cf Tiret & Combes 2009).

CP1241, *Invisible Universe: Proceedings of the Conference,* edited by J.-M. Alimi and A. Füzfa

The motivation for MOND is not only an alternative to dark matter, which particles have not yet been found, but also to solve the problems that the standard CDM model encounters at galaxy scales. Numerical simulations in the standard model predict an over concentration of dark matter in galaxies, and cuspy density profiles, instead of the density cores derived from rotation curves (e.g. de Blok et al 2008, Swaters et al 2009). Also simulations have difficulties to form large galaxy disks, since the angular momentum of baryons is lost against massive dark haloes (e.g. Navarro & Steinmetz 2000), and the missing satellites problem is more acute as spatial resolution is increased (Diemand et al 2007). Possible solutions of these problems may reside in the detailed physics of the baryonic component, that is not presently reproduced realistically enough in simulations: feedback from star formation, supernovae and galaxy-scale winds, or from energy released by the central active nucleus, are currently investigated. In parallel, it is interesting to explore what are the constraints on the MOND phenomenology, due to galaxy dynamics, confronted to observations.

II- CONSTRAINTS ON MOND FROM GALAXY DYNAMICS AND OBSERVATIONS

Reproducing the kinematics of galaxies is not sufficient to test a model. More constraints can be obtained from the stability, evolution and formation of galaxies. In MOND, disks are totally self-gravitating, and not embedded in dark spheroidal halos, which usually do not rotate and are not subject to instabilities. Disks could then be more unstable, however, gravity tends to be proportional to the square root of mass only, which might have a limiting effect. Only simulations can provide an insight on the dominating effect.

II-1. Determination of the interpolation function

In a large number of nearby galaxies, the best interpolating function $\mu(x)$ is the standard form $\mu(x) = x /(1+x^2)^{1/2}$, (e.g. Sanders & Verheijen 1998, Milgrom & Sanders 2007), however, the simple one, $\mu(x) = x /(1+x)$ has also been used, being consistent with the TeVeS (Tensor, Vector, Scalar) relativistic extension of MOND (Bekenstein 2004). The simple function gives also good fits, but with a lower mass-to-light ratios for the stellar component (Sanders & Noordermeer 2007). In the Milky Way, however, it appears that the standard form has some difficulties to fit all the data, and the simple form is better, although another third form would be optimum (Famaey & Binney 2005). It is to be kept in mind, however, that being inside the Galaxy, makes distances and actual rotating velocities highly uncertain. One fact is certain, the Milky Way is dominated by baryons all across the visible disk, contrary to the predictions of the CDM model.

II-2 Escape Velocity

The potential around a galaxy of mass M in the pure MONDian regime is logarithmic $\Phi(r) = (GMa0)^{1/2} \ln r$, and there should not be any escape possible. However, a galaxy is never totally isolated, and the companions provide an external field, which modifies totally the shape of the potential. This is called the External Field Effect (EFE). When solving the Poisson equation for MOND, div ($\mu(x)$ **g**) = $-4\pi G\rho$ (X, Y, Z), with $x=g/a0$, and assuming an external field ge in the X direction, for instance, it is possible to define an equivalent internal potential Φ_{int}, which at large radii, is equivalent to a dilatation Δ in the direction transverse to the external field):

$$\Phi_{int} = -GM/\mu_m \,(X^2 + (1+\Delta)\,(Y^2+Z^2))^{-1/2}$$

Here μ_m is the interpolation function corresponding to the acceleration of the centre of mass. This allows to compute the escape velocity. In the case where $g \ll ge \ll a0$, the shape of the potential has retrieved a Keplerian dependence, with the renormalization of the gravitation constant G into Ga0/ge.

The application has been done to the Milky Way, where recent observations have constrained the escape speed between 498 and 608 km/s (Smith et al 2007). The most important external field effect comes from Andromeda, with ge =a0/100. Simulations have been done with the Besançon mass model of our Galaxy, and provide a good agreement, as shown in Figure 2 (Wu et al 2007).

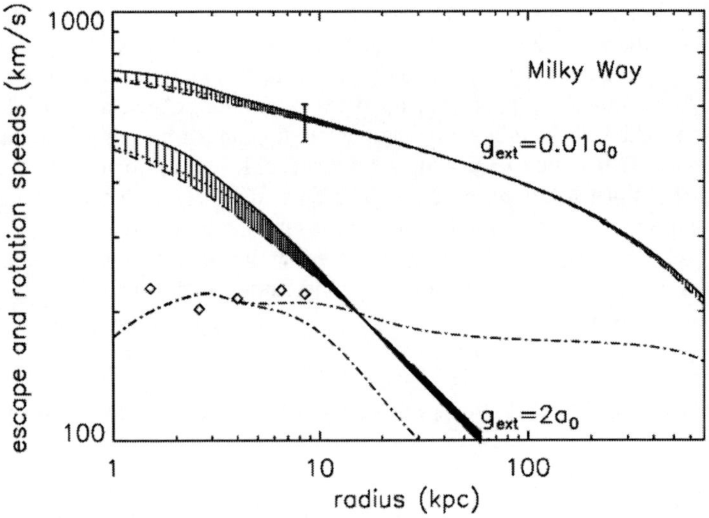

FIGURE 2. The influence of a weak (0.01a0) and strong (2a0) external field is computed on the Milky Way Galaxy. The full lines and hashed zones are the escape speed, for various field directions, compared with the error-bar measured by RAVE (Smith et al 2007). The dash-dot lines are the rotation velocity, compared by the symbols from Caldwell & Ostriker (1981). The weak field provides a good fit to the observations (from Wu et al 2007).

In addition, the EFE has some unexpected consequences, such as disk precession. If the acceleration ge can be considered as constant in amplitude and direction during the rotational period of the central disk, then it implies a gravitational torque and precession around this direction. This is completely absent in the Newton case, where only the tidal effect, and therefore the variation of ge is felt over the disk. This is due to the non-linearity of MOND, which violates the strong equivalence principle. Simulations show that the differential precession over the disk provokes warps (Tiret & Combes, in prep), and this could provide a new interpretation for isolated warped galaxies (see also Brada & Milgrom 2000).

II-3 Stability of galaxy disks: spirals and bars

Spiral waves and bars are the motor of evolution. While CDM dark haloes at first stabilise galaxy disks, they can also provide amplification of bars, since they accept angular momentum: bars are negative momentum waves, they grow through outward transfer of angular momentum. It is therefore interesting to test the stability and bar frequency in both hypotheses.

N-body simulations have resolved the Poisson equation for spiral disks, through a multi-grid algorithm (Tiret & Combes 2007). Initial conditions in phase-space for the visible galaxies were exactly the same, for the Newtonian and MOND cases. Once the baryonic disk had been selected, and provided a realistic rotation curve in MOND, then the dark matter radial distribution was computed, in the Newtonian case, in order to provide exactly the same kinematics, i.e. the same phase-space distribution for the baryons. Figure 3 shows the different results found, in the case of pure stellar systems. With CDM, the bar appears later, and can reform after the first bar is weakened due to the peanut formation (vertical resonance). With MOND, the bar appears quickly through angular momentum transfer to the outer disk, and stays longer, but it does not re-appear, by lack of material to accept more angular momentum. Besides, the pattern speed of the bar behaves very differently: in CDM, the bar is slowed down by dynamical friction against the halo, while the pattern speed stays constant in MOND. The latter is more consistent with observations, that reveal fast bars, through the position of resonances.

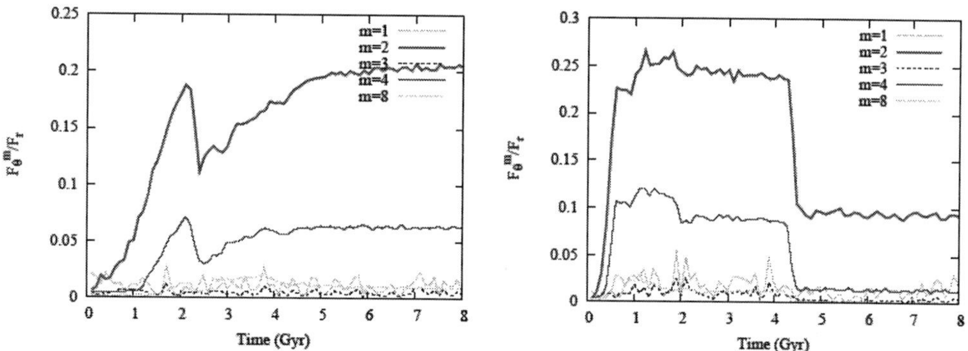

FIGURE 3. Strength of the bar, developed in an Sa galaxy stellar simulation, measured by its Fourier harmonics m=2,3,4 and 8, for the CDM-Newton model (left) and MOND (right). The bar settles earlier in MOND, and stays longer, but after its drop at 4.5 Gyr, it does not develop again, as in the CDM (cf Tiret & Combes 2007).

The slowing down of the bar in the CDM case is accompanied by several peanut formations. The vertical resonance occurs at larger and larger radii, while the bar pattern speed decreases. In MOND, the peanut does not move in radius.

The statistics of the bar strength, taking into account all galaxy models across the Hubble sequence, have been compared with observations. When the gas component is included, a large variety of situations are encountered (gas inflows can destroy bars, external gas accretion can reform them, Bournaud & Combes 2002, Combes 2008), and both models are compatible to the data. Gas radial flows due to the bar gravity torques are responsible to the peculiar morphology of spiral disks: gas piles up at resonances, and form new stars in rings or pseudo-rings. Simulations show that these dynamical phenomena in MOND compatible with observations (cf Figure 4).

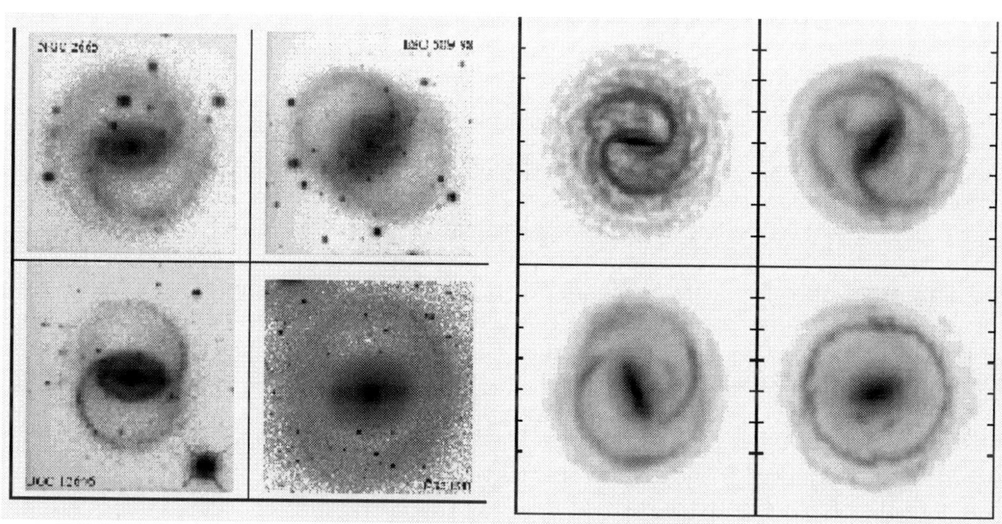

FIGURE 4. Formation of resonant rings through the bar in MOND simulations with gas and star formation. The 4 images at left are particular galaxies to confront with the 4 simulated images at right. MOND can reproduce realistically the morphology of galaxy disks (cf Tiret & Combes 2008a).

II-4 Interactions of galaxies, dynamical friction

Galaxy interactions and mergers are a fundamental element of the hierarchical scenario of galaxy formation. At large distance, the interactions are likely to occur in the MOND regime, and extended dark haloes are essential to absorb the orbital angular momentum for two galaxies to merge. Ciotti & Binney (2004) notice that force fluctuations are larger in the MOND regime, and from a perturbation estimation of the dynamical friction time, conclude that

dynamical friction should be much more efficient in MOND, for instance for stellar clusters spiralling inward or bar slowing down. Mergers should also occur more quickly. However, numerical simulations reveal the opposite effect. Mergers time are much longer in MOND. This has been obtained by merging dissipationless spherical systems with a polar code (Nipoti et al 2007), and also merging spiral galaxies (disks with gas) like the Antennae system, the prototype of major mergers (Tiret & Combes 2008b).
Figure 5 shows that MOND is able to reproduce the Antennae as well as the CDM model. Simulations are done here with an adaptive mesh, easy to implement with multi-grid methods. Dynamical friction is very slow in MOND, since galaxies are not embedded in massive spheres of particles, able to accept the orbital angular momentum. A short merging time-scale, as short as the CDM, is possible, only for nearly radial orbits.

FIGURE 5. Simulations of the Antennae galaxies with CDM-Newton at left, and MOND at right. The morphological details have not been fitted, but the main features and tidal tails are reproduced (cf Tiret & Combes 2008b).

Why then dynamical friction has been predicted analytically to be stronger with MOND than in the ENS (Equivalent Newtonian System) with dark matter (Ciotti & Binney 2004)? Both galaxy interactions or bar slowing down have shown the opposite. This must be due to the approximations done of weak perturbations. Force fluctuations are indeed stronger in MOND, but when a massive body (either a companion, or a galactic bar) has to be slowed down, the particles in the outer disk of galaxies are not sufficient to absorb large amounts of momentum. The reservoir of particles, assumed infinite in the case of a weak perturbation, is insufficient. In the CDM model, the reservoir of dark matter particles is found in the assumed dominating CDM component (cf Nipoti et al 2008).

Merger induced starbursts degeneracy

In the standard model, dynamical friction on DM particles is very efficient, and major mergers between two spiral galaxies in a parabolic orbit are expected to occur in one or two passages. An intense starburst is associated to the final phases, and the number of starbursts is thought to count the number of mergers. In the MOND model, many passages in binary galaxies will be required before the final merging. However, a starburst may be triggered at each passage (e.g. Di Matteo et al 2007). The number of starburst, as a function of redshift, could then be similar, and cannot discriminate the two models. The number of apparently merging galaxy in the sky is also degenerate. There could be a limited number of long-lived mergers, or many more short-live mergers.

II-5 Formation of Tidal Dwarfs

Major mergers between spiral galaxies are frequently observed with young dwarf galaxies at the extremity of their tidal tails, called Tidal Dwarf Galaxies (TDG). In the CDM model, they are difficult to form, and require very extended DM distribution (Bournaud et al 2003). In MOND, the exchange of angular momentum occurs within the disks, which sizes are inflated. It is then much easier with MOND to form TDG. The formation of new galaxies out of tidal debris is the occasion to test the missing mass problem. They are not expected to drag enough CDM, and should present no dark matter problem. The rotation curve of three TDG in the NGC 5291 ring system however, reveals the presence of dark matter (Bournaud et al 2007). A solution with the standard model is to resort to dark baryons in the form of cold molecular gas (e.g. Pfenniger et al 1994). These TDG are well explained in the MOND phenomenology (Gentile et al 2007, Milgrom 2007). However, three TDG are not yet enough statistically to consolidate the result, inclinations on the plane of the sky are uncertain, and selecting a higher inclination eliminates the missing mass. Many other TDG should be observed to randomize the uncertainties.

II-6 MOND and the dark baryons

Since stars and visible gas in galaxies represent only 6% of all baryons, a large fraction of dark baryons must exist outside the galaxies in filaments, and certainly a small fraction of them in galaxies under a cold gas form. Is MOND compatible with the existence of dark gas in galaxies? What fraction of dark gas provides the best fit to the rotation curves? A recent modelling of about 50 rotation curves, introducing dark gas and allowing a0 to vary accordingly, has shown that the optimum gas fraction was c=M(dark)/MHI= 2-3 (Tiret & Combes 2009). Figure 6 shows the quality of the fits for different values of this ratio, for the dwarf Irr NGC 1560. The fact that there is some sort of degeneracy here, comes from the long known observation that dark matter and HI surface densities are proportional in spiral galaxies (e.g. Hoekstra et al 2001).

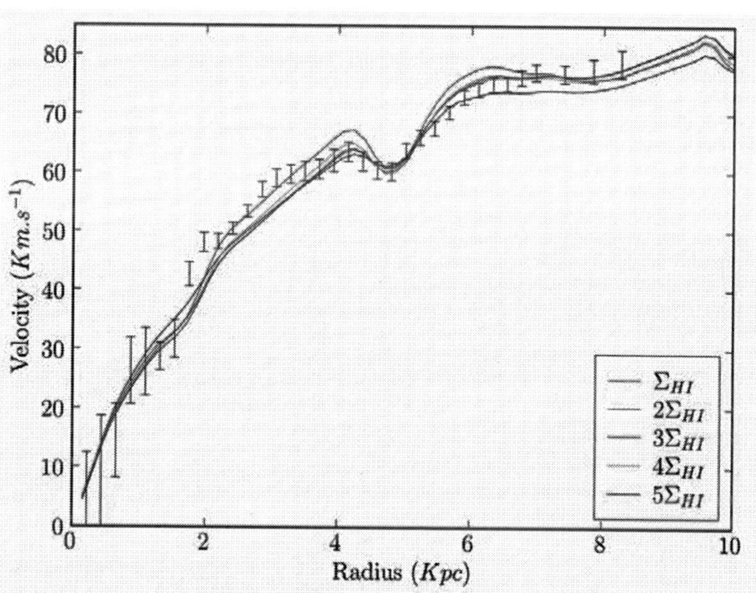

FIGURE 6. Several models for the rotation curve of the dwarf Irregular galaxy NGC 1560. The various curves correspond to the hypothesis that the total gas is either only the HI observed (Σ_{HI}), or a multiple of it, taking into account the dark molecular gas. In each case, the value of a0 is lowered accordingly (cf Tiret & Combes 2009).

II-7 MOND and Ellipticals

Elliptical galaxies are hot systems, with little or no cold gas, and determining the amount of missing mass is much more difficult than for spirals. Several tracers have been used, such as X-ray gas, globular clusters, planetary nebulae, which revealed a possible dearth of dark matter (Romanowsky et al 2003). However, the velocity dispersion could be decreasing only because of radial orbits in the outer parts. In particular, a merger between two spirals is expected to form an elliptical, with radial orbits in the merging direction (Dekel et al 2005). In general, the velocity anisotropy $\beta = 1 - \sigma^2_\theta/\sigma^2_r$ introduces a strong degeneracy, β can take large negative values (circular orbits), =0 (isotropic) to 1 (radial orbits). Figure 7 shows some models in the frame of MOND for a well studied elliptical NGC 3379.

II-8 Missing mass at large scale from satellites in SDSS

There is not a lot of tracers of the dark matter at larger scale. Recently Klypin & Prada (2009) have computed statistically the velocity dispersion of satellites around early-type galaxies, in the Sloan Survey (SDSS). There is only one or zero satellite for each galaxy in average, but thousands of galaxies. Again, the anisotropy of velocities introduce degeneracy in the models, CDM and MOND can both reproduce the data, with different values of β (Figure 7, see Angus et al 2008), however the CDM standard model with cusps is not a good fit.

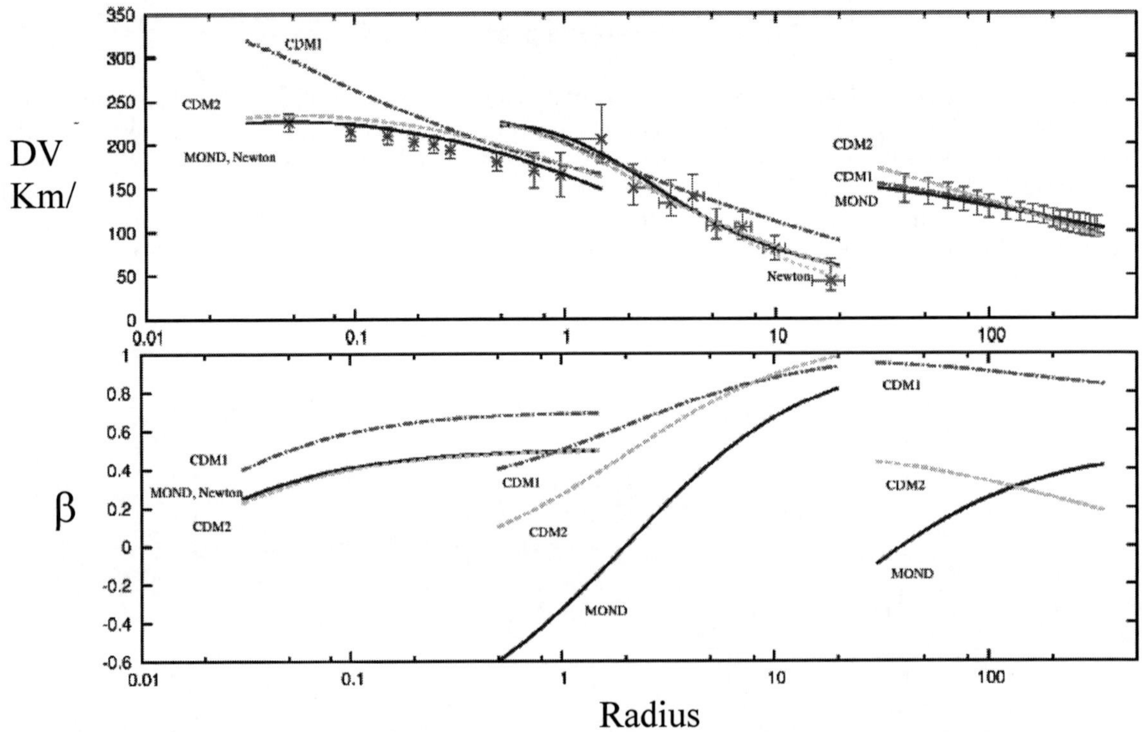

FIGURE 7. Velocity dispersion around the elliptical galaxy NGC 3379 (data from Shapiro et al 2006 in the centre, and Douglas et al 2007 at 10-30kpc), and on the same graph, large-scale mass distribution statistically derived from satellite data around ellipticals (data from Klypin & Prada 2009). The curves correspond to the various models: Newton model without DM, two different radial distributions of CDM (CDM1 with cusps, and CDM2 with cores), and MOND. Fits can be obtained in varying the velocity anisotropy β, shown in the bottom (cf Tiret et al 2007).

II-9 Large scale structure

Cosmological simulations with the MOND gravity have been explored, to test its ability to reproduce the large-scale structure formation, which is one of the best successes of the CDM theory. Previous works had made the strong approximations that the Newton and MOND acceleration fields were parallel (Nusser 2002, Knebe & Gibson 2004). They started from a cosmological Newtonian with CDM universe, assumed that the critical a0 was constant, and found that MOND produced as much clustering. Releasing the parallel approximation, Llinares et al (2009) have carried out new simulations, assuming Newtonian initial state, and a critical acceleration a0 varying with time, proportional to the scale factor of the universe. Their simulations yield comparable results between MOND and the standard model for the large-scale structure, with even more clustering than with the parallel approximation. The cosmological models tested are open with only baryons (without dark energy), and only gravity forces are considered (no hydrodynamics). There is still much freedom and many models to explore, within the MOND frame.

III- CONCLUSION

Recent models and simulations demonstrate that MOND phenomenology is able to reproduce observations of galaxy dynamics, and solve the problems of CDM at galaxy scales. The bar frequency is compatible with what is observed, and the number of mergers/starbursts is degenerate. Dynamical friction is much smaller than in the CDM model, and this could explain the frequency of compact groups.
However, MOND encounters remaining missing mass problems at cluster scale, which are not yet resolved, although neutrinos or dark baryons have been invoked.
The exploration of bar formation and galaxy interactions in MOND has suggested that more observational tests could be carried out to better constrain the various MOND models. In particular the dynamics of tidal debris should be studied statistically.

REFERENCES

Angus, G. W., Famaey, B., Tiret, O., Combes, F., Zhao, H. S. : 2008 MNRAS 383, L1
Bekenstein J.: 2004 PhRvD 70h3509
Bournaud F., Combes F.: 2002 A&A 392, 83
Bournaud F., Duc P-A., Masset F. : 2003, A&A 411, L469
Bournaud F, Duc, P-A., Brinks E. et al: 2007, Science 316, 1166
Brada R., Milgrom M.: 2000 ApJ 531, L21 and 541, 556
Caldwell J.A.R., Ostriker J.P.: 1981, ApJ 251, 61
Ciotti L., Binney J.: 2004, MNRAS 351, 285
Combes F: 2008, in "Pattern Speeds along the Hubble Sequence", ed. E. Corsini and V. Debattista, pub in "Memorie della Societa` Astronomica Italiana" (astro-ph/0811.0153)
Dekel, A., Stoehr, F., Mamon, G. A. et al: 2005 Nature 437, 707
de Blok, W. J. G., Walter, F., Brinks, E.. et al.: 2008 AJ 136, 2648
Diemand, J., Kuhlen, M., Madau, P.: 2007 ApJ 667, 859
Di Matteo P., Combes F., Melchior A-L, Semelin B.: 2007, A&A 468, 61
Douglas N.G., Napolitano N.R., Romanowsky A.J., et al: 2007, ApJ 664, 257
Famaey B., Binney J.: 2005 MNRAS 363, 603
Gentile, G., Famaey B., Combes F., Kroupa P., Zhao, H. S., Tiret O.: 2007A&A 472, L25
Hoekstra H., van Albada T. S., Sancisi R.: 2001 MNRAS 323, 453
Klypin A., Prada F.: 2009 ApJ 690, 1488
Knebe A., Gibson B.K.: 2004 MNRAS 347, 1055
Llinares, C., Zhao, H-S., Knebe, A. 2009 ApJ 695, L145
McGaugh S.S.,Schombert J.M., Bothun G.D., de Blok W.J.G.: 2000, ApJ 533, 99
Milgrom M.: 1983, ApJ 270, 365
Milgrom M.: 2007 ApJ 667, L45
Milgrom M.: 2008 NewAR 51, 906
Milgrom M., Sanders R.H.: 2007 ApJ, 658, 17
Navarro, J.F., Steinmetz, M.: 2000, ApJ 528, 607
Nipoti, C., Londrillo, P., Ciotti, L. 2007 MNRAS 381, L104
Nipoti, C., Ciotti, L., Binney, J., Londrillo, P.: 2008 MNRAS 386, 2194
Nusser A.: 2002 MNRAS 331, 909
Pfenniger D., Combes F., Martinet L.: 1994 A&A 285, 79
Romanowsky, A.J., Douglas, N.G., Arnaboldi, M. et al: 2003 Science, 301, 1696
Sanders R.H., Verheijen M.A.W.: 1998 ApJ 503, 97
Sanders R.H., McGaugh S.: 2002, ARAA 40, 263
Sanders R.H., Noordermeer E.: 2007 MNRAS 379, 702
Shapiro K.L., Cappellari M., de Zeeuw T. et al : 2006, MNRAS 370, 559
Smith M.C., Ruchti, G. R., Helmi, A. et al.: 2007 MNRAS, 379, 755
Swaters R. A., Sancisi, R., van Albada, T. S., van der Hulst, J. M.2009 A&A 493, 871
Tiret O., Combes F.: 2007, A&A 464, 517
Tiret O., Combes F., Angus, G. W., Famaey, B., Zhao, H. S. : 2007 A&A 476, L1
Tiret O., Combes F.: 2008a, A&A 483, 719
Tiret O., Combes F.: 2008b ASPC 396 , 259
Tiret O., Combes F.: 2009 A&A 496, 659
Wu, X., Zhao, H-S., Famaey, B. et al: 2007, ApJ 665, L101

Back Reactions and Dark Energy

Edward W. Kolb

Department of Astronomy and Astrophysics, Enrico Fermi Institute, and Kavli Institute for Cosmological Physics, University of Chicago, Chicago, IL 60637-1433

Abstract. Cosmological observables are used to construct cosmological models and cosmological models are used to interpert cosmological observations. Statements that the expansion of the Universe is accelerating or that dark energy exists are made within the framework of a particular cosmological model. If the cosmological model is inappropriate or incomplete, the interpretation of the observations may be wrong. In particular, I will discuss the possibility that the interpretation of observations within the framework of a homogeneous cosmological model may be inadequate and lead to mistaken conclusion that the universe is accelerating and dark energy exists.

Keywords: Dark energy, backreactions
PACS: 98.80.Cq

INTRODUCTION

In the last decade or so remarkable progress has been made in measuring cosmological parameters to unprecedented accuracy. Parameters such as the Hubble constant, $H_0 = 72 \pm 8$ km s^{-1} Mpc^{-1} [1], the temperature of the cosmic background radiation (CBR), $T_0 = 2.728 \pm 0.004$ K [2], and many other parameters, are now known to impressive precision. Of course, we don't invest so much time and effort determining cosmological parameters because of an occult interest in numerology, but rather, because the parameters are necessary inputs for the task of constructing a standard cosmological model.

In turn, we do not construct cosmological models simply to make predictions to compare with observations, but rather, to guide us to, in the words of Einstein [3], "a deeper and more consistent comprehension of the physical and astronomical facts."

The precision cosmological measurements have lead to the latest cosmological model, usually called the *standard cosmological model,* the *concordance model*, or simply ΛCDM, where Λ indicates the inclusion of Einstein's cosmological constant (or more generally, dark energy), and CDM stands for cold dark matter. The model is but one possible realization of homogeneous and isotropic cosmological solutions to Einstein's equations. (I will refer to homogeneous/isotropic models as Friedmann–Lemaître–Robertson–Walker (FLRW) models.) The question yet to be answered is whether today's standard model will lead us to a deeper and more consistent comprehension of the physical and astronomical facts. Aspects of this cosmological standard model relevant for the purposes of this talk are indicated in Figure 1. The most remarkable feature of the standard cosmological model is that it seems capable of accounting for all cosmological observations; *i.e.,* it seems to work!

There are several interesting features of the ΛCDM model illustrated in Fig. 1. First, while the early universe was radiation dominated, at present the radiation energy density is only 0.005% of the total. The figure also illustrates why chemistry is not very important: the chemical elements (by which I mean elements other than the simplest elements of hydrogen and helium) are only about 0.025% of the total. Neutrinos contribute a much larger fraction of the total mass-energy density, about 0.5%. The neutrino contribution is about as large as the contribution from stars. Most of the baryons in the Universe are not found in stars, but rather they are in the form of a hot intracluster gas of hydrogen and helium. This completes what we "see" (although there is only indirect evidence for the cosmic neutrino background).

The most striking feature of the present composition of the universe in the ΛCDM model is that 95% of the present universe is dark! Of the dark components, roughly 25% of the total mass-energy density is dark matter, associated with galaxies, clusters of galaxies, and other bound structures. The bulk of the mass-energy of the universe, about 70% in the standard ΛCDM model, is dark energy, which drives an accelerated expansion of the Universe.

The existence of dark matter, dark energy, an early epoch of cosmological inflation, and the origin of the baryon asymmetry all require physics beyond the standard model of particle physics. This is one reason cosmology is of such interest to particle physicists. But this is nothing new: on December 9, 1915, Einstein's commented to

CP1241, *Invisible Universe: Proceedings of the Conference,* edited by J.-M. Alimi and A. Füzfa

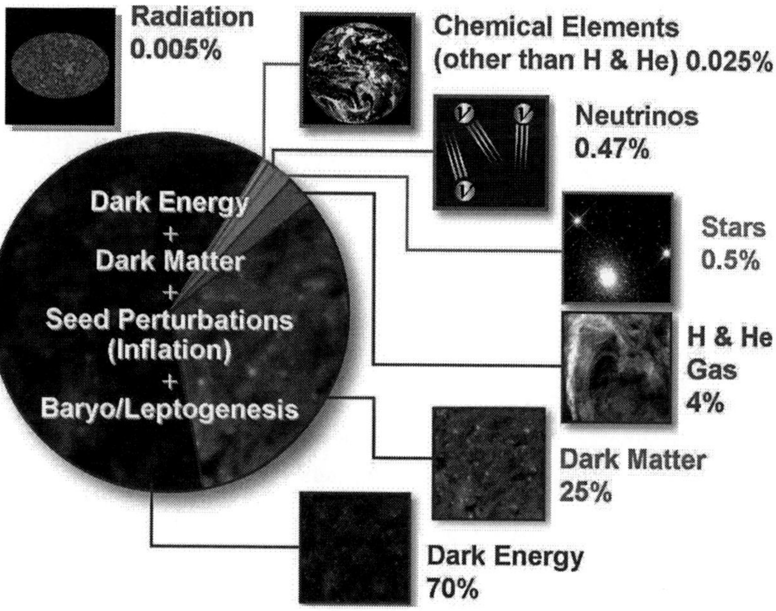

FIGURE 1. Elements of the standard ΛCDM cosmological model. Illustrated by the pie is the fraction of the mass-energy of the Universe in various components. Also noted are some other necessary parts of the standard model, namely seed perturbations and baryo/leptogenesis.

Arnold Sommerfeld, "How helpful to us is astronomy's pedantic accuracy, which I used to secretly ridicule." While Einstein was referring to precision measurements of the advance of the perihelion of Mercury, now we find precision measurements helpful to us because dark energy, dark matter, inflation, and baryo/leptogenesis seem to point to physics beyond the standard model of particle physics.

The observations and phenomena that lead to the assumptions of dark matter and dark energy are real, but that does not necessarily imply that dark matter and dark energy as usually envisioned are real. Dark matter and dark energy are good names for the phenomena. But naming is not explaining!

DARK ENERGY OR A FAILUE OF THE FRIEDMANN EQUATION

The most important point regarding dark energy (and, for that matter, the acceleration of the expansion of the universe), central to the subject of this talk, is that *all* evidence for dark energy is *indirect*, because the only effect of dark energy is on the expansion history of the Universe.

The expansion rate of the Universe is the most fundamental quantity in cosmology. In the standard cosmological model the expansion rate is determined from the 00-component of the Einstein equations, $G_{00} = \kappa T_{00}$, where $\kappa = 8\pi G$. In FLRW models, $G_{00} = 3H^2 + k/a^2$, where $H = \dot{a}/a$ is the expansion rate (here, a is the Robertson-Walker scale factor and $k = \pm 1$ or 0, depending on the geometry). With the assumption of a perfect-fluid stress tensor, $T_{00} = \rho$, where ρ is the mass density.

The stress tensor is assumed to consist of several components with different equations of state. For any component i, the equation of state for that component is defined as $w_i = p_i/\rho_i$. If w_i is constant, the energy density in component i evolves with redshift $z \equiv a_0/a - 1$ as $\rho_i(z) = \rho_i(0)\,(1+z)^{3(1+w_i)}$, where a_0 is the present value of the scale factor. Then the expansion rate as a function of redshift can be expressed as

$$H^2(z) = H_0^2 \left\{ \Omega_\Lambda \exp\left[\int_0^z \frac{dz_1}{1+z_1} 3\left[1 + w_\Lambda(z_1)\right] \right] + \Omega_k (1+z)^2 + \Omega_M (1+z)^3 + \Omega_R (1+z)^4 \right\}. \quad (1)$$

Here, Ω_i is the ratio of the present energy density in component i compared to the critical density $\rho_C = 3H_0^2/\kappa$. The subscript "k" indicates a spatial-curvature contribution ($w_k = -1/3$), "M" indicates a matter component ($w_M = 0$),

TABLE 1. The relationship of observables to the time evolution of the expansion rate of the Universe. The coordinate distance $r(z)$ depends on the time evolution of $H(z)$ through Eq. (3).

Observable	Notation	Definition	Value
luminosity distance	$d_L(z)$	$(\text{Luminosity}/4\pi\text{Flux})^{1/2}$	$d_L(z) \propto r(z)(1+z)$
angular-diameter distance	$d_A(z)$	Physical size/Angular size	$d_A(z) \propto r(z)/(1+z)$
volume element	$dV(z)$	$\sqrt{h}\,drd\Omega$	$dV(z) = \dfrac{r^2(z)}{\sqrt{1-kr^2(z)}}drd\Omega$
age of the Universe	$t(z)$	time from $z=\infty$ to z	$t(z) = \int_z^\infty \dfrac{dz'}{(1+z')H(z')}$

"R" indicates a radiation component ($w_R = 1/3$), and "Λ" represents dark energy. In the event that the function $w_\Lambda(z)$ in the last term is independent of z, the last term in $H^2(z)$ becomes $\Omega_\Lambda(1+z)^{3(1+w_\Lambda)}$. The equation of state for a cosmological constant is $w_\Lambda = -1$.

It appears that the expansion rate as a function of z depends on four constants $\{\Omega_k, \Omega_M, \Omega_R, \Omega_\Lambda\}$ and one function $w_\Lambda(z)$. In reality, the situation is much more predictive. First, since $H^2(0) = H_0^2$, it follows that $\sum_i \Omega_i = 1$, removing one constant.

The value of Ω_k is well determined by WMAP to be small, $\Omega_k = -0.0026^{+0.0066}_{-0.0064}$ at the 68% confidence level [4], effectively removing yet another constant. The value of Ω_R is also well determined by CBR measurements (5×10^{-5}) removing a further constant. One may then make the less secure assumption that Ω_M is "known," or more profitably, use observational constraints to marginalize over the Ω_i's, leaving only the function $w_\Lambda(z)$ undetermined. The function must be determined through its effect on the expansion history of the Universe, which is manifest through various cosmological observables.

Many cosmological observables depend on the expansion history of the Universe. They often depend on the expansion history through the coordinate distance r of a source of redshift z. Recall that the Robertson-Walker metric can be written in the form

$$ds^2 = dt^2 - a^2(t)\left[\frac{dr^2}{1-kr^2} + r^2 d\Omega^2\right], \tag{2}$$

where here $d\Omega$ is the angular differential and r is the comoving "radial" coordinate. The radial coordinate of a source at redshift z is determined by integrating the null geodesic ($ds^2 = 0$) to obtain

$$r(z) = \left.\begin{matrix}\sin\\ 1\\ \sinh\end{matrix}\right\}\left[\int_0^z \frac{dz'}{H(z')}\right], \tag{3}$$

where sin, 1, sinh obtains for $k = +1,\ 0, 1$, respectively.

Now let us turn to the observables defined in Table 1. One can see that they all depend on the time evolution of the expansion rate, hence to the properties of the stress tensor, hence to a new fluid characterized by w.

The best evidence for dark energy comes from luminosity-distance–redshift determinations using Type 1a supernovae as standard candles of known (or at least calibrated) luminosity [5]. A recent large survey is the Supernova Legacy Survey (SNLS) [6]. The results for the observed magnitude of supernovae as a function of redshift is shown in Fig. 2. Without performing a statistical analysis, it is apparent that the data fit the ΛCDM model and the Einstein–de Sitter (EdS) model is "observationally challenged."

It is important to realize that dark energy experiments do not actually measure dark energy directly. They determine the expansion history of the Universe by measuring cosmological observables sensitive to the expansion history. What is determined is that the EdS model does not fit the data. Any information about the new component to the stress tensor depends on the assumptions you put in. You have to input a cosmological model and compare it to the data. An example of such an analysis is shown in Fig. 3. To construct this figure, it was assumed that the expansion history of the Universe is described by a FLRW model with dark energy described by a fluid with $w = -1$, *i.e.,* Einstein's cosmological constant. Also employed are priors on cosmological parameters such as H_0.

If one adopts the model used in Fig. 3, then there is a cosmological constant with an equivalent energy density of $\rho_V \sim 10^{-30}\text{g cm}^{-3}$. Henceforth I will refer to this as the *cosmoillogical* constant. There are two reasons for referring to this cosmo constant as illogical, rather than logical.

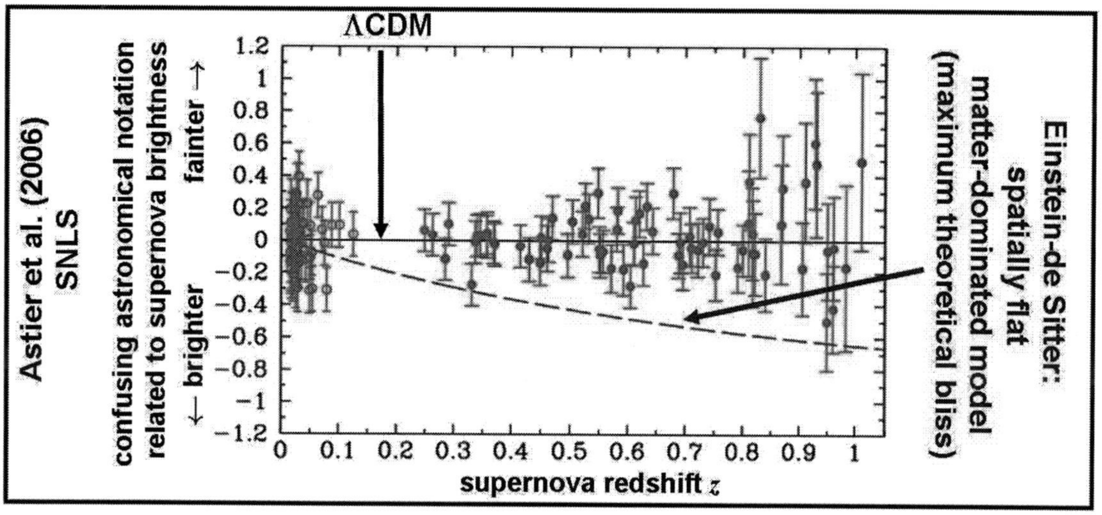

FIGURE 2. The results of the Supernova Legacy Survey [6]. Shown is the apparent magnitude of supernovae normalized to that expected in the standard ΛCDM model.

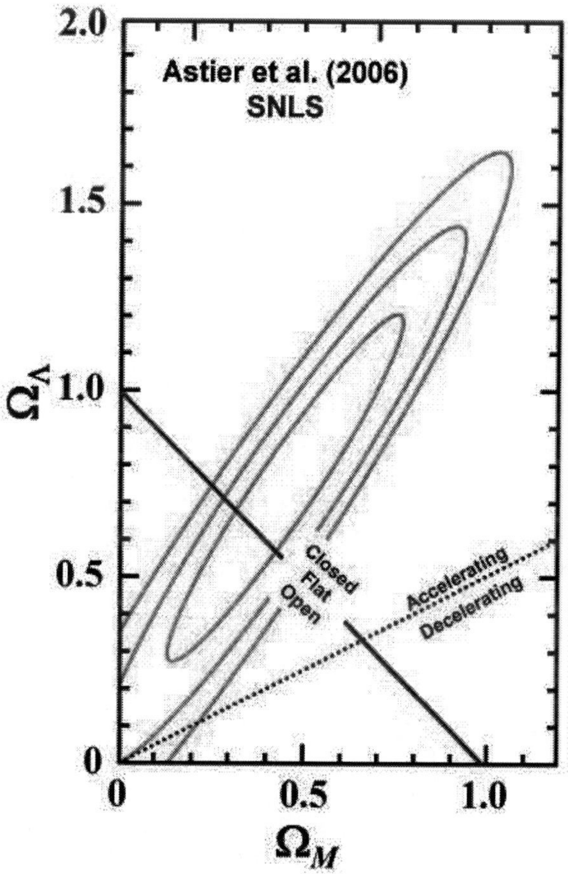

FIGURE 3. Values of Ω_Λ and Ω_M allowed at confidence levels of 68%, 90%, and 95% assuming a cosmological model with $w = -1$. The data used to place the limits are show in Fig. 2.

TABLE 2. The length scales and mass scales implied by a cosmoillogical constant with an energy density of 10^{-30}g cm^{-3}.

	ρ_V	$\Lambda = 8\pi G$
length scale $(\rho_V)^{-1/4}$	10^{-3}cm	10^{29}cm
mass scale $(\rho_V)^{1/4}$	10^{-4}eV	10^{-33}eV

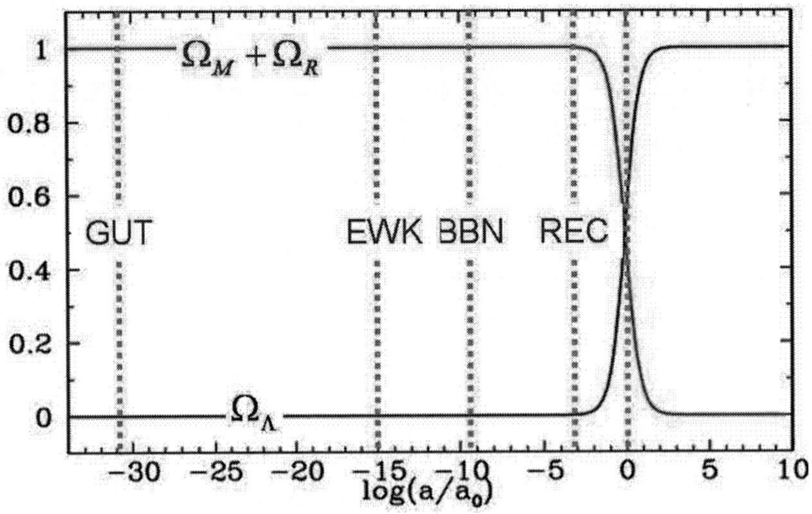

FIGURE 4. The relative contributions of $\Omega_M + \Omega_R$ and Ω_Λ as a function of the scale factor a.

First, the magnitude of the cosmoillogical constant is, well, illogical. Length scales and mass scales of the cosmoillogical constant are given in Table 2. Fundamental length scales and mass scales shown in Table 2 are, as I said, illogical.

A cosmoillogical constant is equivalent to, and indistinguishable from, contributions to the vacuum energy. It is only meaningful to consider them together as a vacuum energy. Among the contributions to vacuum energy is the fact that all fields are harmonic oscillators with a zero point energy. This should contribute a vacuum energy of $\rho_V = \Sigma_{\text{all particles}} \pm \int d^3k\,\sqrt{k^2+m^2}$, where the sign of the contribution depends on the spin-statistics of the particle. The individual integrals have quartic divergences, so formally the answer is infinity. (Unless for some reason–such as supersymmetry (SUSY)–there are exact cancellations between bosons and fermions.) An answer of infinity need not cause alarm because, after all, this is field theory. One might imagine introducing a cutoff Λ_C to the integral so that the integral is finite, and the individual contributions are $\pm\Lambda_C^4$. If you assume Λ_C is related to gravity, then this contribution to ρ_V is $M_{Pl}^4 = (10^{28}\text{ eV})^4 = 10^{112}\text{ eV}^4$. If you conjecture that Λ_C is related to the SUSY breaking scale (assumed here to be 1 TeV), then $\rho_V = M_{\text{SUSY}}^4 = (10^{12}\text{ eV})^4 = 10^{48}\text{ eV}^4$. These scales are far away from the observed value of $\rho_V = (10^{-4}\text{ eV})^4 = 10^{-16}\text{ eV}^4$.

Another contribution to vacuum energy comes from the vacuum energy associated with spontaneous symmetry breaking. Associated with spontaneous symmetry breaking of GUTs, SUSY, electroweak, and chiral symmetries are vacuum energies of 10^{100}eV^4, 10^{48}eV^4, 10^{45}eV^4, and 10^{32}eV^4, respectively. Once again, these scales are far away from the observed value of ρ_V.

A not unrelated issue is the timing of the epoch in the history of the Universe when the dark energy density is comparable to the density of matter and radiation. A graph of the relative contributions of $\Omega_M + \Omega_R$ and Ω_Λ is shown in Fig. 4. It turns out to be very curious that while in the distant past vacuum energy was "in the noise," and in the distant future matter and radiation densities will be "in the noise," while today we live at the very convenient epoch when they are just about the same value. More than curious, this fact is illogical.

Presently, a constant $w_\Lambda(z)$ seems to fit the data, and furthermore $w_\Lambda = -1$, the value of w_Λ for a cosmological constant, is allowed by the data. It is also true that since there is no apparent reason to exclude vacuum energy from the stress tensor, there is no reason *not* to imagine it enters the determination of the expansion rate. As mentioned, the problem is that the magnitude of the expected vacuum energy is very many orders of magnitude larger than expected. If the cosmological constant is part of nature, then some principle must be found to explain the much smaller-than-expected magnitude of Ω_Λ.

This situation has led many to consider alternate explanations of the observations: modified gravity, new long-range forces, new ultra-light scalar fields, extra dimensions, branes, bulk, Lorentz-invariance violation, action of the tooth fairy, *etc.* Some of these approaches are, in many respects, more drastic than a small cosmological constant. For instance, theories of modified gravity typically do not respect the principle of general covariance. Since we should not abandon our principles without good cause, it is tempting to retreat to the position that since a cosmological constant is adequate to account for the data, just assume that it exists and call it a day?

We resist the retreat because our goal in cosmology is more than just explaining the observations. The ingredients of a cosmological model must be deeply grounded in fundamental physics. Dark matter, dark energy, modified gravity, mysterious new forces and particles, *etc.,* unless part of an overarching model of nature, should not be part of a cosmological model. We may plant new ideas and features into cosmological models, but they will ultimately prove barren unless grounded in fundamental physics. No matter how well a concordance model fits the data, its ingredients must not be discordant with fundamental physics.

It is a good time to re-emphasize the important point made previously. The measurements *do not* tell us that there is dark energy. Rather, the observations only provide indirect evidence for dark energy. To elaborate on this, let me review the procedure.

1. Assume a model cosmology. In this case, the assumptions involve:
   (a) The FLRW model, which results in the Friedmann equation for $H(z)$.
   (b) The energy and pressure content of the Universe and how they scale with redshift.
   (c) Input or integrate over cosmological parameters such as H_0, Ω_B, etc.
2. Within the cosmological model, calculate observables such as $d_L(z)$, $d_A(z)$, $H(z)$, etc.
3. Compare to observations.
4. Discover that the model cosmology fits with dark energy, but not without dark energy.

So what the observations seem to tell us is that the observed $H(z)$ is not described by $H(z)$ calculated from the EdS model, but is well described by $H(z)$ described by the ΛCDM model.

Again, all evidence for dark energy is indirect!

So, what should we do? We can no longer just hide from the data. The evidence is overwhelming that $H(z)$ is not described the by the EdS model. Again, this is not just from the supernova projects, but from a variety of techniques.

So we don't know what the answer is, but we do seem to know something very valuable: We know an equation that doesn't work; namely,

$$G_{00}(\text{FLRW}) \neq 8\pi G T_{00}(w=0 \text{ perfect fluid matter}). \tag{4}$$

Let's exploit this non-equation and imagine how we can fix it.

First, we might try to add something to the right-hand side of the equation; a ΔT_{00} if you will. This ΔT_{00} is usually what we mean by dark energy.

Theorists employ two tools in constructing a ΔT_{00}. Now experimentalists believe that theorists don't know anything about tools, but that is not true. In my own case, I owned a British sports car for over 20 years (a 1965 Austin-Healey Sprite) and an Italian economy car (a 1974 Fiat 128) for more than 5 years. Anyone who has owned either of those marvels of engineering has had to use tools very often. In my experience, you only need two tools to fix anything: duct tape and WD-40. There are only two rules:

1. If something moves and it shouldn't, use duct tape.
2. If something doesn't move and it should, squirt it with WD-40.

The equivalent theoretical tools used for dark energy are the anthropic (or Landscape is you speak string) explanation,[1] and scalar fields (known as quintessence for this purpose).

The anthropic/Landscape/duct-tape idea is that there are many sources of vacuum energy. String theory is conjectured to have many (perhaps more than 10^{500}!) vacua. The different vacua have different values of the total vacuum energy density. Presumably most of them have a vacuum energy much, much larger (say, by a factor of 10^{100}) than

[1] The anthropic explanation is sometimes referred to as the anthropic *principle.* Of course, one can have principles instead of ideas.

the observed value. But don't worry, since although exponentially uncommon, vacua with vacuum energy as small as observed are hospitable for life while the more common values result in an inhospitable universe. The anthropic idea is that we should only consider cosmological models with values of the vacuum energy that are hospitable for life. Therefore, there is an anthropic selection effect.

So you see that the anthropic explanation is just like duct tape. There were some loose ends after a calculation. Like duct tape, the anthropic explanation need not be elegant or permanent, but it ties down the loose ends.

Now the quintessence idea borrows a concept from high-energy physics: whenever there is dynamics that is poorly understood, invent a scalar field.[2] One way to visualize the quintessence field is that there are many contributions to ρ_V. Some unknown principle or dynamics results in the minimum value $\rho_V = 0$, but we are not there yet–we are still evolving to the ground state.

Now let's turn to ideas for modifying the left-hand side of non-equation (4), adding a ΔG_{00}. The first class of ideas involves modifying gravity. The first type in this class involves extra dimensions.

1. The braneworld modifies the Friedmann equation [7]. In some models with branes the Friedmann equation does *not* arise from the 00-component of the Einstein equation, but rather from the Israel discontinuity condition as one crosses the brane.
2. In some extra-dimension models the gravitational force law is modified at *large* distances [8].
3. In models with branes the wavefunction of gravitons can leak into the bulk [9, 10]. This is a sort of "tired graviton" explanation for dark energy.
4. There are extra-dimension theories where gravity becomes repulsive at $R \sim$ Gpc [11].
5. All models of extra dimensions have "Kaluza–Klein" (KK) excitations. It could be that the $n = 1$ KK excitation of the graviton is very light, say with a mass $m \sim (\mathrm{Gpc})^{-1}$ [12].

Then there are theories (not necessarily based on extra dimensions) of the type that modify the low-energy gravitational action. The idea is that Einstein and Hilbert got it wrong in 1915, and rather than the Einstein-Hilbert action for gravity, $S = (16\pi G)^{-1}\int d^4x\sqrt{-g}R$, it is more general, say of the form $S = (16\pi G)^{-1}\int d^4x\sqrt{-g}f(R)$, where $f(R)$ is some function of, for instance, the Ricci scalar R [13]. While it is very reasonable to expect terms like R^2 or $R^{\mu\nu}R_{\mu\nu}$, which affect the theory in the ultraviolet, what seems to be required for dark energy are terms that affect the theory in the infrared. Terms like R^{-1} would work, but they lead to problems.

One of the things we have come to understand from all these approaches is how difficult it is to modify gravity on large scales without destroying agreement with observations on solar system scales, or leading to non-linearities, ghosts, or other theoretical unpleasantness.

In this talk I will explore a less traveled path, and consider the possibility that the Friedmann equation is not an adequate description of our inhomogeneous Universe [14, 15, 16, 17, 18, 19, 20, 21, 22, 23, 24, 25]. In this approach there is no need for dark energy, so no need for new long-range forces, modifications of general relativity, new ultralight particles, or anthropic reasoning. The phenomenon of inhomogeneities mimicking dark energy is sometimes referred to as "backreaction." In this scenario (very unpopular in most quarters), the Friedmann equation has large corrections due to inhomogeneities in the Universe. This idea is still under development, but it is my favorite (since I have worked on it a lot). I will turn to it now.

BACK REACTIONS

In cosmology, one usually models the evolution and observables associated with an inhomogeneous universe consisting of density $\rho(\vec{x})$, three-curvature ${}^3\mathscr{R}(\vec{x})$, and expansion rate $H(\vec{x})$ by employing a FLRW model of density $\rho = \langle\rho(\vec{x})\rangle$, and spatial curvature ${}^3\mathscr{R} = \langle{}^3\mathscr{R}\rangle$ (where $\langle\cdots\rangle$ denotes some suitably defined spatial average) and a fluid equation of state given by the local equation of state of the fluid(s) comprising the stress-energy tensor. One then assumes that the expansion history and cosmological observables are those obtained in the corresponding FLRW model. Inhomogeneities are thought of as developing in this fixed background solution according to perturbation theory. In other words, it is implicitely assumed that the FLRW is the only relevant background solution and there are no backreactions.

---

[2] One might think the fact that there is no known fundamental scalar field might temper the enthusiasm for introducing them.

In this approach, inhomogeneities are, therefore, simply neglected: one assumes that inhomogeneities will not affect this long-standing procedure without actually providing a quantitative averaging method to demonstrate it. This approach does not provide any guidance as to the scale upon which such models are supposed to be applicable, nor does it seriously examine the issues arising when we consider the relations between descriptions of the universe at different scales of inhomogeneity [26, 27, 28].

One proposed route to understanding the role of inhomogeneities in mimicking dark energy is by employing an averaging procedure. Our calculations allow us to study this proposal. First, let us review the averaging procedure.

A fundamental quantity in the analysis is the velocity gradient tensor, which is defined as

$$\Theta^i_{\ j} = u^i_{\ ;j} = \frac{1}{2} h^{ik} \dot{h}_{kj}, \tag{5}$$

where h^{ij} is the spatial metric and u^μ is the fluid four velocity. The tensor $\Theta^i_{\ j}$ represents the extrinsic curvature of the spatial hypersurfaces orthogonal to the fluid flow. It may be decomposed in terms of a trace term and a traceless tensor as

$$\Theta^i_{\ j} = \Theta \delta^i_{\ j} + \sigma^i_{\ j}, \tag{6}$$

where Θ is the volume-expansion scalar. The traceless tensor $\sigma^i_{\ j}$ is the shear.

The evolution equations for the expansion and the shear come from the space-space components of Einstein's equations (see *e.g.*, Ref. [29]). Combining the expansion evolution equation with the energy constraint gives the *Raychaudhuri equation*,

$$\dot{\Theta} + \frac{1}{3}\Theta^2 + 2\sigma^2 + \frac{1}{2}\kappa\rho = 0. \tag{7}$$

From the Raychaudhuri equation it is straightforward to verify that local fluid elements cannot undergo accelerated expansion. (This point was emphasized by Hirata and Seljak [30].) But, of course, as now appreciated, that point is irrelevant. Locally the expansion does not accelerate, but it is incorrect to assume that acceleration does not occur when the fluid is coarse-grained over a finite domain. The reason is trivial: the time derivative of the average of Θ and the average of the time derivative of Θ are not the same because of the time dependence of the coarse-graining volume.

There has been a lot of work recently regarding the averaging procedure. First, we take the original proposal and average at a fixed time. We will then average over the light cone.

Let us denote the coarse-grained value of a quantity $\mathscr{F}$ by its average over a spatial domain D:

$$\langle \mathscr{F} \rangle_D = \frac{\int_D \sqrt{h}\mathscr{F}\, d^3x}{\int_D \sqrt{h}\, d^3x}. \tag{8}$$

Following the work of Buchert [31, 32], we define a dimensionless scale factor

$$a_D(t) \equiv \left(\frac{V_D}{V_{D_0}}\right)^{1/3} ; \qquad V_D = \int_D \sqrt{h}\, d^3x, \tag{9}$$

where V_D is the volume of our coarse-graining domain (the subscript "0" denotes the present time). In the LTB model $\sqrt{h} = R^2(r,t_0)R'(r,t_0)/\sqrt{1+\beta(r)}$.

The coarse-grained Hubble rate H_D will be

$$H_D = \frac{\dot{a}_D}{a_D} = \frac{1}{3}\langle\Theta\rangle_D. \tag{10}$$

The smoothing procedure leads to the evolution equations for the coarse-grained scale factor [31, 32]:

$$\frac{\ddot{a}_D}{a_D} = -\frac{\kappa}{6}\left(\rho_{\text{eff}} + 3p_{\text{eff}}\right), \tag{11}$$

$$\left(\frac{\dot{a}_D}{a_D}\right)^2 = \frac{\kappa}{3}\rho_{\text{eff}}, \tag{12}$$

where the effective energy density and pressure terms are ($^3\mathscr{R}$ is the spatial curvature)

$$\rho_{\text{eff}} = \langle\rho\rangle_D - \frac{Q_D}{2\kappa} - \frac{\langle^3\mathscr{R}\rangle_{\mathscr{D}}}{2\kappa} \tag{13}$$

$$p_{\text{eff}} = -\frac{Q_D}{2\kappa} + \frac{\langle^3\mathscr{R}\rangle_{\mathscr{D}}}{6\kappa}, \tag{14}$$

and we have introduced the *kinematical backreaction*

$$Q_D = \frac{2}{3}\left(\langle\Theta^2\rangle_D - \langle\Theta\rangle_D^2\right) - 2\langle\sigma^2\rangle_D. \tag{15}$$

Finally, we can define an effective equation of state, $w_{\rm eff}$, to be $w_{\rm eff} = p_{\rm eff}/\rho_{\rm eff}$.

Here's the crucial point: While it is true that locally the expansion does not accelerate, it is incorrect to assume that acceleration can not occur when the fluid is coarse-grained over a finite domain. The reason is trivial: the time derivative of the average of Θ and the average of the time derivative of Θ are not the same because of the time dependence of the coarse-graining volume:

$$\langle\Theta\rangle_D^{\cdot} = \langle\dot{\Theta}\rangle_D + \langle\Theta^2\rangle_D - \langle\Theta\rangle_D^2 \geq \langle\dot{\Theta}\rangle_D\,. \tag{16}$$

Although $\langle\dot{\Theta}\rangle_D \leq -\frac{1}{3}\langle\Theta^2\rangle_D \leq 0$, the coarse-grained deceleration parameter $q_D \equiv -3\langle\Theta\rangle_D^{\cdot}/\langle\Theta\rangle_D^2 - 1$ is related to $\langle\Theta\rangle_D^{\cdot}$, which is not the same as $\langle\dot{\Theta}\rangle_D$.

While it would be naïve to assume that the averaging procedure outlined is directly related to observations, it would be equally naïve to assume that the averaging procedure is completely unrelated to observations. In the past couple of years there has been a lot of activity studying different averaging procedures. In the procedure discussed above the averaging was performed on a spatial hypersurface at constant time. Perhaps a better procedure would be light-cone averaging, or perhaps a covariant averaging prescription must be developed. Some ideas along these lines were presented in the parallel session of this conference. The reader is referred there.

It has often been stated that backreactions must have a small effect because the effect must be related to density perturbations on large scales, which we know to be small.

First of all, it has been claimed that there is an instability in the perturbative expansion [19, 20], so results based on perturnation theory are suspect. Secondly, the arguments do not seem to hold in Lemaître-Tolman-Bondi (LTB) models [33, 34, 35].

There have been several attempts to construct LTB models to reproduce the luminosity density–redshift observations [14, 36, 37, 38, 39, 40, 41, 42, 43, 44, 45, 46, 47, 48, 49, 50]. Here I will just remark that while it is clear that such models can be constructed, they require our location to be at or very near the center of a spherically symmetric Universe. I regard them to be toy models to study how an fully inhomogeneous universe might mimic dark energy.

Here, I would just like to extract but one of the interesting results from such studies. Recently, in a work reported at this meeting, Célérier, Bolejko, Krasinski, and Hellaby reported on an LTB model that reproduces *exactly* both the luminosity distance as a function of redshift and the mass density as a function of redshift of ΛCDM models [51]. I would like to use this to address the argument that purports to prove that inhomogeneous models cannot mimic the effects of dark energy (see, *e.g.*, Refs. [52, 53, 54]). The argument might be paraphrased as follows: Perform an expansion of the FLRW metric to Newtonian order in potential and velocity but take into account the fully nonlinear density inhomogeneities. In the conformal Newtonian gauge, the metric will be of the form

$$ds^2 = a^2(\eta)[-(1+2\psi)d\eta^2 + (1-2\phi)d\mathbf{x}^2], \tag{17}$$

where η is conformal time and $\phi \simeq \psi$. The volume average of the equation then yields [53]

$$\left(\frac{\dot{a}}{a}\right)^2 = \frac{\kappa}{3}\langle\rho\rangle(1-5W+2K), \tag{18}$$

where W and K are the Newtonian potential and kinetic energy (per unit mass),

$$W = \frac{1}{2}\langle(1+\delta)\phi\rangle, \qquad K = \frac{1}{2}\langle(1+\delta)v^2\rangle. \tag{19}$$

Here, $\delta = \delta\rho/\langle\rho\rangle$ is the density perturbation in the matter rest frame and v is the peculiar velocity. The argument is that observationally both W and K are small, hence, inhomogeneities cannot mimic dark energy.

However the model constructed by Célérier *et al.* seems to belie this simple and compelling argument. If one interprets the observations of $\rho(z)$ of the LTB model in terms of the fiducial ΛCDM phenomenological background solution, one would conclude that the perturbations are very small (*viz.* zero), hence W and K zero, and inhomogeneities could not mimic dark energy. But, at least for $d_L(z)$ and $\rho(z)$, inhomogeneities do exactly that!

We point out that the very notion of peculiar velocities requires a background solution (a norm) about which to define "peculiarity" [55]. Also, Enqvist, Mattsson, and Rigopoulos [56] explicitly demonstrate how spherically symmetric perturbations of a dust Universe can lead to acceleration while the perturbations of the gravitational potential remain small on all scales. Perhaps the general lesson is that there are highly non-Gaussian inhomogeneities in the late Universe, and the coherence of structures causes small deviations in observables to sum to a large deviation.

We can venture a different way to frame the argument: If the observed density fluctuations, $\delta\rho(z)$, originated from the growth of perturbations in an FLRW cosmology, and the nonlinear perturbations on one scale did not affect the growth of structure or the evolution of the background solution on any other scale, and inhomogeneities do not show coherence or non-Gaussianity, and $\phi \simeq \psi$, then one can say that the interpretation of observations would yield a value of W and K that are too small to mimic dark energy.

This way of framing the argument makes manifest the fact that there are dynamical assumptions that enter, and it cannot simply be a matter that small peculiar velocities rule out the backreaction proposal. If any of the dynamical assumptions fail, the argument fails.

CONCLUSIONS

This talk has developed and discussed issues related to the interpretation of cosmological observables in constructing cosmological models and theories. Observations of $d_L(z)$ and $\rho(z)$ do not, by and of themselves, prove that there is dark energy or that the Universe is accelerating in the usual sense; they do so only if one assumes that the Universe is homogeneous and isotropic and the dynamics of the Universe are governed by general relativity. This talk discussed aspects of the program to develop understanding of how an inhomogeneous Universe might mimic dark energy. While consideration of the results of this investigation might lead one to conclude that there are no arguments that conclusively exclude the backreaction proposal, we are no closer to developing an inhomogeneous cosmology alternative to dark energy.

REFERENCES

1. W. L. Freedman et al., Ap. J. 553, 47 (2001).
2. D. J. Fixen, et al., Ap. J. 473, 576 (1996).
3. A. Einstein, in the foreword to *Dialogue Concerning the Two Chief World Sysyems*, translated by S. Drake (University of California Press, Berkeley, 1967, second edition).
4. E. Komatsu *et al.*, Ap. J. Suppl. 180, 330 (2009).
5. A. Riess et al., Astron. J. 116 1009 (1998); S. Perlmutter et al., Ap. J. 517, 565 (1999).
6. P. Astier et al. [The SNLS Collaboration], Astron. Astrophys. 447, 31 (2006).
7. P. Binetruy, C. Deffayet, U. Ellwanger and D. Langlois, Phys. Lett. B 477, 285 (2000).
8. C. Deffayet, G. R. Dvali and G. Gabadadze, Phys. Rev. D 65, 044023 (2002).
9. R. Gregory, V. A. Rubakov and S. M. Sibiryakov, Phys. Lett. B 489, 203 (2000).
10. G. R. Dvali, G. Gabadadze and M. Porrati, Phys. Lett. B 485, 208 (2000).
11. C. Csaki, J. Erlich, T. J. Hollowood and J. Terning, Phys. Rev. D 63, 065019 (2001).
12. I. I. Kogan, S. Mouslopoulos, A. Papazoglou, G. G. Ross and J. Santiago, Nucl. Phys. B 584, 313 (2000).
13. S. M. Carroll, V. Duvvuri, M. Trodden and M. S. Turner, Phys. Rev. D 70, 043528 (2004).
14. M.-N. Célérier, Astron. Astrophys. **353**, 63 (2000).
15. T. Buchert, JGRG **9**, 306, (2000).
16. D. J. Schwarz, "Accelerated expansion without dark energy," arXiv:astro-ph/0209584.
17. S. Rasanen, JCAP **0402**, 003 (2004).
18. E. W. Kolb, S. Matarrese, A. Notari, and A. Riotto, Phys. Rev. D **71**, 023524 (2005).
19. E. W. Kolb, S. Matarrese and A. Riotto, New J. Phys. **8**, 322 (2006).
20. A. Notari, Mod. Phys. Lett. A **21**, 2997 (2006).
21. S. Rasanen, JCAP **0411**, 010 (2004).
22. E. W. Kolb, S. Matarrese, A. Notari, and A. Riotto, "Primordial inflation explains why the universe is accelerating today," arXiv:hep-th/0503117; Mod. Phys. Lett. A **20**, 2705 (2005),
23. S. Rasanen, Class. Quant. Grav. **23**, 1823 (2006).
24. E. Barausse, S. Matarrese, and A. Riotto, Phys. Rev. D **71**, 063537, (2005).
25. S. Rasanen, JCAP **0611**, 003 (2006).
26. G. F. R. Ellis, Relativistic cosmology – its nature, aims and problems. In *General Relativity and Gravitation* (D. Reidel Publishing Co., Dordrecht), ed. B. Bertotti, F. de Felice and A. Pascolini, pp. 215–288 (1984).
27. G. F. R. Ellis and W. Stoeger, Class. Quant. Grav. **4**, 1697 (1987).
28. G. Ellis, Nature **452**, 158 (2008).
29. S. Matarrese and D. Terranova, Mon. Not. Roy. Astron. Soc. **283**, 400 (1996).

30. C. M. Hirata and U. Seljak, Phys. Rev. D **72**, 083501 (2005).
31. T. Buchert, Gen. Rel. Grav. **32**, 105 (2000).
32. T. Buchert, Gen. Rel. Grav. **33**, 1381 (2001).
33. G. Lemaitre, Gen. Rel. Grav. **29**, 641 (1997) [Annales Soc. Sci. Brux. Ser. I Sci. Math. Astron. Phys. A **53**, 51 (1933)].
34. R. C. Tolman, Proc. Nat. Acad. Sci. **20**, 169 (1934).
35. H. Bondi, Mon. Not. Roy. Astron. Soc. **107**, 410 (1947).
36. K. Tomita, Prog. Theor. Phys. **105**, 419 (2001); Mon. Not. Roy. Astron. Soc. **326**, 287 (2001); Prog. Theor. Phys. **106**, 929 (2001); Astrophys. J. **584**, 580 580 (2003).
37. H. Iguchi, T. Nakamura, and K.-I. Nakao, Prog. Theor. Phys. **108**, 809 (2002).
38. J. W. Moffat, JCAP **0510** 012 (2005); "Large scale cosmological inhomogeneities, inflation and acceleration without dark energy," arXiv:astro-ph/0504004; JCAP **0605** 001 (2006).
39. H. Alnes, M. Amarzguioui, and O. Gron, Phys. Rev. **D73** 083519 (2006).
40. R. Mansouri, "Structured FRW universe leads to acceleration: A non-perturbative approach," arXiv:astro-ph/0512605.
41. R. A. Vanderveld, E. E. Flanagan, and I. Wasserman, Phys. Rev. **D74** 023506 (2006).
42. D. Garfinkle, Class. Quant. Grav. **23** 4811 (2006).
43. T. Biswas, R. Mansouri and A. Notari, JCAP **0712**, 017 (2007).
44. H. Alnes and M. Amarzguioui, Phys. Rev. **D75** 023506 (2007).
45. R. R. Caldwell and A. Stebbins, Phys. Rev. Lett. **100** 191302 (2008).
46. S. Alexander, T. Biswas, A. Notari, and D. Vaid, JCAP **0909**, 025 (2009).
47. J. Garcia-Bellido and T. Haugboelle, JCAP **0804** 003 (2008), JCAP **0809** 016, (2008).
48. T. Clifton, P. G. Ferreira, and K. Land, Phys. Rev. Lett. **101** 131302 (2008).
49. P. Hunt and S. Sarkar, "Constraints on large scale voids from WMAP-5 and SDSS," arXiv:0807.4508 [astro-ph].
50. W. Valkenburg, JCAP **0906**, 010 (2009).
51. M.-N. Célérier, K. Bolejko, A. Krasinski, and C. Hellaby, "A (giant) void is not mandatory to explain away dark energy with a Lemaitre–Tolman model," arXiv:0906.0905 [astro-ph.CO].
52. L. Hui and U. Seljak, in *Proceedings of the 173rd IAU Symposium*, edited by C. S. Kochanek and J. N. Hewitt (Kluwer Academic Publishers, Dordrecht, 1995); U. Seljak and L. Hui, in *Proceedings of Clusters, Lensing and the Future of the Universe*, ASP Conference Series, edited by V. Trimble and A. Reisenegger (IAU Publications, Provo, 1995) Vol. 88, p. 267.
53. E. R. Siegel and J. N. Fry, Astrophys. J. **628**, L1 (2005).
54. A. Ishibashi and R. M. Wald, Class. Quant. Grav. **23**, 235 (2006).
55. E. W. Kolb, V. Marra and S. Matarrese, Phys. Rev. D **78**, 103002 (2008).
56. K. Enqvist, M. Mattsson and G. Rigopoulos, JCAP **0909**, 022 (2009).

Dynamics and properties of the Invisible Universe

THE ABNORMALLY WEIGHTING ENERGY HYPOTHESIS: THE ORIGIN OF THE COSMIC ACCELERATION (ABSTRACT)

J.-M. Alimi[1,2] and A. Füzfa[1,2,3] and

[1] *CNRS, Laboratoire Univers et Théories (LUTh), UMR 8102 CNRS, Observatoire de Paris, Université Paris Diderot; 5 Place Jules Janssen, 92190 Meudon, France*

[2] *Department of Mathematics, University of Namur (FUNDP); Rue de Bruxelles, 61, B-5000 Namur, Belgium*

[3] *Center for Particle Physics and Phenomenology (CP3), Université catholique de Louvain, Chemin du Cyclotron, 2, B-1348 Louvain-la-Neuve, Belgium*

We generalize tensor-scalar theories of gravitation by the introduction of an abnormally weighting type of energy. This theory of tensor-scalar anomalous gravity is based on a relaxation of the weak equivalence principle that is now restricted to ordinary visible matter only. As a consequence, the convergence mechanism toward general relativity is modified and produces naturally cosmic acceleration as an inescapable gravitational feedback induced by the mass-variation of some invisible sector. The cosmological implications of this new theoretical framework are studied. This glimpses at an enticing new symmetry between the visible and invisible sectors, namely that the scalar charges of visible and invisible matter are exactly opposite.

Keywords: Dark Matter, Dark Energy, Tensor-Scalar theories, Equivalence Principle, Cosmology
PACS: 95.35.+d-95.36.+x-98.80.-k-04.

THE UNIFYING DARK FLUID MODEL (ABSTRACT)

Alexandre Arbey

Université de Lyon, Lyon, F-69000, France ; Université Lyon 1, Villeurbanne, F-69622, France

Centre de Recherche Astrophysique de Lyon, Observatoire de Lyon, 9 avenue Charles André, Saint-Genis Laval cedex, F-69561, France; CNRS, UMR 5574

Ecole Normale Supérieure de Lyon, Lyon, France.

The standard model of cosmology relies on the existence of two components, "dark matter" and "dark energy", which dominate the expansion of the Universe. There is no direct proof of their existence, and their nature is still unknown. Many alternative models suggest other cosmological scenarios, and in particular the *dark fluid* model replace the dark matter and dark energy components by a unique dark component able to mimic the behaviour of both components. The current cosmological constraints on the unifying dark fluid model is discussed, and a dark fluid model based on a complex scalar field is presented. Finally the consequences of quantum corrections on the scalar field potential are investigated.

CP1241, *Invisible Universe: Proceedings of the Conference,* edited by J.-M. Alimi and A. Füzfa

Keywords: Dark Fluid; Dark Energy; Dark Matter; Quantum Corrections
PACS: 98.80.-k, 11.10.-z, 95.35.+d, 95.36.+x

LAGRANGE PERTURBATION AND AVERAGING OF A GENERAL RELATIVISTIC INHOMOGENEOUS UNIVERSE(ABSTRACT)

Hideki Asada

Faculty of Science and Technology, Hirosaki University, Hirosaki 036-8561, Japan

In the framework of general relativity, we discuss Lagrange perturbation approach to describe locally nonlinear inhomogeneities in a globally FLRW-like universe, in which we assume two cases: only the dust matter and a matter accompanied with pressure. Next, we discuss apparent discrepancies among a lot of works on effects of the nonlinear backreaction. By defining the spatially averaged matter energy density as a conserved quantity in the large comoving volume, it is shown that the nonlinear backreaction neither accelerates nor decelerates the cosmic expansion in a matter-dominated universe.

Keywords: Lagrange, averaging, relativity, cosmology
PACS: 98.80.Hw, 04.25.Nx

THE EVENT HORIZON THERMODYNAMICAL (EHT-) MODEL OF THE DARK ENERGY(ABSTRACT)

C. Barbachoux, J. Gariel and G. Le Denmat

LERMA-UPMC Université P. et M. Curie, (Observatoire de Paris, ENS, UCP, CNRS) B.C. 142, 3, Rue Galilée, 94 200 Ivry, France

We present the main results of a model, namely the event-horizon model, based on a thermodynamical approach for the law of $\Lambda(t)$. We consider a phenomenological model where we give to the effective cosmological "constant" Λ the physical meaning of a "vacuum" energy density, as usually admitted. In the same vein as the generalization of the black holes and of the de Sitter event-horizon thermodynamics to the FLRW space-time, we associated to this "vacuum" energy density a full thermodynamics for cosmological event horizons without further microscopic precision about the fields describing the vacuum. Without any ansatz and using thermodynamics arguments, we then deduce the law $\Lambda \approx r^{-2}$ where r is the proper radius of the event horizon in quasi-de Sitter FLRW spatially flat spacetime. With this relation, the coincidence problem is solved by the decay of $\Lambda(t)$ due to the expansion of the universe. We apply this model to the study of the recent accelerated expansion phase and to the coincidence problem. Finally, we introduce a logarithmic corrective term in the entropy associated to the DE, which leads to a new term in the DE density. This term was derived for any thermodynamical system encountering fluctuations around an equilibrium state. This corrective term was first applied to black hole. Because of the analogy between the cosmological FLRW event horizon and the Black hole event horizon, it seems legitimate to introduce this corrective term in the entropy of Λ. After a numerical resolution, we compute the evolutions of the main cosmological parameters: $\Lambda(t)$, the Hubble parameter H(t), the vacuum density $\Omega_\Lambda(t)$. The computed values are compatible with an initial inflation, and the current observations of the "concordance" model. It also predicts a graceful exit of early inflation and the present acceleration and solves in the same time the age and the coincidence problem.

Keywords: dark energy theory, thermodynamics, Supernova Type I a, cosmological parameters
PACS: 95.36.+x,98.80.-k,98.80.Jk

INTERACTING DARK MATTER AS AN ALTERNATIVE TO DARK ENERGY (ABSTRACT)

Spyros Basilakos[1] **and Manolis Plionis**[2,3]

[1]*Academy of Athens, Research Center for Astronomy and Applied Mathematics, Soranou Efesiou 4, 11527, Athens, Greece*

[2]*Institute of Astronomy & Astrophysics, National Observatory of Athens, Palaia Penteli 152 36, Athens, Greece*

[3]*Instituto Nacional de Astrofísica, Óptica y Electronica, Puebla, Mexico*

We investigate the global dynamics of the universe within the framework of the Interacting Dark Matter (IDM) scenario. Considering that the dark matter obeys the collisional Boltzmann equation, we can obtain analytical solutions of the global density evolution, which can accommodate an accelerated expansion, equivalent to either the *quintessence* or the standard Λ models. This is possible if there is a disequilibrium between the DM particle creation and annihilation processes with the former process dominating, which creates an effective source term with negative pressure. Comparing the predicted Hubble expansion of one of the IDM models (the simplest) with observational data, we find that the effective annihilation term is quite small, as suggested by various experiments.

Keywords: Cosmology; dark matter; dark energy
PACS: 98.80.-k, 95.35.+d,95.36.+x

MODIFIED THEORIES OF GRAVITY WITH NON-MINIMAL CURVATURE-MATTER COUPLING (ABSTRACT)

Orfeu Bertolami[1,2,3] **and Miguel Carvalho Sequeira**[1]

[1]*Instituto Superior Técnico, Departamento de Física, Av. Rovisco Pais, 1049-001 Lisboa, Portugal*

[2]*Instituto de Plasmas e Fusão Nuclear, Instituto Superior Técnico, Lisboa*

[3]*orfeu@cosmos.ist.utl.pt*

In this contribution one examines the generalization of the $f(R)$ theories of gravity where one introduces a non-minimal coupling between curvature and matter. This model has new and interesting features. However, as any modified gravity theories, these may give origin to anomalies which might turn the theory physically meaningless. In this respect, one undertakes a study on the energy conditions and the Dolgov-Kawasaki criterion specific of this new model.

Keywords: Modified theories of gravitation, Energy Conditions, Dolgov-Kawasaki criterion.
PACS: 04.50.Kd, 98.80.-k, 98.80.Jk

DO WMAP DATA FAVOR NEUTRINO MASS AND A COUPLING BETWEEN COLD DARK MATTER AND DARK ENERGY ? (ABSTRACT)

S. A. Bonometto[1], G. La Vacca[1], J. R. Kristiansen[2], R. Mainini[2] and L. P. L. Colombo[3]

[1]*Department of Physics G. Occhialini – Milano – Bicocca University, Piazza della Scienza 3, 20126 Milano, I.N.F.N., Sezione di Milano*

[2]*Institute of Theoretical Astrophysics, University of Oslo, Box 1029, 0315 Oslo, Norway*

[3]*Department of Physics & Astronomy, University of Southern California, Los Angeles, CA 90089-0484*

We fit WMAP5 and related data by allowing for a CDM–DE coupling and non–zero neutrino masses, simultaneously. We find a significant correlation between these parameters, so that simultaneous higher coupling and ν–masses are allowed. Furthermore, models with a significant coupling and ν–mass are statistically favoured in respect to a cosmology with no coupling and negligible neutrino mass (our best fits are: $C \sim 1/2m_p$, $m_\nu \sim 0.12\,\text{eV}$ per flavor). We use a standard Monte Carlo Markov Chain approach, by assuming DE to be a scalar field self–interacting through Ratra–Peebles or SUGRA potentials.

Keywords: dynamical dark energy, massive neutrinos
PACS: 14.60.Pq,95.36.+x,95.35.+d

FIRST ORDER EXTENDED GRAVITY AND THE DARK SIDE OF THE UNIVERSE: MATCHING WITH OBSERVATIONS (ABSTRACT)

S. Capozziello[1], M. De Laurentis[1], M. Francaviglia[2] and S. Mercadante[2]

[1]*Dipartimento di Scienze Fisiche, Università di Napoli "Federico II", and INFN Sez. di Napoli, Compl. Univ. di Monte S. Angelo, Edificio G, Via Cinthia, I-80126, Napoli, Italy*

[2]*Dipartimento di Matematica, Università di Torino, and INFN Sez. di Torino, Via Carlo Alberto 10, 10123 Torino, Italy*

The so called f (R)-gravity could be, in principle, able to explain the accelerated expansion of the Universe without adding unknown forms of dark energy/dark matter but, more simply, extending the General Relativity by generic functions of the Ricci scalar. However, a part several phenomenological models, there is no final f (R)-theory capable of fitting all the observations and addressing all the issues related to the presence of dark energy and dark matter. Astrophysical observations are pointing out huge amounts of "dark matter" and "dark energy" needed to explain the observed large scale structures and cosmic accelerating expansion. Up to now, no experimental evidence has been found, at fundamental level, to explain such mysterious components. The problem could be completely reversed considering dark matter and dark energy as "shortcomings" of General Relativity and considering the concept of Dark Metric introduced in the companion paper of these Proceedings by the same authors.

Keywords: Dark matter, Dark energy, Dark metric, Gravitation, Observations
PACS: 95.30.Sf-98.80.Jk

HIGH ACCURACY SPECTRA AT HIGH Z IN DYNAMICAL DARK ENERGY SIMULATIONS (ABSTRACT)

L. Casarini[1,2], A. V. Macciò[3] and S. A. Bonometto[1,2]

[1] *Department of Physics G. Occhialini – Milano – Bicocca University, Piazza della Scienza 3, 20126 Milano*

[2]*I.N.F.N., Sezione di Milano*

[3] *Max-Planck-Institut für Astronomie, Königstuhl 17, 69117 Heidelberg, Germany*

Non-linear power spectra, at various redshift z, will be measured by the next generation mass probes, and this will allow to investigate the nature of DE. Seemingly, to exploit such data, we however need high precision spectral predictions, spanning the functional space of w(a) DE state equations, for the admitted cosmological parameter ranges. In this paper we show that data analysis can be simplified, by comparing them with w = const. models only, although allowing for different w's at different z. The point we test is that, at a any redshift z, spectra depend just on the r.m.s. fluctuation amplitude $\sigma_8(z)$, the reduced density parameters $\omega_{b,c}$ and the distance $d_{LSB}(z)$ from the Last Scattering Band. We test this through N-body simulations, by comparing dynamical DE models and auxiliary w = const. models. We find spectral discrepancies steadily < 1%, but even smaller at higher redshift, where spectra approach a permil precision. An implicit conclusion, however, is that any w(z) law, deduced from data at various z, by testing them against constant-*w* models, is not the actual *z*-dependence of the DE state parameter.

Keywords: dynamical dark energy, non linear power spectrum, N-body simulations
PACS:

INHOMOGENEITIES IN THE UNIVERSE WITH EXACT SOLUTIONS OF GENERAL RELATIVITY (ABSTRACT)

Marie-Noëlle Célérier

Laboratoire Univers et Théories (LUTH), Observatoire de Paris, CNRS, Université Paris-Diderot, 5 place Jules Janssen, 92190 Meudon, France

It is commonly stated that we have entered the era of precision cosmology in which a number of important observations have reached a degree of precision, and a level of agreement with theory, that is comparable with many Earth-based physics experiments. One of the consequences is the need to examine at what point our usual, well-worn assumption of homogeneity associated to the use of perturbation theory begins to compromise the accuracy of our models. It is now a widely accepted fact that the effect of the inhomogeneities observed in the Universe cannot be ignored when one wants to construct an accurate cosmological model. Well-established physics can explain several of the observed phenomena without introducing highly speculative elements, like dark matter, dark energy, exponential expansion at densities never attained in any experiment (i.e. inflation), and the like. Two main classes of methods are currently used to deal with these issues. Averaging, sometimes linked to fitting procedures a la Stoegger and Ellis, provide us with one promising way of solving the problem. Another approach is the use of exact inhomogeneous solutions of General Relativity. This will be developed here.

Keywords: Cosmology; Large scale structures formation; Large scale structure evolution; Dark energy; Local void
PACS: 98.80.-k; 98.65.Dx; 95.36.+x

ON THE DETERMINATION OF DARK ENERGY (ABSTRACT)

Chris Clarkson

Cosmology & Gravity Group, Department of Mathematics and Applied Mathematics, University of Cape Town, Rondebosch 7701, South Africa.

I consider some of the issues we face in trying to understand dark energy. Huge fluctuations in the unknown dark energy equation of state can be hidden in distance data, so I argue that model-independent tests which signal if the cosmological constant is wrong are valuable. These can be constructed to remove degeneracies with the cosmological parameters. Gravitational effects can play an important role. Even small inhomogeneity clouds our ability to say something definite about dark energy. I discuss how the averaging problem confuses our potential understanding of dark energy by considering the backreaction from density perturbations to second-order in the concordance model: this effect leads to at least a 10value of the deceleration parameter, and could be significantly higher. Large Hubble-scale inhomogeneity has not been investigated in detail, and could conceivably be the cause of apparent cosmic acceleration. I discuss void models which defy the Copernican principle in our Hubble patch, and describe how we can potentially rule out these models. This article is a summary of two talks given at the Invisible Universe Conference, Paris, 2009.

THE INNOCUOUSNESS OF ADIABATIC INSTABILITIES IN COUPLED SCALAR FIELD-DARK MATTER MODELS (ABSTRACT)

P.-S. Corasaniti

LUTH, Observatoire de Paris, CNRS UMR 8102, Université Paris Diderot, 5 Place Jules Janssen, 92195 Meudon Cedex, France

Non-minimally coupled scalar field models suffer of unstable growing modes at the linear perturbation level. The nature of these instabilities depends on the dynamical state of the scalar field. In particular in systems which admit adiabatic solutions, large scale instabilities are suppressed by the slow-roll dynamics of the field. Here we review these results and present a preliminary likelihood data analysis suggesting that along adiabatic solutions coupled models with coupling of order of gravitational strength can provide viable cosmological scenarios satisfying constraints from SN Ia, CMB and large scale structure data.

Keywords: Cosmology-Structure Formation-Linear Perturbations
PACS: 95.35.+d,95.36+x

IMPRINTS OF DARK ENERGY ON STRUCTURE FORMATION : NO UNIVERSALITY IN MASS FUNCTIONS ? (ABSTRACT)

J. Courtin[1], J-M. Alimi[1], Y. Rasera[1], P-S. Corasaniti[1], A. Fuzfa[1,2,3] and V. Boucher[3]

[1]*CNRS, Laboratoire Univers et Théories (LUTh), UMR 8102 CNRS, Observatoire de Paris, Université Paris Diderot; 5 Place Jules Janssen, 92190 Meudon, France*

[2]*Groupe d'Application des MAthématiques aux Sciences du COsmos (GAMASCO), University of Namur (FUNDP), Belgium*

[3]*Center for Particle Physics and Phenomenology (CP3), Université catholique de Louvain, Chemin du Cyclotron, 2, B-1348 Louvain-la-Neuve, Belgium*

In this proceedings for the invisible universe conference, we present our results on the imprints of dark energy on structure formation. This article follows the one of Rasera et al. 2010 which features the imprints of dark energy on the non linear power spectrum. In this proceedings, we focus on the non linear imprints of dark energy on the mass functions and question the Jenkins et al. approximation of the universality of mass functions in the context of dark energies [6]. We first calculate the critical density formation threshold $\delta_c(z)$ and the overdensity detection parameter $\Delta(z)$ for dark energy cosmologies, using the spherical collapse model. We show that these non linear parameters keep a record of the history of the universe expansion and in particular that these parameters are quantitatively modified by the recent acceleration, driven by the dark energy. We show that the consequences on the mass functions should be detected in numerical simulations. In the second section, we present our dark energy models, the numerical setup and discuss the numerical checks that were necessary to guarantee a relevant precision for the purpose of detecting dark imprints on the mass functions. In the third section, we present our numerical results. We show that a different nature of the dark energy, or a variation of the quantity of dark energy in the universe, can lead to more than 20% deviations of mass functions to a universal behavior.We then quantitatively explain these deviations on the basis of the non linear parameters and give a parameterisation for the detection parameter that permits a unique prediction of the dark energy cosmologies mass functions up to a precision of 5% (Courtin et al [4]). We finally conclude.

Keywords: N-body simulations, dark energy, mass function, large-scale structures, quintessence, cosmology
PACS: 98.80.-k,95.36.+x,95.35.+d, 98.62.Ai,98.62.Py,98.65.Cw,98.65.Dx

DARK ENERGY DUE TO LATE TIME QUANTUM DECAY PROCESS (ABSTRACT)

A. de la Macorra and F. Briscese

Instituto de Física, Universidad Nacional Autonoma de Mexico, Apdo. Postal 20-364, 01000 México D.F., México

We show that the dark energy field ϕ can be dynamically obtained at a low scale, e.g. at $E = O(eV)$, via a quantum transitions process. This is achieved if the ϕ field is coupled with a relativistic scalar field which we call φ. The $2 \leftrightarrow 2$ coupling has a transition rate $\Gamma \propto g^2 T_\varphi$ and the fields are coupled at low energies when $\Gamma/H \equiv T_{gen}/T_\varphi \propto g^2/T_\varphi$ is larger than one, i.e. at $T < T_{gen} \propto g^2$. After the ϕ particles have been produced we solve the dynamical equations and show that the energy of ϕ is transferred to its scalar classical potential $V(\phi)$ The classical evolution of ϕ is then given by the form of $V(\phi)$.

Keywords: Dark Energy, Inflation, Cosmology
PACS: 98.90.Cq,95.36.+x,98.80.-k

COSMOLOGICAL SIMULATIONS ON A GRID OF COMPUTERS (ABSTRACT)

Benjamin Depardon[1], Eddy Caron[1], Frédéric Desprez[1], Jérémy Blaizot[2] and Hélène Courtois[3]

[1]*University of Lyon; ENS-Lyon/INRIA/CNRS/UCBL; LIP Laboratory. France.*

[2]*University of Lyon; UCB Lyon 1/CNRS/INSU; CRAL. France.*

[3]*University of Lyon; UCB Lyon 1/CNRS/IN2P3/INSU; IPN Lyon. France.*

The work presented in this paper aims at restricting the input parameter values of the semi-analytical model used in GALICS and MOMAF, so as to derive which parameters influence the most the results, e.g., star formation, feedback and halo recycling efficiencies, etc. Our approach is to proceed empirically: we run lots of simulations and derive the correct ranges of values. The computation time needed is so large, that we need to run on a grid of computers. Hence, we model GALICS and MOMAF execution time and output files size, and run the simulation using a grid middleware: DIET. All the complexity of accessing resources, scheduling simulations and managing data is harnessed by DIET and hidden behind a web portal accessible to the users.

Keywords: Cosmology, N-body simulations, Parallel computing, Local universe.
PACS: 98.80.Es, 97.75.Pq

EFFECT OF PRIMORDIAL NON-GAUSSIANITY ON HALO BIAS AND MASS FUNCTION (ABSTRACT)

Vincent Desjacques[1], Uroš Seljak[1,2] and Ilian T. Iliev[1,3]

[1]*Institute for Theoretical Physics, University of Zürich, Winterthurerstrasse 190, CH-8057 Zürich, Switzerland*

[2]*Physics and Astronomy Department, University of California, and Lawrence Berkeley National Laboratory, Berkeley, California 94720, USA*

[3]*Astronomy Centre, Department of Physics & Astronomy, University of Sussex, Brighton BN1 9QH, UK*

We discuss the effect of quadratic and cubic local non-Gaussianity on the mass function and bias of dark matter halos extracted from cosmological N-body simulations. This type of non-Gaussianity induces a k-dependent bias in the large-scale clustering of rare objects. While we find that at low wavenumbers $k < 0.03\ hMpc^{-1}$ the theory and the simulations agree well with each other for biased halos with $b(M) > 1.5$, including a scale independent correction to the non-Gaussian bias improves the agreement on small scales where the k-dependent effect becomes rapidly negligible. Using available large-scale structure data, we obtain a limit on the size of the cubic nonlinear parameter of $-3.5 \times 10^5 < g_{NL} < 8.2 \times 10^5$. Future observations shall improve this bound by 1-2 orders of magnitude.

Keywords: large-scale structures, dark matter, primordial non-Gaussianity
PACS: 98.80.-k, 98.80.Cq, 98.80.Es, 98.70.Vc

AN EXPLORATION OF THE PHYSICS OF F (R) GRAVITY THEORIES (ABSTRACT)

Peter Dunsby

Cosmology & Gravity Group, Department of Mathematics and Applied Mathematics, University of Cape Town, Rondebosch 7701, South Africa.

We present a review of progress made in the study of structure growth and the properties of spherically symmetric solutions in f (R) theories of gravity. This article is a summary of a talk given at the Invisible Universe Conference, Paris, 2009.

FIRST ORDER EXTENDED GRAVITY AND THE DARK SIDE OF THE UNIVERSE: THE GENERAL THEORY (ABSTRACT)

S. Capozziello[1], M. De Laurentis[1], M. Francaviglia[2] and S. Mercadante[2]

[1]*Dipartimento di Scienze Fisiche, Università di Napoli "Federico II", and INFN Sez. di Napoli, Compl. Univ. di Monte S. Angelo, Edificio G, Via Cinthia, I-80126, Napoli, Italy*

[2]*Dipartimento di Matematica, Università di Torino, and INFN Sez. di Torino, Via Carlo Alberto 10, 10123 Torino, Italy*

General Relativity is not the definitive theory of Gravitation due to several shortcomings which are coming out both from theoretical and experimental viewpoints. At large scales (astrophysical and cosmological scales) the attempts to match it with the today observational data lead to invoke Dark Energy and Dark Matter as the bulk components of the cosmic fluid. Since no final evidence, at fundamental level, exists for such ingredients, it is clear that General Relativity presents shortcomings at infrared scales. On the other hand, the attempts to formulate theories more general than the Einstein one give rise to mathematical difficulties that need workarounds which, in turn, generate problems from the interpretative viewpoint.We present here a completely new approach to the mathematical objects in terms of which a theory of Gravitation may be written in a first-order (à la Palatini) formalism, and introduce the concept of Dark Metric which could completely bypass the introduction of disturbing concepts as Dark Energy and Dark Matter.

Keywords: Dark matter, Dark energy, Dark metric, Gravitation
PACS: 95.30.Sf-98.80.Jk

DARK MATTER AND DARK ENERGY FROM GRAVITATIONAL SYMMETRY BREAKING (ABSTRACT)

A. Füzfa[1,2,3] and J.-M. Alimi[3]

[1]*Department of Mathematics, University of Namur (FUNDP); Rue de Bruxelles, 61, B-5000 Namur, Belgium*

[2]*Center for Particle Physics and Phenomenology (CP3), Université catholique de Louvain, Chemin du Cyclotron, 2, B-1348 Louvain-la-Neuve, Belgium*

[3]*CNRS, Laboratoire Univers et Théories (LUTh), UMR 8102 CNRS, Observatoire de Paris, Université Paris Diderot; 5 Place Jules Janssen, 92190 Meudon, France*

We build a mechanism of gravitational symmetry breaking (GSB) of a global U(1) symmetry based on the relaxation of the equivalence principle due to the mass variation of pseudo Nambu-Goldstone dark matter (DM) particles. This GSB process is described by the modified cosmological convergence mechanism of the Abnormally Weighting Energy (AWE) Hypothesis previously introduced by the authors. Several remarkable constraints from the Hubble diagram of faraway supernovae are derived, notably on the explicit and gravitational symmetry breaking energy scales of the model. We then briefly present some consequences on neutrino masses when this mechanism is applied to the particular case of the breaking of lepton number symmetry.

Keywords: Mass-varying Dark Matter, Axions, Dark Energy, Symmetry Breaking, Tensor-Scalar theories, Neutrino masses
PACS: 98.80.-k,95.35.+d,95.36.+x,14.60.Pq,11.30.Fs

RUNNING COSMOLOGICAL CONSTANT AND RUNNING NEWTON CONSTANT IN MODIFIED GRAVITY THEORIES (ABSTRACT)

Remo Garattini

Università degli Studi di Bergamo, Facoltà di Ingegneria, Viale Marconi 5, 24044 Dalmine (Bergamo) Italy and I.N.F.N. - sezione di Milano, Milan, Italy. E-mail:remo.garattini@unibg.it

We discuss how to extract information about the cosmological constant from the Wheeler-DeWitt equation, considered as an eigenvalue of a Sturm-Liouville problem in a de Sitter and Anti-de Sitter background. The equation is approximated to one loop with the help of a variational approach with Gaussian trial wave functionals. A canonical decomposition of modes is used to separate transverse-traceless tensors (graviton) from ghosts and scalar. We show that no ghosts appear in the final evaluation of the cosmological constant. A zeta function regularization is used to handle with divergences. A renormalization procedure is introduced to remove the infinities together with a renormalization group equation. We apply this procedure on the induced cosmological constant Λ and, as an alternative, on the Newton constant G. A brief discussion on the extension to a $f(R)$ theory is considered.

Keywords: Cosmological Constant, Quantum Cosmology, Quantum Gravity, Renormalization Group Equations
PACS: 04.60.-m, 04.62+v, 05.10.Cc

COSMIC ACCELERATION FROM TOPOLOGICAL CONSIDERATIONS (ABSTRACT)

Miguel Angel Garcia-Aspeitia and Tonatiuh Matos

Departamento de Fisica, Centro de Investigacion y de Estudios Avanzados del I.P.N. Apdo. Post. 14-740 07000, D.F., Mexico

In this work we explore the possibility that the dynamics of the universe can be reproduced choosing appropriately the global topology of the cosmos. We explore two concentric three-dimensional spherical branes immersed in a fivedimensional space-time. Before to the collision, in the interior sphere there exist only a spin-zero fundamental field (scalar field), in the exterior one there exist only fundamental spin-one interactions and spin-two interactions in the bulk. In this model, like in the Epkyrotic, the Big Bang is caused for the collision of the branes and generate all the fields predicted by the standard model in the exterior brane (our universe). In the interior brane the scalar field behaves like scalar field dark matter. We discuss two different regimens where the energy density and the brane tension are compared, with the aim to obtain the dynamics of the universe after and before the collision. Finally we discuse the perturbations in the modified Einstein equations of the scalar field dark matter in the inner brane and the consequence in the high energy universe dynamics and the corrections in the standard general relativity.

DO SUPERNOVAE FAVOR TACHYONIC BIG BRAKE INSTEAD DE SITTER ? (ABSTRACT)

L. A. Gergely[1,2], Z. Keresztes[1,2], A. Yu. Kamenshchik[3,4], V. Gorini[5,6] and U. Moschella[5,6]

[1]*Department of Theoretical Physics, University of Szeged, Tisza Lajos krt 84-86, Szeged 6720, Hungary*

[2]*Department of Experimental Physics, University of Szeged, Dóm Tér 9, Szeged 6720, Hungary*

[3]*Dipartimento di Fisica and INFN, via Irnerio 46, 40126 Bologna, Italy*

[4]*L.D. Landau Institute for Theoretical Physics, Russian Academy of Sciences, Kosygin street 2, 119334 Moscow, Russia*

[5]*Dipartimento di Scienze Fisiche e Mathematiche, Università dell'Insubria, Via Valleggio 11, 22100 Como, Italy*

[6]*INFN, sez. di Milano, Via Celoria 16, 20133 Milano, Italy*

We investigate whether a tachyonic scalar field, encompassing both dark energy and dark matter-like features will drive our universe towards a Big Brake singularity or a de Sitter expansion. In doing this it is crucial to establish the parameter domain of the model, which is compatible with type Ia supernovae data. We find the 1σ contours and evolve the tachyonic sytem into the future.We conclude, that both future evolutions are allowed by observations, Big Brake becoming increasingly likely with the increase of the positive model parameter k.

Keywords: dark energy, tachyons, supernovae, cosmological singularities
PACS: 98.80.Cq, 98.80.Jk, 98.80.Es, 95.36.+x

ACCRETION HISTORY OF SUBHALO POPULATION NOW AND THEN (ABSTRACT)

Carlo Giocoli

Zentrum für Astronomie, ITA, Universität Heidelberg, Albert-Ueberle-Str. 2, 69120 Heidelberg, Germany

In the standard model of structure formation galaxies reside in virialized dark matter haloes which extend much beyond the observational radius of the central system. The dark matter halo formation process is hierarchical, small systems collapse at high redshift and then merge together forming larger ones. In this work we study the mass assembly history of host haloes at different observation redshifts and the mass function of accreted satellites (haloes that merge directly on the main halo progenitor). We show that the satellite mass function is universal, both independent on the host halo mass and observation redshift. The satellite mass function also turn out to be universal once only satellites before or after the host halo formation redshift (time at which the main halo progenitor assembles half of its final mass) are considered. We show that the normalizations of these distributions are directly related to the main halo progenitor mass distributions before and after its formation, while their slope and the exponential high mass cut-off remain unchanged.

Keywords: Dark Matter, Haloes, Subhaloes, Mass Function, Galaxies
PACS: 98.80-k

ON THE CO-EXISTENCE OF MATTER DOMINATED AND ACCELERATING SOLUTIONS IN F(G)-GRAVITY AND THE CASE OF F(R)-GRAVITY (ABSTRACT)

Naureen Goheer[1], Rituparno Goswami[1], Julien Larena[1], Peter K. S. Dunsby[1,2] and Kishore Ananda[1]

[1]*Department of Mathematics and Applied Mathematics, University of Cape Town, 7701 Rondebosch, Cape Town, South Africa*

[2]*South African Astronomical Observatory, Observatory 7925, Cape Town, South Africa*

Working within the theory of modified Gauss-Bonnet gravity, we show that FLRW-like power-law solutions only exist for a very special class of f (G) theories. Furthermore, we point out that any transition from decelerated to accelerated expansion must pass through G = 0, and no function f (G) that is differentiable at this point can admit both a decelerating power-law solution and any accelerating solution. This strongly constrains the cosmological viability of f (G)-gravity, since it may not be possible to obtain an expansion history of the universe which is compatible with observations. We compare to the case of f (R)-gravity, and discuss possible caveats for the case of f (G)-gravity.

Keywords: Cosmology, modified gravity, expansion history
PACS: 98.80.Cq

NATURAL NEUTRINO DARK ENERGY (ABSTRACT)

Ilya Gurwich

Physics Department, Ben-Gurion University, Beer-Sheva 84105, Israel

I construct a general description for neutrino dark energy models, that do not require exotic particles or strange couplings. With the help of the above, this class of models is reduced to a single function with several constraints. It is shown that these models lead to some concrete predictions that can be verified (or disproved) within the next decade, using results from PLANK, EUCLID and JDEM.

Keywords: Dark Energy, Neutrino masses, Neutrino dark energy, Cosmic Neutrino Background
PACS: 14.60.Pq,95.36.+x,98.80.Es

BLACK HOLES AND COLLISIONLESS BULGES (ABSTRACT)

R.N. Henriksen[1], M. LeDelliou[2] and J.D. MacMillan[3]

[1]*Department of Physics, Engineering Physics and Astronomy, Queen's University, Kingston, Ontario,Canada, K7L 3N6*

[2]*Facultad de Ciencias, C-XI, Universidad Autonoma de Madrid, 28049 Madrid, Spain*

[3]*Faculty of Science, University of Ontario Institute of Technology, Oshawa, Ontario, Canada, L1H 7K4*

Currently central black holes in collisionless self-gravitating systems such as the bulges of galaxies are abundant. One of the most definite cases lies in the center of the milky way. Simulations have studied surrounding cusp profiles as produced by binary black hole mergers. However there seems to be no unified picture that places a black hole in a surrounding bulge and distant halo in a self-consistent state. Moreover the time and nature of the growth of the black hole, together with the reason for the correlation between its mass and that of the distant galactic bulge, remain unidentified and unexplained. In this paper we study the problem analytically. We seek distribution functions that have a "memory" of previous dynamical evolution. We justify these by fits to numerical simulations. One simple choice of distribution function is able to give a "zero flux" outer boundary region, followed by a $1/r$ inner shell that gives way to an $1/r^2$ cusp surrounding the black hole. The black hole e-folding time is a few million years. It is likely to apply during the rapid growth phase of the dark matter halo since it is very anisotropic.

Keywords: Dark Matter, Black Holes, Density Cusps
PACS: 95.35.+d, 95.75.Pq

SOME LIGHT ON THE DARK MATTER VS. MODIFIED GRAVITY DEBATE FROM DWARF SPHEROIDAL SCALINGS (ABSTRACT)

X. Hernandez, S. Mendoza, T. Suarez and T. Bernal

Instituto de Astronomía, Universidad Nacional Autónoma de México, AP 70-264, Ciudad Universitaria, D. F. 04510, México

Dwarf spheroidal (dSph) galaxies are the smallest galactic systems known. Also, under standard gravity, they are the most dark matter dominated systems known in astrophysics. The local sample of dSph satellites of the Milky Way comprise some two dozen systems, ranging from a few million, down to just a few thousand solar masses in stellar content, for the recently discovered ones. High quality observations of their stellar kinematics now allow detailed inferences of their structural and dynamical properties. We perform a modeling of the gravitational equilibrium configurations for a sample of the best studied dSphs under MOND, and find that a satisfactory description is obtained, without the need to invoke the presence of dark matter or any tidal effects, for a particular interpolation function, valid at this regime. We find also that the now well established scalings of the structural properties of these systems are naturally accounted for, while they require fine tuning under the dark matter hypothesis. In particular we find a previously unrecognised scaling between the inferred mass to light ratio of a particular dSph, with the age of the most recent star formation episode for the system. Again, a trait which emerges naturally under any modified gravity scheme, and which appears as an unexplained coincidence under the dark matter hypothesis.

Keywords: gravitation – galaxies: dwarf — galaxies: kinematics and dynamics – dark matter — Local Group
PACS: 95.35.+d, 98.10.+z,98.58.Ay

TURBULENCE MODELING AND THE PHYSICS OF THE INTRA-CLUSTER MEDIUM (ABSTRACT)

L. Iapichino[1], A. Maier[2], W. Schmidt[3] and J. C. Niemeyer[3]

[1]*Zentrum für Astronomie der Universität Heidelberg, Institut für Theoretische Astrophysik, Albert-Ueberle-Str. 2, D-69120 Heidelberg, Germany*

[2]*Lehrstuhl für Astronomie, Universität Würzburg, Am Hubland, D-97074 Würzburg, Germany*

[3]*Institut für Astrophysik, Universität Göttingen, Friedrich-Hund-Platz 1, D-37077 Göttingen, Germany*

FEARLESS (Fluid mEchanics with Adaptively Refined Large Eddy SimulationS) is a new numerical scheme arising from the combined use of subgrid scale (SGS) model for turbulence at the unresolved length scales and adaptive mesh refinement (AMR) for resolving the large scales. This tool is especially suitable for the study of turbulent flows in strongly clumped media. In this contribution, the main features of FEARLESS are briefly outlined. We then summarize the main results of FEARLESS cosmological simulations of galaxy cluster evolution. In clusters, the production of turbulence is closely correlated with merger events; for minor mergers, we find that turbulent dissipation affects the cluster energy budget only locally. The level of entropy in the cluster core is enhanced in FEARLESS simulations, in accord with a better modeling of the unresolved flow, and with its feedback on the resolved mixing in the ICM.

Keywords: galaxies: clusters: general hydrodynamics methods: numerical turbulence
PACS: 95.30.Lz,98.65.Cw,98.65.Fz

THE IMPACT OF THE EXTERNAL FIELD EFFECT IN THE MODIFIED NEWTONIAN DYNAMICS ON SOLAR SYSTEMŠS ORBITS (ABSTRACT)

L. Iorio

Viale Unità di Italia 68, 70125, Bari (BA), Italy. INFN-Sezione di Pisa

We looked at the orbital motions of test particles according to the External Field Effect (EFE) predicted by the MOdified Newtonian Dynamics (MOND) in the Oort cloud which falls in the deep MONDian regime ($r \approx 50 - 150$ kAU). Concerning the interpolating function $\mu(x)$, we extensively used the forms $\mu_1 = 1/(1+x), \mu_2 = x/(1+x^2)^{1/2}, \mu_{3/2} = x/(1+x^{3/2})^{2/3}$. We integrated both the MOND and the Newtonian equations of motion in Cartesian coordinates sharing the same initial conditions. We considered both ecliptic and nearly polar trajectories, all with high eccentricities ($e > 0.1$). In order to evaluate the characteristic MOND parameters μ_g and L_g entering the problem, we used two different values ($V = 220$ km s^{-1} and $V = 254$ km s^{-1}) of the circular speed of the solar system's motion through the Galaxy; V allows to evaluate the Milky Way's gravitational field at the Sun's location. It turns out that EFE induces strong distortions of the Newtonian ellipses, especially in the ecliptic plane yielding more involved paths which span less extended spatial regions.

Keywords: Modified theories of gravity; Experimental tests of gravitational theories; Celestial mechanics; Ephemerides, almanacs, and calendars; Oort cloud
PACS: 04.50.Kd; 04.80.Cc; 95.10.Ce; 95.10.Km; 96.50.Hp

HOW BAO MEASUREMENTS CAN FAIL TO DETECT QUINTESSENCE (ABSTRACT)

E. Jennings[1,2], C. M. Baugh[1], R. E. Angulo[3] and S. Pascoli[2]

[1]*Institute for Computational Cosmology, Department of Physics, University of Durham, South Road, Durham, DH1 3LE, U.K.*

[2]*Institute for Particle Physics Phenomenology, Department of Physics, University of Durham, South Road, Durham, DH1 3LE, U.K.*

[3]*Max Planck Intitute fur Astrophysik, D-85741 Garching, Germany.*

We model the nonlinear growth of cosmic structure in different dark energy models, using large volume N-body simulations.We consider a range of quintessence models which feature both rapidly and slowly varying dark energy equations of state, and compare the growth of structure to that in a universe with a cosmological constant. We use a four parameter equation of state for the dark energy which accurately reproduces the quintessence dynamics over a wide range of redshifts. The adoption of a quintessence model changes the expansion history of the universe, the form of the linear theory power spectrum and can alter key observables, such as the horizon scale and the distance to last scattering. The difference in structure formation can be explained to first order by the difference in growth factor at a given epoch; this scaling also accounts for the nonlinear growth at the 15% level. We find that quintessence models which feature late ($z < 2$), rapid transitions towards $w = -1$ in the equation of state, can have identical baryonic acoustic oscillation (BAO) peak positions to those in LCDM, despite being very different from LCDM both today and at high redshifts ($z \sim 1000$). We find that a second class of models which feature non-negligible amounts of dark energy at early times cannot be distinguished from LCDM using measurements of the mass function or the BAO. These results highlight the need to accurately model quintessence dark energy in N-body simulations when testing cosmological probes of dynamical dark energy.

Keywords: N-body simulations - Cosmology: theory - large-scale structure of the Universe

COLD UNIFORM SPHERICAL COLLAPSE REVISITED (ABSTRACT)

M. Joyce[1], B. Marcos[2] and F. Sylos Labini[3,4]

[1]*Laboratoire de Physique Nucléaire et de Hautes Energies, UMR 7585, Université Pierre et Marie Curie – Paris 6, 75252 Paris Cedex 05, France*

[2]*Laboratoire J.-A. Dieudonné, UMR 6621, Université de Nice – Sophia Antipolis, Parc Valrose 06108 Nice Cedex 02, France*

[3]*"E. Fermi" Center, Via Panisperna 89 A, Compendio del Viminale, I-00184 Rome, Italy*

[4]*ISC-CNR, Via dei Taurini 19, I-00185 Rome, Italy*

We report results of a study of the Newtonian dynamics of N self-gravitating particles which start in a quasiuniform spherical configuration, without initial velocities. These initial conditions would lead to a density singularity at the origin at a finite time when $N \to \infty$, but this singularity is regulated at any finite N (by the associated density fluctuations). While previous studies have focussed on the behaviour as a function of N of the minimal size reached during the contracting phase, we examine in particular the size and energy of the virialized halo which results. We find the unexpected result that the structure decreases in size as N increases, scaling in proportion to $N^{-1/3}$, a behaviour which is associated with an ejection of kinetic energy during violent relaxation which grows in propertion to $N^{1/3}$. This latter scaling may be qualitatively understood, and if it represents the asymptotic behaviour in N implies that this ejected energy is unbounded above. We discuss also tests we have performed which indicate that this ejection is a mean-field phenomenon (i.e. a result of collisionless dynamics).

Keywords: N-body simulations; virialization; spherical collapse
PACS: 95.30.-k, 95.30.Sf, 05.20.-y

BRIEF INTRODUCTION TO Q-THEORY (ABSTRACT)

F.R. Klinkhamer

Institute for Theoretical Physics, University of Karlsruhe, KIT, 76128 Karlsruhe, Germany

A special type of vacuum variable q has been introduced to describe the gravitational effects of the vacuum energy density (cosmological constant). The basic logic of q-theory is explained and the main results are discussed.

Keywords: Lorentz invariance, general relativity, cosmology, dark energy
PACS: 11.30.Cp, 04.20.Cv, 98.80.Jk, 95.36.+x

DYNAMICS DRIVEN BY THE TRACE ANOMALY IN FLRW UNIVERSES (ABSTRACT)

Jurjen F. Koksma

Institute for Theoretical Physics (ITP) & Spinoza Institute, Utrecht University, Postbus 80195, 3508 TD Utrecht, The Netherlands

By means of a semiclassical analysis we show that the trace anomaly does not affect the cosmological constant. We calculate the evolution of the Hubble parameter in quasi de Sitter spacetime, where the Hubble parameter varies slowly in time, and in FLRWspacetimes.We show dynamically that a Universe consisting of matter with a constant equation of state, a cosmological constant and the quantum trace anomaly evolves either to the classical de Sitter attractor or to a quantum trace anomaly driven one. There is no dynamical effect that influences the effective value of the cosmological constant.

Keywords: Trace Anomaly, Quantum Backreaction, Cosmological Constant, Dark Energy
PACS: 98.80.-k, 04.62.+v, 95.36.+x

COARSE-GRAINING OF INHOMOGENEOUS DUST FLOW IN GENERAL RELATIVITY VIA ISOMETRIC EMBEDDINGS (ABSTRACT)

Mikolaj Korzynski

Max-Planck-Institut für Gravitationsphysik (Albert-Einstein-Institut) 14476 Golm, Germany

We present a new approach to coarse-graining of variables describing dust flow in GR. It is based on assigning quasi-local shear, twist and expansion to 2-dimensional surfaces with the help of isometric embeddings into the 3-dimensional Euclidean space and deriving the time evolution equations for them. In contrast to the popular Buchert's scheme it allows to coarse-grain tensorial quantities in a coordinate-independent way. The framework can be used to estimate backreaction in inhomogeneous cosmological models.

Keywords: relativistic cosmology, cosmological inhomogeneities, backreaction problem
PACS: 04.20.Cv, 95.30.Sf, 95.36.+x, 98.80.Jk

ON AVERAGED COSMOLOGIES (ABSTRACT)

Julien Larena

Department of Mathematics and Applied Mathematics, University of Cape Town, Rondebosch 7701, South Africa

We review some aspects of the backreaction problem. In a first part, using an averaging scheme generalized to arbitrary frames, we quantify the effect in the Poisson gauge of perturbation theory up to second order. The main result is the definition of a scale of homogeneity above which the backreaction effect becomes scale invariant. In a second part we present results about the structure of the effective lightcone in averaged cosmologies and show that a non-trivial structure due to the backreaction may allow to test the relevance of the backreaction effect as a solution to the dark energy conundrum with future data.

Keywords: Inhomogeneous cosmologies; Averaging problem; Dark Energy
PACS: 04.20.-q, 95.30.Sf, 98.80.-k, 98.80.Jk, 95.30.-k, 95.36.+x, 98.62.En, 98.62.Py

DYNAMICS OF THE LOCAL UNIVERSE: COSMIC VELOCITY FLOWS AND VOIDS (ABSTRACT)

Guilhem Lavaux

University of Illinois at Urbana-Champaign, 1110 W Green St, 61801 Urbana, IL, USA

A valuable amount of information is available in peculiar velocities of galaxies. Peculiar velocity surveys have recently allowed the discovery of potential problems with ΛCDM. Nonetheless, their direct observation through distance measurements remains a daunting task. Another way of considering the problem is to use orbit reconstruction methods assuming some mass-to-light assignment for galaxies. We give here two applications of this procedure for the study of large-scale bulk flows and the dynamics of voids in the Local Universe.We concentrate our study on the use of the Monge- Ampère-Kantorovitch reconstruction technique. Using peculiar velocities reconstructed from the 2MASS Galaxy Redshift Survey, and after comparison with the NGB-3k peculiar velocity catalog, we look in the details of these peculiar velocities. More particularly, we estimate the constraints that the peculiar velocities put on the cosmology. The information on dynamics that is included in reconstructed orbits of galaxies also allows us to have a much better prescription for defining and identifying voids in simulations and redshift catalogs. We present this new technique and how voids may give us additional constraints on cosmology with current and future surveys.

Keywords: Peculiar velocities, method of inversion, N-body simulation, large-scale structures, statistical methods, dark energy
PACS: 95.35.+d, 95.36.+x, 98.65.Dx, 98.80.-k, 02.30.Zz, 05.10.Ln, 98.80.Es

SEPARATING EXPANSION FROM CONTRACTION: GENERALIZED TOV CONDITION, LTB MODELS WITH PRESSURE AND ΛCDM (ABSTRACT)

Morgan Le Delliou[1,2], Filipe C. Mena[3] and José P. Mimoso[2,4]

[1]*Instituto de Física Teórica UAM/CSIC, Facultad de Ciencias, C-XI, Universidad Autónoma de Madrid, Cantoblanco, 28049 Madrid SPAIN. Email: Morgan.LeDelliou@uam.es*

[2]*Centro de Física Teórica e Computacional, Universidade de Lisboa, Av. Gama Pinto 2, 1649-003 Lisboa, Portugal.*

[3]*Centro de Matemática, Universidade do Minho, Campus de Gualtar, 4710-057 Braga, Portugal Email: fmena@math.uminho.pt*

[4]*Departamento de Física, Faculdade de Ciências, Edifício C8, Campo Grande, P-1749-016 Lisboa, Portugal Email: jpmimoso@cii.fc.ul.pt*

We discuss the existence of a dividing shell separating expanding and collapsing regions in spherically symmetric solutions with pressure. We obtain gauge invariant conditions relating not only the intrinsic spatial curvature of the shells to the ADM mass, but also a function of the pressure which we introduce that generalises the Tolman-Oppenheimer-Volkoff equilibrium condition, in the framework of a 3+1 spacetime splitting. We consider the particular case of a Lemaître-Tolman- Bondi dust models with a cosmological constant (a Λ-CDM model) as an example of our results.

Keywords: Cosmology, trapped matter surfaces, Tolman-Oppenheimer-Volkoff, General Relativity and Gravitation
PACS: 98.80.-k, 98.80.Cq, 98.80.Jk, 95.30.Sf , 04.40.Nr, 04.20.Jb

INSTABILITIES IN DARK COUPLED MODELS AND CONSTRAINTS FROM COSMOLOGICAL DATA (ABSTRACT)

Laura Lopez Honorez[1,2] **and Olga Mena**[3]

[1]*Departamento de Física Teórica & Instituto de Física Teórica, Universidad Autónoma de Madrid, 28049 Cantoblanco,Madrid, Spain*

[2]*Service de Physique Théorique, Université Libre de Bruxelles, 1050 Brussels, Belgium*

[3]*Instituto de Física Corpuscular, IFIC, CSIC and Universidad de Valencia, Spain*

Coupled dark matter-dark energy systems can suffer from non-adiabatic instabilities at early times and large scales. In these proceedings, we consider two parameterizations of the dark sector interaction. In the first one the energy-momentum transfer 4-vector is parallel to the dark matter 4-velocity and in the second one to the dark energy 4-velocity. In these cases, coupled models which suffer from non-adiabatic instabilities can be identified as a function of a generic coupling Q and of the dark energy equation of state w. In our analysis, we do not refer to any particular cosmic field. We confront then a viable class of models in which the interaction is directly proportional to the dark energy density and to the Hubble rate parameter to recent cosmological data. In that framework, we show that correlations between the dark coupling and several cosmological parameters allow for a larger neutrino mass than in uncoupled models.

Keywords: dark energy, dark matter, perturbation theory,cosmology of theories beyond the SM
PACS: 98.80.Cq,98.80.Es

PROBING LONG RANGE SCALAR DARK FORCES IN TERRESTRIAL EXPERIMENTS (ABSTRACT)

Sonny Mantry

University of Wisconsin at Madison, WI, 53705,USA

A long range Weak Equivalence Principle (WEP) violating force between Dark Matter (DM) particles, mediated by an ultralight scalar, is tightly constrained by galactic dynamics and large scale structure formation. We examine the implications of such a "dark force" for several terrestrial experiments, including Eötvös tests of the WEP and direct-detection DM searches. The presence of a dark force implies a non-vanishing effect in Eötvös tests that could be probed by current and future experiments depending on the DM model. For scalar singlet DM scenarios, a dark force of astrophysically relevant magnitude is ruled out in large regions of parameter space by the DM relic density and WEP constraints. WEP tests also imply constraints on the Higgs-exchange contributions to the spin-independent (SI) DM-nucleus direct detection cross-section. For WIMP scenarios, these considerations constrain Higgs-exchange contributions to the SI cross-section to be subleading compared to gauge-boson mediated contributions. The combination of observations from galactic dynamics, large scale structure formation, Eötvös experiments, DM-direct-detection experiments, and colliders can further constrain the size of new long range forces in the dark sector.

Keywords: Dark forces, Dark Matter, Weak Equivalence Principle, Eotvos
PACS: 95.35.+d, 98.80.Cq

ELECTROMAGNETIC NATURE OF DARK ENERGY (ABSTRACT)

Jose Beltrán Jiménez and Antonio L. Maroto

Departamento de Física Teórica, Universidad Complutense de Madrid, 28040 Madrid, Spain

Out of the four components of the electromagnetic field, Maxwell's theory only contains two physical degrees of freedom. However, in an expanding universe, consistently eliminating one of the "unphysical" states in the covariant (Gupta-Bleuler) formalism turns out to be difficult to realize. In this work we explore the possibility of quantization without subsidiary conditions. This implies that the theory would contain a third physical state. The presence of such a new (temporal) electromagnetic mode on cosmological scales is shown to generate an effective cosmological constant which can account for the accelerated expansion of the universe. This new polarization state is completely decoupled from charged matter, but can be excited gravitationally. In fact, primordial electromagnetic quantum fluctuations produced during electroweak scale inflation could naturally explain the presence of this mode and also the measured value of the cosmological constant. The theory is compatible with all the local gravity tests, it is free from classical or quantum instabilities and reduces to standard QED in the flat space-time limit. Thus we see that, not only the true nature of dark energy can be established without resorting to new physics, but also the value of the cosmological constant finds a natural explanation in the context of standard inflationary cosmology. Possible signals, discriminating this model from LCDM, are also discussed.

Keywords: Dark energy, cosmological electromagnetic fields
PACS: 95.36.+x, 04.62.+v

IMPACT OF COSMIC INHOMOGENEITIES ON SNE OBSERVATIONS (ABSTRACT)

Valerio Marra and Kimmo Kainulainen

Department of Physics, University of Jyväskylä, PL 35 (YFL), FIN-40014 Jyväskylä, Finland and Helsinki Institute of Physics, University of Helsinki, PL 64, FIN-00014 Helsinki, Finland

We study the impact of cosmic inhomogeneities on the interpretation of SNe observations. We build an inhomogeneous universe model that can confront supernova data and yet is reasonably well compatible with the Copernican Principle. Our model combines a relatively small local void, that gives apparent acceleration at low redshifts, with a meatball model that gives sizeable lensing (dimming) at high redshifts. Together these two elements, which focus on different effects of voids on the data, allow the model to mimic the concordance model.

Keywords: Inhomogeneous Universe Models, Dark Energy, Backreaction, Observational Cosmology
PACS: 95.36.+x, 98.62.Sb, 98.65.Dx, 98.80.Es

ANISOTROPIC COSMOLOGICAL MODELS WITH A GENERALIZED CHAPLYGIN GAS (ABSTRACT)

N. Mebarki

Laboratoire de Physique Mathématique et Subatomique, Faculty of Science, Mentouri University, Constantine, Algeria

Anisotropic cosmological models with a generalized Chaplygin gas in a Finsler space-time geometry are considered and a class of exact solutions as well as cosmological parameters behaviors are studied. Moreover, the viability and stability criteria for a general solution are also discussed.

Keywords: Cosmology, Chaplygin gas, Finsler geometry.
PACS: 96. 10. +i, 04. 50. kd, 95. 36. +x

EXPANDING UNIVERSES IN THE CONFORMAL FRAME OF $f(R)$ GRAVITY (ABSTRACT)

John Miritzis[1] and Roberto Giambò[2]

[1]*Department of Marine Sciences, University of the Aegean, University Hill, Mytilene 81100, Greece, e-mail: imyr@aegean.gr*

[2]*Department of Mathematics and Computer Science, University of Camerino 62032 Camerino (MC) Italy*

The late time evolution of Friedmann-Robertson-Walker (FRW) models with a perfect fluid matter source is studied in the conformal frame of $f(R)$ gravity. We assume that the corresponding scalar field, nonminimally coupled to matter, has an arbitrary non-negative potential function $V(\phi)$. We prove that equilibria corresponding to non-negative local minima for V are asymptotically stable. We investigate all cases where one of the matter components eventually dominates. The results are valid for a large class of non-negative potentials without any particular assumptions about the behavior of the potential at infinity. In particular for a nondegenerate minimum of the potential with zero critical value we show that if γ, the parameter of the equation of state is larger than one, then there is a transfer of energy from the fluid to the scalar field and the later eventually dominates.

Keywords: higher-order gravity, cosmological dynamical systems.
PACS: 98.80.Jk, 98.80.Cq

OBSERVATIONALLY VERIFIABLE PREDICTIONS OF MODIFIED GRAVITY (ABSTRACT)

J. W. Moffat[1,2] and V. T. Toth[1]

[1]*Perimeter Institute for Theoretical Physics, Waterloo, Ontario N2L 2Y5, Canada*

[2]*Department of Physics, University of Waterloo, Waterloo, Ontario N2L 3G1, Canada*

MOG is a fully relativistic modified theory of gravity based on an action principle. The MOG field equations are exactly solvable numerically in two important cases. In the spherically symmetric, static case of a gravitating mass, the equations also admit an approximate solution that closely resembles the Reissner-Nordström metric. Furthermore, for weak gravitational fields, a Yukawa-type modification to the Newtonian acceleration law can be obtained, which can be used to model a range of astronomical observations. Without nonbaryonic dark matter, MOG provides good agreement with the data for galaxy rotation curves, galaxy cluster masses, and gravitational lensing, while predicting no appreciable deviation from EinsteinŠs predictions on the scale of the solar system. Another solution of the field equations is obtained for the case of a a spatially homogeneous, isotropic cosmology. MOG predicts an accelerating universe without introducing EinsteinŠs cosmological constant; it also predicts a CMB acoustic power spectrum and a mass power spectrum that are consistent with observations without relying on non-baryonic dark matter. Increased sensitivity in future observations or space-based experiments may be sufficient to distinguish MOG from other theories, notably the LCDM "standard model" of cosmology.

Keywords: Modified gravity, Galaxy rotation curves, Gravitational lensing, Bullet Cluster 1E0657-56, Cosmology
PACS: 04.20.Cv,04.50.Kd,04.80.Cc,98.80.-k

RELATIVE INFORMATION ENTROPY OF AN INHOMOGENEOUS UNIVERSE (ABSTRACT)

Masaaki Morita[1], Thomas Buchert[2], Akio Hosoya[3] and Nan Li[4]

[1]*Department of Material Science and Technology, Nagaoka University of Technology, 1603-1 Kamitomioka, Nagaoka, Niigata 940-2188, Japan*

[2]*Université Lyon 1, Centre de Recherche Astrophysique de Lyon, CNRS UMR 5574, 9 avenue Charles André, F-69230 Saint-Genis-Laval, France*

[3]*Department of Physics, Tokyo Institute of Technology, Oh-Okayama,Meguro, Tokyo 152-8551, Japan*

[4]*Center for Particle Physics and Phenomenology, Université catholique de Louvain, 2 Chemin du Cyclotron, B-1348 Louvain-la-Neuve, Belgium*

In the context of averaging an inhomogeneous cosmological model, we propose a natural measure identical to the Kullback-Leibler relative information entropy, which expresses the distinguishability of the local inhomogeneous density field from its spatial average on arbitrary compact domains. This measure is expected to be an increasing function in time and thus to play a significant role in studying gravitational entropy. To verify this conjecture, we explore the time evolution of the measure using the linear perturbation theory of a spatially flat FLRW model and a spherically symmetric nonlinear solution. We discuss the generality and conditions for the time-increasing nature of the measure, and also the connection to the backreaction effect caused by inhomogeneities.

Keywords: inhomogeneous cosmology, gravitational instability, large-scale structure of the universe, information entropy
PACS: 98.80.Jk, 95.30.Sf, 89.70.Cf, 98.65.Dx

LUMINOSITY DISTANCE $\neq$ PROPER DISTANCE: A COSMOLOGICAL DISSIMILITUDE INDUCED BY NONLINEAR ELECTRODYNAMICS (ABSTRACT)

Herman J. Mosquera Cuesta

Instituto de Cosmologia, Relatividade e Astrofísica (ICRA-BR), Centro Brasileiro de Pesquisas Físicas Rua Dr. Xavier Sigaud 150, CEP 22290-180 Urca, Rio de Janeiro, RJ, Brazil Instituto de Física, Universidade Federal do Ceará (UFC), Fortaleza, Ceará, Brazil

The current understanding of our universe is built upon the information that we extract from light coming from cosmological sources like far-away galaxies, quasars, supernovae, gamma-ray bursts, etc, including the emission in radio wavelengths of those sources. The particular analysis of supernovae type Ia (SNIa) observations has led to the idea that the universe is undergoing a late-time accelerate phase which started when it was at redshift $z \sim 1$. The redshift z is a cosmological parameter inferred from observations of emission (or absorption) lines from the expanding SNIa debris or from the supernova host galaxy, presuming the light properties and interactions, as described by Maxwell's theory, do not change while it travels through the intervening intergalactic magnetic fields. In this paper we demonstrate that the nonlinear electrodynamics (NLED) description of photon propagation through the weak background intergalactic magnetic fields introduces a fundamental modification of the cosmological redshift, as compared to the one that a direct computation within a specific cosmological model ascribes to a distant source. It is shown that independently of the class of NLED Lagrangian, the effective redshift turns out to be $(1+z)|_{\rm eff} = (1+z)\,\Delta$, where $\Delta \equiv (1+\Phi_e)/(1+\Phi_o)$, with $\Phi \equiv 8/3(L_{FF}/L_F)B^2$, being $L_F = dL/dF$, $L_{FF} = d^2L/dF^2$, the field $F \equiv F_{\alpha\beta}F^{\alpha\beta}$, and B the magnetic field strength. It comes out that for field strengths as those estimated over the intergalactic space the effective redshift is always much lower than the standard redshift, and it recovers such limit when the NLED correction $\Delta(\Phi_e,\Phi_o) \longrightarrow 1$. Since *we do not actually ever observe proper distances*, then one can argue that for a particular redshift the *observed luminosity distance* of the light-emitting far-away source *is different*. The implications of this result for cosmography studies using SNIa (and gamma-ray bursts) are discussed.

Keywords: Cosmology: standard model :: proper distance - luminosity distance - redshift - astronomical observables :: Nonlinear electrodynamics
PACS: 98.80.-k, 98.62.Ra, 95.30.Dr, 95.30.Sf

NON-SINGULAR MODIFIED GRAVITY: THE UNIFICATION OF THE INFLATION, DARK ENERGY AND DARK MATTER (ABSTRACT)

Shin' ichi Nojiri[1] and Sergei D. Odintsov[2,3]

[1]*Department of Physics, Nagoya University, Nagoya 464-8602, Japan*

[2]*Instituciò Catalana de Recerca i Estudis Avançats (ICREA), Barcelona*

[3]*Institut de Ciencies de lŠEspai (IEEC-CSIC), Campus UAB, Facultat de Ciencies, Torre C5-Par-2a pl, E-08193 Bellaterra (Barcelona), Spain*

Using the fluid representation, we formulate the conditions for the appearance of all four types finite-time future singularity in modified gravity in accelerating FRW universe. It stressed that number of standard quintessence/phantom DE theories (including scalar, fluid, DBI ones, etc) brings the accelerating cosmology to future singularity precisely in the same way as singular modified gravity DE. The viable non-singular modified gravity unifying the early-time inflation with late-time acceleration is considered. It is shown that adding such non-singular theory to another realistic singular modified gravity which has the accelerating solution with future singularity may cure the singularity of resulting combined model. This universal scenario may be naturally applied to standard singular DE models as well as to inflationary theories with initial singularity. This suggests the additional fundamental reasoning for modification of General Relativity.

Keywords: dark energy, modified gravity
PACS: 95.36.+x, 98.80.Cq, 04.50.Kd, 11.10.Kk, 11.25.-w

NEW PHENOMENOLOGY FOR PALATINI F (R) THEORIES: NON-SINGULAR UNIVERSES (ABSTRACT)

Gonzalo J. Olmo

Instituto de Estructura de la Materia, CSIC, Serrano 121, 28006 Madrid, Spain

We study modified theories of gravity of the f (R) type in Palatini formalism. We first consider the stability of atoms when the Palatini gravitational interaction is taken into account in the derivation of the non-relativistic Schrödinger equation. We show that theories with infrared curvature corrections are ruled out by the mere existence of atoms. In particular, we carry out fully perturbative calculations that, for the first time, convincingly rule out the $1/R$ model of Carroll et al. in its Palatini version. We then study the Planck scale corrected quadratic model $f(R) = R + R^2/R_P$ and show that it can avoid the big bang singularity for matter sources which satisfy all the energy conditions. We comment on the mechanisms that cure this singularity and point out that they are closely related to non-perturbative terms also present in the atomic Hamiltonian of infrared corrected models.

Keywords: Modified Gravity, Bouncing Cosmology, Backreaction
PACS: 04.50.Kd, 98.80.-k

THE SKELETON: CONNECTING LARGE SCALE STRUCTURES TO GALAXY FORMATION (ABSTRACT)

Christophe Pichon[1,2], Christophe Gay[2], Dmitry Pogosyan[2,3], Simon Prunet[2], Thierry Sousbie[2,4], Stephane Colombi[2], Adrianne Slyz[1] and Julien Devriendt[1]

[1]*Department of Physics, Denys Wilkinson Building, Keble Road, Oxford, OX1 3RH, United Kingdom*

[2]*Institut d'Astrophysique de Paris, UPMC Univ Paris 06, CNRS, UMR7095, F-75014, Paris, France*

[3]*Department of Physics, University of Alberta, 11322-89 Avenue, Edmonton, Alberta, T6G 2G7, Canada*

[4]*Tokyo University, Physics Dept 7-3-1 Hongo Bunkyo-ku, JP Tokyo, 113-0033, Japan*

We report on two quantitative, morphological estimators of the filamentary structure of the Cosmic Web, the socalled global and local skeletons. The first, based on a global study of the matter density gradient flow, allows us to study the connectivity between a density peak and its surroundings, with direct relevance to the anisotropic accretion via cold flows on galactic halos. From the second, based on a local constraint equation involving the derivatives of the field, we can derive predictions for powerful statistics, such as the differential length and the relative saddle to extrema counts of the Cosmic web as a function of density threshold (with application to percolation of structures and connectivity), as well as a theoretical framework to study their cosmic evolution through the onset of gravity-induced non-linearities.

Keywords: cosmology, larges scales structures, topology
PACS: 95.35.+d, 95.36.+x 98.80.-k, 98.80.-k 98.80.Jk

COSMOLOGICAL MODEL DISCRIMINATION FROM WEAK LENSING DATA (ABSTRACT)

S. Pires[1], J.-L. Starck[1], A. Amara[2], A. Réfrégier[1] and R. Teyssier[1]

[1]*Laboratoire AIM, CEA/DSM-CNRS-Université Paris Diderot, IRFU/SEDI, Service d'Astrophysique, CEA Saclay, Orme de Merisiers, 91191 Gif-sur-Yvette, France*

[2]*Department of Physics, ETH Zurich, Wolfgang-Pauli-Strasse 16, CH-8093 Zurich, Switzerland*

Weak gravitational lensing provides a unique way of mapping directly the dark matter in the Universe. The majority of lensing analyses use the two-point statistics of the cosmic shear field to constrain the cosmological model, a method that is affected by degeneracies, such as that between σ_8 and Ω_m which are respectively the rms of the mass fluctuations on a scale of $8Mpc/h$ and the matter density parameter, both at $z = 0$. However, the two-point statistics only measure the Gaussian properties of the field, and the weak lensing field is non-Gaussian. It has been shown that the estimation of non-Gaussian statistics for weak lensing data can improve the constraints on cosmological parameters. In this paper, we systematically compare a wide range of non-Gaussian estimators to determine which one provides tighter constraints on the cosmological parameters. These statistical methods include skewness, kurtosis, and the higher criticism test, in several sparse representations such as wavelet and curvelet; as well as the bispectrum, peak counting, and a newly introduced statistic called wavelet peak counting (WPC). Comparisons based on sparse representations indicate that the wavelet transform is the most sensitive to non-Gaussian cosmological structures. It also appears that the most helpful statistic for non-Gaussian characterization in weak lensing mass maps is the WPC. Finally, we show that the $\sigma_8 - \Omega_m$ degeneracy could be even better broken if the WPC estimation is performed on weak lensing mass maps filtered by the wavelet method, MRLens.

Keywords: Cosmology: Weak Lensing, Methods : Statistics, Data Analysis

MASSIVE GRAVITATIONAL WAVES FROM THE COSMIC DEFECT THEORY (ABSTRACT)

N. Radicella and A. Tartaglia

Dipartimento di Fisica, Politecnico di Torino and INFN sez. Torino

The Cosmic Defect theory (CD), which is presented elsewhere in this conference, introduces in the standard Einstein-Hilbert Lagrangian an elastic term accounting for the strain of space-time viewed as a four-dimensional physical continuum. In this framework the Ricci scalar acts as the kinetical term of the strain field whose potential is represented by the additional terms. Here we are presenting the linearised version of the theory in order to analyze its implications in the weak field limit. First we discuss the recovery of the Newtonian limit. We find that the typical static weak field limit imposes a constraint on the values of the two parameters (Lamé coefficients) of the theory. Once the constraint has been implemented, the typical gravitational potential turns out to be Yukawa-like. The value for the Yukawa parameter is consistent with the constraints coming from the experimental data at the Solar system and galactic scales. We then come to the propagating solutions of the linearised Einstein equations in vacuo, i.e. to gravitational waves. Here, analogously with other alternative or extended theories of gravity, the presence of the strain field produces massive waves, where massive (in this completely classical context) means subluminal. Furthermore longitudinal polarization modes are allowed too, thus lending, in principle, a way for discriminating these waves from the plane GR ones.

Keywords: Alternative models of gravity, gravitational waves
PACS: 04.30.-w, 04.50.Kd

INTRODUCING THE DARK ENERGY UNIVERSE SIMULATION SERIES (DEUSS) (ABSTRACT)

Y. Rasera[1], J.-M. Alimi[1,2], J. Courtin[1], F. Roy[1], P-S. Corasaniti[1], A. Füzfa[1,2,3] and V. Boucher[3]

[1]*CNRS, Laboratoire Univers et Théories (LUTh), UMR 8102 CNRS, Observatoire de Paris, Université Paris Diderot; 5 Place Jules Janssen, 92190 Meudon, France*

[2]*Groupe d'Application des MAthématiques aux Sciences du COsmos (GAMASCO), University of Namur (FUNDP), Belgium*

[3]*Center for Particle Physics and Phenomenology (CP3), Université catholique de Louvain, Chemin du Cyclotron, 2, B-1348 Louvain-la-Neuve, Belgium*

In this "Invisible Universe" proceedings, we introduce the Dark Energy Universe Simulation Series (DEUSS) which aim at investigating the imprints of realistic dark energy models on cosmic structure formation. It represents the largest dynamical dark energy simulation suite to date in term of spatial dynamics. We first present the 3 realistic dark energy models (calibrated on latest SNIa and CMB data): ΛCDM, quintessence with Ratra-Peebles potential, and quintessence with Sugra potential. We then isolate various contributions for non-linear matter power spectra from a series of pre-DEUSS highresolution simulations (130 million particles). Finally, we introduce DEUSS which consist in 9 Grand Challenge runs with 1 billion particles each thus probing scales from 4 Gpc down to 3 kpc at z=0. Our goal is to make these simulations available to the community through the "Dark Energy Universe Virtual Observatory" (DEUVO), and the "Dark Energy Universe Simulations" (DEUS) consortium.

Keywords: N-body simulations, dark energy, power spectrum, large-scale structures, quintessence, cosmology
PACS: 98.80.-k,95.36.+x,95.35.+d, 98.62.Ai,98.62.Py,98.65.Cw,98.65.Dx

A RESIDUAL ACCELERATION EFFECT DUE TO AN INHOMOGENEITY (ABSTRACT)

Boudewijn F. Roukema

Torun Centre for Astronomy, Nicolaus Copernicus University, ul. Gagarina 11, 87-100 Torun, Poland

The perturbed Friedmann-Lemaître-Robertson-Walker models allow many different possibilities for the 3- manifold of the comoving spatial section of the Universe. It used to be thought that global properties of the spatial section, including the topology of the space, have no feedback effect on dynamics. However, an elementary, weak-limit calculation shows that in the presence of a density perturbation, a gravitational feedback effect that is algebraically similar to dark energy does exist. Moreover, the effect differs between different 3-spaces. Among the well-proportioned spaces, the effect disappears down to third order in several cases, and down to fifth order for the Poincaré dodecahedral space $S^3/I*$. The Poincaré space, that which also is preferred in many observational analyses, is better-balanced than the other spaces.

Keywords: cosmology: observations Ű cosmic microwave background
PACS: 98.80.-k, 98.80.Es, 98.80.Jk

DBI MODELS FOR THE UNIFICATION OF DARK MATTER AND DARK ENERGY (ABSTRACT)

Luis P. Chimento[1], Ruth Lazkoz[2] and Irene Sendra[2]

[1]*Departamento de Física, Universidad de Buenos Aires, 1428 Buenos Aires, Argentina*

[2]*Fisika Teorikoa, Zientzia eta Teknologia Fakultatea, Euskal Herriko Unibertsitatea, 644 Posta Kutxatila, 48080 Bilbao, Spain*

We propose a model based on a DBI action for the unification of dark matter and dark energy. This is supported by the results of the study of its background behavior at early and late times, and reinforced by the analysis of the evolution of perturbations. We also perform a Bayesian analysis to set observational constraints on the parameters of the model using type Ia SN, CMB shift and BAO data. Finally, to complete the study we investigate its kinematics aspects, such as the effective equation of state parameter, acceleration parameter and transition redshift. Particularizing them for the best fit one appreciates that an effective phantom behavior is preferred.

Keywords: dark energy, dark matter, cosmological perturbations, geometrical constraints, DBI Lagrangians
PACS: 98.80.Jk, 98.80.Es, 95.36.+x, 95.35.+d

QUASI-LOCAL VARIABLES AND SCALAR AVERAGING IN LTB DUST MODELS (ABSTRACT)

Roberto A. Sussman

Instituto de Ciencias Nucleares, UNAM, México D.F. 04510, México

We introduce quasi-local (QL) scalar variables in spherically symmetric LTB models. If the QL scalars are defined as functionals, they become weighed averages that generalize the standard proper volume averages on space slices orthogonal to the 4-velocity. We examine the connection between QL functions and functionals and the "back-reaction" term $\mathscr{Q}$ in the context of Buchert's scalar averaging formalism.With the help of the QL scalars we provide rigorous proof that back-reaction is positive for (i) all LTB models with negative and asymptotically negative spatial curvature, and (ii) models with positive curvature decaying to zero asymptotically in the radial direction. We show by means of qualitative, but robust, arguments that generic LTB models exist, either with clump or void profiles, for which an "effective" acceleration associated with Buchert's formalism can mimic the effects of dark energy.

Keywords: Theoretical Cosmology, backŰreaction, inhomogeneous cosmological models, quasiŰlocal massŰenergy
PACS: 98.80.-Jk, 04.20.-q, 95.36.+x, 95.35.+d

CHARACTERIZING THE LARGE SCALE INHOMOGENEITY OF THE GALAXY DISTRIBUTION (ABSTRACT)

Francesco Sylos Labini

Museo Storico della Fisica e Centro Studi e Ricerche Enrico Fermi, - Piazzale del Viminale 1, 00184 Rome, Italy & Istituto dei Sistemi Complessi CNR, - Via dei Taurini 19, 00185 Rome, Italy

In order to investigate whether galaxy structures are compatible with the predictions of the standard LCDM cosmology, we focus here on the analysis of several simple and basic statistical properties of the galaxy density field.

Namely, we test whether, on large enough scales (i.e., $r > 10Mpc/h$), this is self-averaging, uniform and characterized by a Gaussian probability density function of fluctuations. These are three different and clear predictions of the LCDMcosmology which are fulfilled in mock galaxy catalogs generated from cosmological N-body simulations representing this model. We consider some simple statistical measurements able to tests these properties in a finite sample. We discuss that the analysis of several samples of the Two Degree Field Galaxy Redshift Survey and of the Sloan Digital Sky Survey show that galaxy structures are non self-averaging and inhomogeneous on scales of $\approx 100Mpc/h$, and are thus intrinsically different from LCDM model predictions. Correspondingly the probability density function of fluctuations shows a "fat tail" and it is thus different from the Gaussian prediction. Finally we discuss other recent observations which are odds with LCDM predictions and which are , at least theoretically, compatible with the highly inhomogeneous nature of galaxy distribution. We point out that inhomogeneous structures can be fully compatible with statistical isotropy and homogeneity, and thus with a relaxed version of the Cosmological Principle.

Keywords: Cosmology,Fluctuations phenomena in random processes
PACS: 98.80.-k,05.40.-a,02.50.-r

FROM ELASTIC CONTINUA TO SPACE-TIME (ABSTRACT)

Angelo Tartaglia and Ninfa Radicella

Dipartimento di Fisica, Politecnico, Corso Duca degli Abruzzi 24, 10129 Torino, Italy and INFN, Sezione di Torino, Via Pietro Giuria 1, 10126 Torino Italy

Since the early days of the theory of electromagnetism and of gravity the idea of space, then space-time, as a sort of physical continuum hovered the scientific community. Actually general relativity shows the strong similarity that exists between the geometrical properties of space-time and the ones of a strained elastic continuum. The bridge between geometry and the elastic potential, as well in three as in three plus one dimensions, is the strain tensor, read as the nontrivial part of the metric tensor. On the basis of this remark and exploiting appropriate multidimensional embeddings, it is possible to build a full theory of space-time, allowing to account for the accelerated expansion of the universe. How this can be obtained is the content of the paper. The theory fits the cosmic accelerated expansion data from type Ia supernovae better than the ΛCDM model.

Keywords: Elasticity, Strain, Defects, Space-time, Gravity
PACS: 04.50.Kd, 98.80.-k, 46.25.-y, 81.40.Jj

DYNAMICS OF INTERACTING DARK ENERGY (ABSTRACT)

Gabriela Caldera-Cabral[1], Roy Maartens[1], L. Arturo Ureña-López[2]

[1]*Insitute of Cosmology & Gravitation, University of Portsmouth, Portsmouth PO1 2EG, UK*

[2]*Departamento de Física, Campus León, Universidad de Guanajuato, C.P. 37150, León, Guanajuato, México*

It is possible that there is an exchange of energy within the dark sector of Cosmology. We consider two broad classes of interacting models where the energy exchange is a linear combination of the dark sector densities. We give general conditions on the parameters in both classes to avoid unphysical behavior, such as negative energy densities and the appearance of instabilities in their linear perturbations.

Keywords: Cosmology, Dark Energy, Dark Matter
PACS: 98.80.-k, 95.35.+d, 98.80.Jk

MASS FUNCTIONS AND BIAS OF DARK MATTER HALOS (ABSTRACT)

P. Valageas

Institut de Physique Théorique, CEA Saclay, 91191 Gif-sur-Yvette, France

We report on some results on the mass function and the bias of dark matter halos. Focusing on the limit of rare massive halos, we point out that exact analytical results can be obtained for the large-mass tail of the halo mass function. This is most easily seen from a steepest-descent approach, that becomes asymptotically exact for rare events. We also revisit the traditional derivation of the bias of massive halos, associated with overdense regions in the primordial density field, paying attention to the Lagrangian-Eulerian mapping (i.e. corrections associated with the motions of halos). This allows us to improve the standard analytical formula for the bias of massive halos without introducing any free parameter. Next, we briefly note that this approach can be extended to non-Gaussian initial conditions. We again obtain a good agreement with numerical simulations. Moreover, the nonlinear real-space expression we obtain for the halo correlation clearly shows how the baryon acoustic oscillation at $\sim 100h^{-1}Mpc$ is amplified by the bias of massive halos and modified by primordial non-Gaussianity. At smaller scales, $30 < x < 90h^{-1}Mpc$, we find that the correction to the real-space bias roughly scales as $f_{NL}b_M(f_{NL}=0)x^2$.

Keywords: Cosmology: large-scale structure of Universe; gravitation; Methods: analytical
PACS: 98.65.-r, 98.65.Cw, 98.65.Dx, 98.80.-k, 95.35.+d

THE SWISS-CHEESE TOY MODEL IN THE LIGHT OF THE CMB (ABSTRACT)

Wessel Valkenburg

LAPTH, CNRS & Université de Savoie, 9 chemin de Bellevue, 74941 Annecy-Le-Vieux, France

Keywords: Cosmology, Inhomogeneous Universe, Dark Energy, Large Scale Structure
PACS: 98.65.-r, 98.80.-k, 98.80.-CQ, 04.25.D

GRAVITATIONAL ENERGY AS DARK ENERGY: AVERAGE OBSERVATIONAL QUANTITIES (ABSTRACT)

David L. Wiltshire

Department of Physics & Astronomy, University of Canterbury, Private Bag 4800, Christchurch 8140, New Zealand; International Center for Relativistic Astrophysics Network (ICRANet), P.le della Repubblica 10, Pescara 65121,Italy

In the timescape scenario cosmic acceleration is understand as an apparent effect, due to gravitational energy gradients that grow when spatial curvature gradients become significant with the nonlinear growth of cosmic structure. This affects the calibratation of local geometry to the solutions of the volumeŰaverage evolution equations corrected by backreaction. In this paper I discuss recent work on defining observational tests for average geometric quantities which can distinguish the timescape model from a cosmological constant or other models of dark energy.

Keywords: dark energy, theoretical cosmology, observational cosmology
PACS: 98.80.-k 98.80.Es 95.36.+x 98.80.Jk

CONCLUSION

Phenomenology of the Invisible Universe

P. J. E. Peebles

Joseph Henry Laboratories, Princeton University, Princeton NJ USA

Abstract. Cosmology is operating now on a well established and tightly constraining empirical basis. The relativistic ΛCDM hot big bang theory is consistent with all the present tests; it has become the benchmark. But the many open issues in this subject make it reasonable to expect that a more accurate cosmology will have more interesting physics in the invisible sector of the universe, and maybe also in the visible part.

Keywords: cosmology, dark sector
PACS: 98.80.-k

THE COSMOLOGICAL TESTS

Lest the exciting debates over the great open issues of cosmology cause us to forget, I offer in Figure 1 and Table 1 reminders that we have a firm and substantial empirical basis for our subject. The figure shows the spectrum of the cosmic microwave radiation that nearly uniformly fills space: energy density as a function of frequency. The measurements are diverse and well cross-checked, from the ground, balloons, rocket and satellite missions, and indirectly from spin temperatures derived from interstellar molecular absorption lines. Within the measurement uncertainties, which over much of the range of frequencies in this figure are smaller than the symbols, this wealth of data fits the Planck thermal spectrum at temperature $T = 2.725$ K plotted as the solid curve.

We know the universe as it is now is close to transparent at the frequencies plotted here, because active galaxies at redshifts exceeding unity are detected at these frequencies. Relaxation to this thermal spectrum thus had to have happened earlier in the expansion of the universe, when the density of absorbing matter was great enough make the universe opaque across the Hubble length then. That is, the figure presents us with almost tangible evidence that the universe really did evolve from a very different state. We should take a moment to admire this wonderfully simple and deeply significant phenomenon.

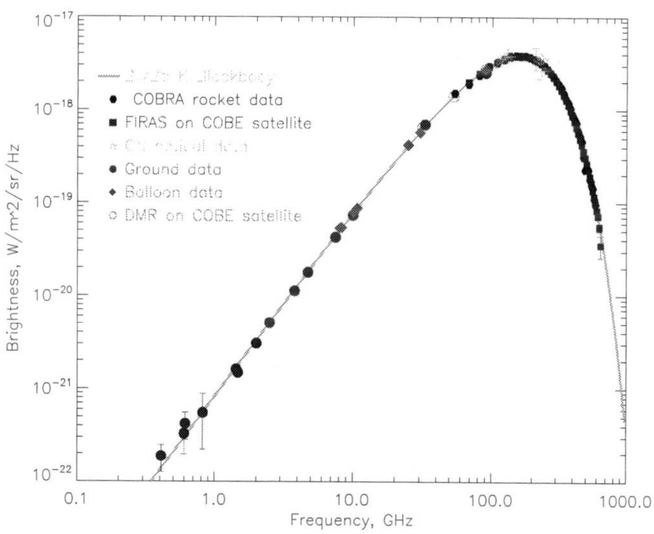

FIGURE 1. Energy spectrum of the cosmic microwave background, from Alan Kogut.

CP1241, *Invisible Universe: Proceedings of the Conference,* edited by J.-M. Alimi and A. Füzfa

TABLE 1. Cosmological Tests

1. fossil CMB radiation	6. cosmic mass density
(a) energy spectrum	(a) galaxy peculiar velocities
(b) temperature acoustic oscillation	(b) gravitational lensing
(c) temperature-polarization cross spectrum	7. large-scale structure
(d) integral Sach-Wolfe effect	(a) baryon acoustic oscillation
2. light element abundances	(b) galaxy N-point functions
(a) deuterium in Lyα absorbers at redshift ~ 3	8. clusters of galaxies
(b) helium in dwarf low-metallicity galaxies	(a) mass function as a function of mass & redshift
(c) observed baryon budget	(b) baryon mass fraction
(d) count of neutrino families	9. small-scale structure
3. redshift-magnitude relation	(a) galaxy structure and evolution
4. stellar evolution & radioactive decay ages	(b) Lyman-α forest
5. distance scale	
(a) distance ladders	
(b) gravitational lensing	

The figure also shows that expansion and cooling to the present transparent state of the universe had to have preserved the thermal radiation spectrum formed when the universe was opaque. Sufficient conditions include standard local physics and a metric description of spacetime. It does not require general relativity theory. It does require that, despite the distinctly inhomogeneous distribution of matter in galaxies and concentrations of galaxies, spacetime is close enough to homogeneous and isotropic not to have produced a sigificant spread of redshifts along different lines of sight (as could have happened in a homogeneous anisotropic universe, for example). Space has to be close to what Einstein envisioned in 1917.

The ΛCDM cosmology adopts Einstein's near homogeneous space, his general relativity theory, and his cosmological constant, and it adds the general expansion first discussed by Friedmann and Lemaître, Tolman's thermal radiation, Gamow's thermonuclear element formation, and the more recent notions of nonbaryonic dark matter and adiabatic near scale-invariant Gaussian initial departures from homogeneity. The application of general relativity is a long extrapolation from its precision tests, but it was natural to try this theory first. Other of these assumptions were not so obvious first guesses; they were adopted because they were seen to offer promising fits to the improving tests. We would be poor theorists indeed if ΛCDM did not at least approximate what is now measured. But this cosmology has gone beyond that; it is predictive.

This situation is illustrated in Table 1 (adapted from the book, *Finding the Big Bang* [1]; details are there and in other reviews of the state of the cosmological tests). Each measurement named in the table is an independent (or close to it) probe of the universe. Each is capable of falsifying ΛCDM. This cosmology passes the tests so far. The measurements are difficult, and issues of interpretation and systematic errors are well worth continued close examination. But the diversity of these probes, and the consistency of independent constraints on the relevant parameters in ΛCDM, make a close to compelling case that we have not been seriously misled. That certainly does not mean the ΛCDM cosmology is the final word on the structure and physics of the universe that will be needed to interpret coming generations of measurements. But we can be sure that if ΛCDM is replaced by something better the new theory will predict a universe that looks much like ΛCDM, because that is what is observed.

Precision Cosmology and Accurate Cosmology

We should take another moment to admire the abundance of probes of the large-scale nature of the universe represented in Table 1, and to recall the importance of the diversity of these measurements.

It has been said that we are moving toward precision cosmology, which is correct but misses the point. A digital scale may read out the weight of an object to many significant figures, in a precise measurement. But if the scale is not well calibrated the measurement may not be very accurate. That is, accuracy is what remains of precision after discounting systematic errors. And our goal is an accurate cosmology.

In the pursuit of precision cosmology it is natural to concentrate on the relatively few observations that lend themselves to the most precise measurements. This is good science, provided one bears in mind that if the precision measurements are not more numerous than the parameters that may be adjusted to fit them we are getting a cosmology of dubious accuracy. I offer one example, the properties of galaxies.

The mass distributions in the outer parts of galaxies were an apparent anomaly within the old cosmology, and a hint of something new: dark matter. Recent research proceeds in the opposite direction: take ΛCDM, dark matter and all, as given, to be used as the basis for analyses of how galaxies acquired their observed properties. Since galaxies are complicated — our understanding of star formation and its impact on the diffuse baryons is particularly schematic — the program must add parameters that are physically well motivated but adjustable because we don't understand how the physics actually is expressed. The freedom of adjustment makes it difficult to judge the significance of the notable advances in this program. The pursuit of more accurate cosmology may be aided by a return to the earlier direction. There are apparent anomalies in the properties of galaxies in terms of what might have been expected in a straightforward reading of numerical simulations of structure formation in ΛCDM. Instead of asking how the parameters needed to describe the histories of the baryons may be adjusted to remedy this situation consider instead whether apparent anomalies might be hints to better fundamental physics. It is not difficult to invent alternative physics that fits the tests on the scale of the Hubble length about as well as ΛCDM, but on the scale of galaxies does something different, interesting, and just possibly better. An example of an alternative theory on the scale of galaxies is discussed next; others are in [2] – [4].

THE INVISIBLE UNIVERSE

Dark Matter

Milgrom's MOdified Newtonian Dynamics, MOND, is a very interesting alternative, an adjustment to gravity physics that allows dark matter to be subdominant to baryons almost everywhere. MOND offers a natural account of the scaling relation $v \propto L^{1/4}$ known as the Tully-Fisher relation between spiral galaxy circular velocity v and luminosity L, and the Faber-Jackson relation between stellar velocity dispersion and luminosity of ellipticals. The relation is even tighter when one uses baryon mass instead of luminosity, and it extends to baryon masses well below what was known when MOND was introduced. That is, on the scales of its intended application MOND is predictive. In these Proceedings Hammer describes observations that indicate spiral galaxies arrive at the $v \propto L^{1/4}$ relation as they form, whether early or late in the expansion of the universe. This is to be expected in MOND, but curious in ΛCDM. Ellipticals follow a scaling relation among velocity dispersion, size and luminosity (or baryon mass) that is even tighter than $v \propto L^{1/4}$, and the relation is little affected by environment. I recommend Figure 3 in Bernardi *et al.* [5] for a striking — even startling — illustration of the general insensitivity to what is happening around the galaxy.[1]

How do galaxies "know" to join these near universal relations? It is tempting to think the relations might be "wired into" the physics, as in MOND. But we know that complex processes operating on smaller scales are capable of producing wonderful scaling relations, and maybe the same is true of galaxy formation.

MOND is not ruled out by the cosmological tests because it does not make agreed upon predictions on these scales. But MOND faces what I count as a daunting challenge: show how a generalization can fit the tests on larger scales that on the face of it quite systematically point to the dominance of nonbaryonic over baryonic matter.

I remain persuaded by the empirical case for nonbaryonic dark matter. But the cautionary example of MOND shows why we should be open to possibly better ideas and watching for phenomena such as the galaxy scaling relations that may be pointing to better physics than the woefully simple dark sector of ΛCDM. The universe is large, and quite capable of continuing to instruct us.

Dark Energy

I have not seen any way to avoid the quantum physics prediction of significant contributions to the vacuum energy density. We recall that if the vacuum does not define a preferred frame of motion then its energy appears in the stress-energy tensor in the form of Einstein's Λ, in elegant agreement with the evidence for Λ from cosmology. But, as widely lamented, the value of Λ in cosmology seems absurdly small compared to natural estimates of contributions from quantum physics. Maybe some symmetry forces the net quantum energy to vanish, another arguably natural

[1] The central generally most luminous galaxy in a rich cluster does systematically differ from the general population, but I am impressed at how small the differences are. For discussions of the evidence and the challenges of its interpretation see [6].

value. But if so how do we understand the cosmological tests? Is dark energy really present? If so is it the result of a miraculously near but not exact cancellation of quantum contributions, a parameter of the theory – Einstein's Λ, or a dynamical entity? If dynamical what is it and how do we understand its apparently curious value?

Departures from Homogeneity and the General Expansion of the Universe

Could the accelerated cosmic expansion usually attributed to Λ be instead an effect of the departure from an exactly homogeneous mass distribution? Analyses in these Proceedings illustrate the richly complicated physics of the departure from homogeneity in the classical chaos of matter collecting into galaxies. But I think it is clear that this physics has negligibly small effect on the general expansion. I explain at some length because the issue is important and the physics interesting.

The idea that the general expansion of the universe is significantly affected by the observed clustering of matter is inspired by analyses of variants of Raychaudhuri's equations. They relate the evolution of the expansion, density, shear and vorticity along the world line attached to a fluid element. Within the assumption of a pressureless continuous fluid with a differentiable fluid velocity, which is a sensible approximation for this purpose, and apart from the effect of orbit crossings, these relations are identities in general relativity. But deriving the general expansion by averaging the expansions of fluid elements that generally have ended up in the complexity of strongly nonlinear mass concentrations in or around galaxies is an exceedingly challenging problem.

There is a simpler way [7] – [10]. The mass concentrations in galaxies have characteristic Newtonian gravitational potentials on the order of $\phi \sim v^2 \sim 10^{-6}$, where $v \sim 300$ km s^{-1} is a typical velocity between galaxies and within galaxies. Since ϕ is small the departure from perfectly homogeneous and isotropic spacetime is small and readily analyzed in perturbation theory.

Consider, for example, that the cosmic microwave radiation has passed through the strongly nonlinear mass concentrations in galaxies. But because the gravitational potentials belonging to these departures from homogeneity are small, and they vary slowly with time, the effect on the radiation is exceedingly small. It is not detected in the radiation energy spectrum shown in Figure 1. The effect is observed [11] in the cross-correlation of the variation of the cosmic microwave radiation temperature across the sky (at $\delta T/T \sim 10^{-5}$) with the large-scale galaxy distribution, which is a reasonably useful tracer of mass. This cross-correlation — the integrated Sachs-Wolfe effect — is computed to all needed precision in linear perturbation theory of the departure from a homogeneous and isotropic spacetime produced by the observed clustering of matter.

Computation of the effect of mass clustering on the general expansion of the universe requires some elements of second-order perturbation theory. For simplicity let us model the material content of the universe as a gas of noninteracting dark matter particles with mass m and mean number density $\bar{n}$. The matter is gravitationally bound in dark halos — galaxies — with internal velocities $v \sim 300$ km s^{-1} and sizes $r_g \sim 10$ kpc. Close to equivalent parameters that measure the effect of the clustering of mass in dark halos on the spacetime geometry are

$$\varepsilon \sim \phi \sim (v/c)^2 \sim r_g/ct \sim 10^{-6}, \tag{1}$$

where t is the Hubble time. This is a reasonable approximation to our cosmologically flat universe at the present epoch, apart from the subdominant mass in baryons and radiation. Galaxies are nonrelativistic; ε is small. This suggests the line element

$$ds^2 = a(\tau)^2[(1+2\phi)d\tau^2 - (1-2\phi)\delta_{ij}dx^i dx^j], \tag{2}$$

which reduces to the nonrelativistic Newtonian form on the scale of galaxies, and describes the general expansion of the universe in the average over much larger scales.

In this line element the time-time part of Einstein's field equation to zeroth order in ε is

$$R_{00} = -3\frac{d}{d\tau}\frac{1}{a}\frac{da}{d\tau} + \nabla^2\phi = 4\pi Gm\bar{n}a^2(1+\delta), \tag{3}$$

where $\delta(\vec{x},t) = n(\vec{x},t)/\bar{n}(t) - 1$ is the dimensionless measure of the departure from a homogeneous mass distribution. In the central parts of galaxies δ is much greater than unity, but that does not concern us; we are interested in ϕ, which we know is small (apart from neutron stars and black holes, an interpretation of which is offered below). The standard and sensible practice is to separate this relation into the two parts

$$\frac{\ddot{a}}{a} = -\frac{4}{3}\pi Gm\bar{n}, \qquad \nabla^2\phi = 4\pi Gm\bar{n}a^2\delta, \tag{4}$$

where the dots are derivatives with respect to proper time ($dt = ad\tau$), and the boundary condition is that ϕ has no source-free part. This means the spatial average of ϕ in the coordinates of Eq. 2 vanishes along with the spatial average of δ.

To find the effect of mass clustering on the Friedmann-Lamaître equations to order ε we must compute spatial averages of Einstein's field equation to order ϕ^2 for the dominant term,

$$t^2\langle\nabla\phi\cdot\nabla\phi\rangle \sim \varepsilon. \tag{5}$$

To order ε we can evaluate Eq. 5 using the zeroth order equation for ϕ in Eq. 4. It is an interesting exercise to complete the calculation (which requires modest extensions of the computations in [7] – [10]) and check that the spatial averages of the time and space parts of Einstein's equation yield the usual two Friedmann-Lemaître equations in which the effective mean mass density and pressure terms are

$$\bar{\rho} = m\bar{n}(1+K+W), \quad \bar{p} = m\bar{n}(2K+W)/3, \quad K = \langle v^2\rangle/2, \quad W = -\frac{1}{2}Gm\bar{n}a^2\int\frac{d^3x}{x}\langle\delta(\mathbf{r})\delta(\mathbf{x}-\mathbf{r})\rangle. \tag{6}$$

These are in the coordinates of Eq. 2, which to order ε are what an astronomer uses to measure δ and v. The mean kinetic energy per unit mass, K, comes from the stress-energy tensor for the collisionless particle model. The gravitational energy per unit mass, W, has contributions from the stress-energy tensor and from the term proportional to Eq. 5 in the left-hand side of the field equation.

For an intuitive understanding of Eq. 6 consider the stress-energy tensor of a galaxy. We know from the nonrelativistic kinetic theory of gasses that in the galaxy rest frame an isotropic dark matter particle velocity dispersion has pressure $m\bar{n}\langle v^2\rangle/3$, which appears in the diagonal space part of the stress-energy tensor. It is not surprising that the space average of this expression is the term $2m\bar{n}K/3 = m\bar{n}\langle v^2\rangle/3$ in Eq. 6. Now consider the spatial part of the total stress-energy tensor. In a statistically isotropic solid with a free boundary the space parts of the stress-energy tensor have zero mean. That comes about because the interaction holding the solid together adds to the stress-energy tensor tensor a term that just cancels the contribution by the velocity dispersion of the ions. In a gravitationally bound and stationary galaxy the virial theorem says the gravitational energy per unit mass, W, and the kinetic energy per unit mass, K, satisfy $W = -2K$, meaning $\bar{p}$ vanishes in Eq. 6, in analogy to the solid. A non-rotating neutron star has a large internal pressure, again just balanced by its internal gravity: the galaxy sees it as a particle, to excellent accuracy. The orbits of stars around the massive black hole in the center of a galaxy with a well-developed bulge show that the stars see the black hole as an ordinary point mass, to excellent accuracy, and it is scarcely surprising that the rest of the universe sees it in the same way. The gravity holding a galaxy together sees a considerable mass deficit in a neutron star or black hole. And, as expected, the effective mass density $\bar{\rho}$ in Eq. 6 is lower than the rest mass density $m\bar{n}$ by the amount of the binding energy per unit mass, $-(K+W)$. Finally, I offer another interesting exercise: following [8] check consistency of Eq. 6 with the relation between K and W that Irvine [12] derived in 1961,

$$\frac{d}{dt}(K+W) = -\frac{1}{a}\frac{da}{dt}(2K+W). \tag{7}$$

To avoid confusion I repeat that this paragraph is not a derivation of Eq. 6 but rather an explanation of why I find the result persuasive.

The large-scale growing mass fluctuations in the real universe are not in statistical equilibrium, so $2K+W$ does not vanish. These fluctuations contribute to $\bar{p}$, in an analog of sorts to the integrated Sachs-Wolfe effect. Also, Eq. 6 is a spatial average, which is appropriate for the effect of spacetime variations along a line of sight but not for the potential ϕ at the observer. But these corrections to the Friedmann-Lemaître equations are on the order of ε, roughly a part in 10^6, well below what would needed to vitiate the evidence for detection of Λ. I respect the deep analyses that start from Raychaudhuri's equations, but I think the situation is clear. Within general relativity theory we must learn to live with Λ.

Modified Gravity

One can avoid that, eliminate Λ, by adjusting the gravity physics, as also is well discussed in these Proceedings. One approach is to choose a Lagrangian density for gravity operating four-dimensional spacetime that is close enough to the Ricci scalar R of general relativity theory in regions of high density and on relatively small scales to pass the precision tests of gravity physics, and on large scales acts as a close enough surrogate for R plus a constant to pass the cosmological tests. One can choose a function $f(R)$ that fits these conditions and passes the present tests of gravity and cosmology about as well as general relativity with a cosmological constant — provided quantum physics has no net effect on the vacuum. If more precise and accurate tests prove to be inconsistent with R plus a constant and agree with an $f(R)$ theory it will be a really profound advance.

Other ways to modify gravity are discussed in these Proceedings. I depart from my charge, the phenomenology of the invisible universe, to offer my supposition that the provenance within some deeper unified theory of everything of a successful modified gravity would be an even greater puzzle than the standard R plus a constant. I expect the community by and large will continue to work with R plus a constant in ΛCDM until driven from it. And it is right and proper that others will continue to prepare the way for the eventuality that something better than R plus a constant proves to be needed.

Learning to Live with Λ

In the ΛCDM cosmology what might be the nature of this mysterious dark energy component, and what might we make of its curious value?

The evidence is that our universe now is becoming dominated by dark energy, in effect approaching another epoch of inflation but at really low energy. It is natural to consider the possibility that the dark energy manifest in the cosmological tests is a remnant of the dark energy that drove inflation; maybe the present dark energy density is small because it has been rolling toward zero for a long time. These Proceedings document the rich considerations in research on the physics of inflation that may also shed light on the nature of the present Invisible Universe.

A demonstration that the dark energy density actually is evolving, as opposed to Einstein's constant Λ, certainly would stimulate thinking about the nature of dark energy, though what the community would think about is less clear; we are short of manifestly promising ideas. A detection of variation of the strength of the electromagnetic interaction, or of other of the parameters of physics, also would be a deeply influential advance. The same would be true of detection of HI radiation (redshifted 21-cm emission from atomic hydrogen) from reionization, or of a characterization of the dark matter, or of a third generation all-sky position-magnitude-redshift survey, or of a really searching and complete study of how stars form as a function of environment. Research aimed at the first of these problems, detecting or better constraining evolution of the dark energy density, certainly merits community support. But I don't think our subject is ready for a dark energy mission in the big science mode of the search for the Higgs boson; there is still too much other work to be done in straightening the foundations of our subject.

What are we to make of the value of Λ? Most widely discussed is the anthropic argument, as well framed by Weinberg [13]: among the ensemble of all possible universes, we have to have found ourselves in one that allows the formation of galaxies of stars that last some 10^{10} years. This is an elegant answer, but at the expense of a new question: how do we determine whether the anthropic argument is more than a "just so" story invented to save the phenomenon?

On the positive side, the anthropic argument offers a possible window into the deeper physics that allows many universes and, in universes in which dark energy is a meaningful concept, some hint to the range of values the deeper physics allows this parameter. It is noteworthy that, at a time when Λ was not at all popular, Weinberg's considerations along these lines led him to conclude that the dark energy density might be expected to be larger than the present matter density, which has proved to be the case. Similar arguments have been applied to other of the many parameters in our physics that are so very well and finely tuned to allow our existence, from the strength of the strong interaction to the expansion of water on freezing. This offers fascinating lines of investigation, but I would be more comfortable with a less anthropocentric way to put it, as follows.

In another universe different physics may allow development of systems that become so deeply complex as to "wonder why" they are in a universe that is so wonderfully well suited to the requirements of their existence. (The phrase, "wonder why," is anthropocentric, of course, but I don't have a better way to put it.) The answer many of us would be happy with is that that is where these systems evolved. Maybe somewhat fewer would be happy with the similar argument that all the details of all the physics of our universe are exquisitely well suited for our existence because this is where we evolved, but it seems logical to me. In this way of putting it the challenge is to determine the bounds on the values of Λ and on all the rest of physics in the class of all possible universes in which systems can get complex enough to start "wondering why" conditions are what they are.

SUMMARY REMARKS

Cosmology has changed from the mode of operation the older of us remember without nostalgia. We are now tightly constrained by a rich observational basis that makes it exceedingly delicate — though certainly not impossible — to find and demonstrate viability of alternatives to the ΛCDM cosmology. We each have chosen a mode of research appropriate for this relatively new situation in cosmology. It is worth listing strategies, to remind ourselves and funding agencies that we do have choices.

It is entirely reasonable to take the conservative position that ΛCDM, maybe with slow evolution of the dark energy density and fine tuning of initial conditions, includes all the physics that is going to be relevant to analyses of phenomenological cosmology. Then the goal becomes to understand how this given physics is expressed in the processes that make the universe we observe, from the Hubble length down to galaxies and stars and planets, and back in time to light element production and maybe baryogenesis. This strategy has proven its worth by defining a host of productive research problems.

If ΛCDM does differ from reality enough to matter it means this conservative strategy will reveal it through identification of phenomena that cannot be reconciled with the cosmology. But beware of precision cosmology, because the observations that lend themselves precision measurements reduce the tests in Table 1 to a number perilously close to the parameters we are willing to adjust in ΛCDM. (Currently discussed additions to the more traditional cosmological parameters include evolution and maybe spatial gradients of the dark energy density, a soupçon of annihilating or collisional or warm dark matter, or maybe even a new long-range interaction in the dark sector. A competent theorist likely could offer quite a few other ways to save the phenomena within ΛCDM.) I expect that the better strategy for finding a more accurate cosmology will be to examine all measurements, however inaccurate, that probe different aspects of the universe and may challenge standard ideas.

A more proactive strategy for discovering a possibly more accurate cosmology is to search for apparent anomalies in ΛCDM, situations that seem particularly challenging for this theory. My list commences with a pronounced difference between the physics of the visible and invisible universes. Physics in the visible sector is wonderfully simple, in the appropriate sense, but its expression is spectacularly complicated. Physics in the invisible sector is so exceedingly simple that its expression is simple: the dark matter just piles up in halos and the dark energy sits there or maybe slowly evolves. It is wishful thinking, but also sensible, to suspect that this apparently anomalous situation is telling us that we have settled for the simplest physics we can get away with at the present crude levels of probes of this sector. It presents no hints of phenomena to look for; I read it an invitation to look broadly.

Other apparent anomalies may offer more direct hints of better physics. Examples include preliminary results from underground dark matter particle detectors and the search for signatures of dark matter annihilation. At the time of writing the experimental/observational situation is confused, maybe in part a result of systematic errors in difficult measurements, maybe in part because we are seeing hints to something new. Other examples are the more curious properties of galaxies. How did galaxies arrive at their scaling relations that show such little regard for environment? Why do the voids defined by normal galaxies look so very empty? Why do most of the nearest spiral galaxies edge up to the very empty Local Void? How did massive black holes in active galactic nuclei form so early in the expansion of the universe? Why is large-scale structure in the galaxy distribution so very large? If such curiosities are expressions of standard physics then understanding how they came about will teach us something of value. And maybe some are hints to better physics, which would be a good thing too.

Research in the proactive direction may accept the standard physics of the visible universe and seek something more interesting in the poorly explored invisible part. As one sees in these Proceedings, there is no shortage of ideas here. The adoption of standard visible sector physics is conservative, but maybe overly so. For example, the precision tests of general relativity are on scales some fifteen orders of magnitude smaller than the Hubble length. It is sensible to continue to question this enormous extrapolation, to consider modifications of gravity physics. This strategy has led to viable alternatives that to date seem to me to be less logically compelling than general relativity with ΛCDM. This is not a very convincing argument, but I expect it will lead most in the community to continue to work within the physics of ΛCDM unless/until some new phenomenology or really attractive idea drives us from it.

I refer to Wigner's [14] essay, *The Unreasonable Effectiveness of Mathematics in the Natural Sciences*, for the arguments that lead me to expect that the community will not arrive at the one true cosmology, if such a thing exists. We will instead end up with the best approximation allowed by the limited ability to observe. We cannot rule out the possibility that something will turn up that invalidates our present cosmology, but from all experience that is exceedingly unlikely. Much easier to imagine is fine tuning to fit new measurements will lead to a more accurate cosmology that contains elements of ΛCDM along with new and interesting adjustments.

REFERENCES

1. P. J. E Peebles, L. A. Page and R. B. Partridge, *Finding the Big Bang*, Cambridge University Press, Cambridge, 2009.
2. N. Afshordi, G. Geshnizjani and J. Khoury, *Journal of Cosmology and Astro-Particle Physics* **8**, 30 (2009).
3. J.A. Keselman, A. Nusser and P.J.E. Peebles, arXiv:0902.3452 (2009).
4. S.M. Carroll, S. Mantry, M.J. Ramsey-Musolf and C.W. Stubbs, *Physical Review Letters* **103**, 011301 (2009).
5. M. Bernardi, R.C. Nichol, R.K. Sheth, C.J. Miller and J. Brinkmann, *Astronomical Journal* **131**, 1288 (2006).
6. M. Bernardi, J.B. Hyde, R.K. Sheth, C.J. Miller and R.C. Nichol, R. C. *Astronomical Journal* **133**, 1741 (2007).
7. U. Seljak and L. Hui, in *Clusters, Lensing, and the Future of the Universe*, edited by V. Trimble and A. Reisenegger, Astronomical Society of the Pacific, San Francisco, 1996, p. 267.
8. E. R. Siegel and J. N. Fry, *Astrophysical Journal Letters* **268**, L1 (2005).
9. M. Kasai, H. Asada and T. Futamase, *Progress of Theoretical Physics* **115**, 827 (2006).
10. A. Paranjape and T.P. Singh, *Physical Review Letters* **101**, 181101(2008).
11. T. Giannantonio, R. Scranton, R.G. Crittenden, R.C. Nichol, S.P. Boughn, A.D. Myers and G.T. Richards, *Physical Review D* **77** 123520 (2008).
12. Irvine, W. M. Ph.D. Thesis, Harvard University (1961).
13. S. Weinberg, *Physical Review Letters* **59**, 2607 (1987).
14. E. P. Wigner, *Communications in Pure and Applied Mathematics* **13**, 1 (1960),

POSTERS

POSTERS

HIGHER SPIN DARK MATTER (ABSTRACT)

M. Asorey[1] and D. García-Álvarez[2]

[1]*Departamento de Física Teórica. Facultad de Ciencias. Universidad de Zaragoza. 50009 Zaragoza. Spain*

[2]*Department of Earth and Environmental Sciences. The Open University. Milton Keynes MK7 6AA, United Kingdom*

We analyze the role of spin three gauge fields as dark matter candidates. The relevant feature of these higher spin fields is that they do not couple directly to Standard Model particles. Their only direct coupling is to the gravitational field which explains the very weak interactions with other fields and enhances their qualities as realistic candidates for dark matter. The mass of these particles is provided by a Higgs-like mechanism with the gravitational field playing the role of Higgs field.

Keywords: High spin gauge fields, Dark matter, High spin Higgs fields
PACS: 11.15.-q, 95.35.+d, 12.10.-g, 14.80.Fd

STAR MODELS WITH DARK ENERGY (ABSTRACT)

R. Chan[1], M. A. F. da Silva[2] and Jaime F. Villas da Rocha[3]

[1]*Coordenação de Astronomia e Astrofísica, Observatório Nacional, Rua General José Cristino 77, São Cristóvão, CEP 20921-400, Rio de Janeiro, RJ, Brazil*

[2]*Departamento de Física Teórica, Universidade do Estado do Rio de Janeiro, Rua São Francisco Xavier* 524, *Maracanã, CEP 20550-013, Rio de Janeiro, RJ, Brazil.*

[3]*Universidade Federal do Estado do Rio de Janeiro, Instituto de Biociências, Departamento de Ciências Naturais, Av. Pasteur 458, Urca, CEP 22290-240, Rio de Janeiro, RJ, Brazil.*

Inspired by the gravastars models, we have constructed star models consisting of a homogeneous inner core with anisotropic pressure, an infinitesimal thin shell separating the core and the envelope, an envelope of inhomogeneous density and isotropic pressure and an infinitesimal thin shell matching the envelope boundary and the exterior Schwarzschild spacetime. Analyzing all the energy conditions for the core, envelope and the two thin shells, we have found that static solutions are possible only if at least one of the regions must be constituted by dark energy. Our results show that there is not a superior limit for the mass of these objects, but for the ratio of mass and radius.

Keywords: Dark energy, Gravastars, Star models
PACS: 04.20.Jb, 04.40.Dg, 97.10.q

CP1241, *Invisible Universe: Proceedings of the Conference*, edited by J.-M. Alimi and A. Füzfa

PRIMORDIAL NON-GAUSSIANITY IN DENSITY FLUCTUATIONS (ABSTRACT)

F. Fraschetti[1,2], J.-M. Alimi[2], J. Courtin [2] and P.-S. Corasaniti[2]

[1]*Lunar and Planetary Lab & Dept. Physics, 1629 E University Bvd, University of Arizona, 85721-0092, Tucson, AZ, USA*

[2]*LUTh, Observatoire de Paris, CNRS-UMR8102 and Université Paris VII, 5 Place Jules Janssen, F-92195 Meudon Cédex, France.*

We present N-body numerical simulations including a primordial non-Gaussianity in the density fluctuation field quantified by the non-linear parameter f_{NL}. We have used MPGRAFIC code to produce initial conditions and the Adaptive Mesh Refinement (AMR) code RAMSES to evolve the scale structure formation. We estimated the higher order momenta of the initial distribution of density fluctuations, investigated the redshift evolution of the non-linear power spectrum and estimated the discrepancy introduced by the primordial non-Gaussianity in the non-linear power spectrum.

Keywords: Cosmology; Large-scale structure of Universe
PACS: 95.30.Sf; 98, 98.65 Dx, 98.80Bp

AUGER, HIRES, RELATIVITY, COSMIC RAYS, SUPERBRADYONS, DARK MATTER, DARK ENERGY (ABSTRACT)

Luis Gonzalez-Mestres

LAPP, CNRS/IN2P3 - Université de Savoie, B.P. 110, 74941 Annecy-le-Vieux Cedex, France.

Taking into account the experimental results of the HiRes and AUGER collaborations, the present status of bounds on Lorentz symmetry violation (LSV) patterns is discussed. Although significant constraints will emerge, a wide range of models and values of parameters will still be left open. Cosmological implications of allowed LSV patterns are discussed focusing on the origin of our Universe, the cosmological constant, dark matter and dark energy. Superbradyons (superluminal preons) may be the actual constituents of vacuum and of standard particles, and form equally a cosmological sea leading to new forms of dark matter and dark energy.

Keywords: Auger, HiRes, relativity, cosmic rays, superbradyons, dark matter, dark energy
PACS: 11.30.Cp, 12.60.Rc, 95.30.Cq, 95.35.+d, 95.36.+x, 98.70.Sa

COSMOLOGY OF WEYL'S GAUGE GRAVITY (ABSTRACT)

T.Maki[1] and K. Shiraishi[2]

[1]*Japan Women's College of Physical Education, Setagaya, Tokyo 157-8565, Japan*

[2]*Faculty of Science, Yamaguchi University, Yoshida, Yamaguchi-shi, Yamaguchi 753-8512, Japan*

In order to solve cosmological problems, many authors have proposed various cosmo- logical models, e.g. chaotic inflation, k-inflation, and vector inflation. While these models would be successful in solving the problems, these models are not supported by fundamental physical the- ories. At the early stage of the universe, it can be expected that the gravity theory had a hidden symmetry near the plank scale, which Einstein theory of gravity does not. Weyl's gauge gravity is a scale invariant and can naturally introduce scalar and vector bosons. We investigate Weyl's gauge gravity in arbitrary dimensions and apply the early universe. The behaviors of scale factors are demonstrated in 4 and 5-dimensions.

Keywords: Weyl's gauge gravity, Cosmology
PACS: 04.50.Kd

SCALAR FIELD IN THE JORDAN-BRANS-DICKE THEORY AND DARK ENERGY (ABSTRACT)

E.V. Chubaryan, R.M. Avagyan, G.G. Harutunyan and A.S. Piloyan

Yerevan State University, Faculty of Physics, Alex Manoogian 1, Yerevan 0025, Armenia

The article considers cosmological models in the framework of the Jordan-Brans-Dicke (JBD) theory in the Einstein frame. It separately discusses the cases of scalar field domination, as well as the presence of cosmological constant Λ, with matter, which is described by the barometric equation of state $P = \alpha\varepsilon$ (P-pressure, ε–energy density of matter). The analysis of obtained results is carried out in the light of modern observational data. It is shown that the contribution of the scalar field and Λ-term ($\Lambda > 0$) in the case of $q = -\frac{1}{2}$ ($q = \frac{\ddot{a}a}{\dot{a}^2}$ decelerating parameter) compensate each other, and the situation becomes like in Einstein's theory.

Keywords: Relativity and gravitation, Dark energy, Cosmology, Particle-theory and field-theory models of the early Universe (including cosmic pancakes, cosmic strings, chaotic phenomena, inflationary universe, etc), Observational cosmology (including Hubble constant, distance scale, cosmological constant, early Universe, etc), Mathematical and relativistic aspects of cosmology.
PACS: 95.30.Sf, 95.36.+x, 98.80.-k, 98.80.Cq, 98.80.Es, 98.80.Jk

A NEW MODEL FOR CLAUSIUS'VIRIAL MAXIMUM THEORY: EXPLANATION OF GALAXY FUNDAMENTAL PLANE *TILT* (ABSTRACT)

D. Bindoni, L. Secco and E. Contini

Department of Astronomy, University of Padova, Padova, Italy

The theory of the Clausius' virial maximum proposed by Secco (2000, 2001, 2005) is based on the existence of a maximum in the Clausius' Virial potential energy of a baryonic component when it is completely embedded inside a dark matter (DM) halo. When this maximum is reached the baryonic inner component acquires a scale length in its own gravity field due to the induced effect on it from the dark matter gravitation. The existence of this special configuration among the infinity number of those an inner component may assume inside an outer DM halo is able to explain the typical scale relationships observed for the early type galaxies (ETGs) which derive from their lying on the Fundamental Plane (FP). To prove that we model an ETG in the following way: an inner baryonic component of stellar kind built-up by a King-model with a cut-off and a DM halo with a cored power law density profile. The two components, which are gravitationally interacting, settle on virial equilibrium after violent relaxation. The new improved model allows us to understand more deeply the dynamical reason of the FP existence and in particular why it has a *tilt*. The main puzzle may be also fully understood that is why the FP is degenerated with respect to the initial density perturbation spectrum in a CDM hierarchical clustering cosmological scenario.

Keywords: Galaxies: Clusters; Cosmology; Celestial Mechanics, Stellar Dynamics.
PACS: 98.65.Cw; 98.80.-k; 95.10.Ce; 98.10.+z.

ASTROPHYSICAL APPLICATION OF STANDARD MODEL PARAMETERS (ABSTRACT)

Sergey Sukhoruchkin

Petersburg Nuclear Physics Institute 188300 Gatchina Russia

We discuss relations in particle mass spectrum and consider results of analysis of spacing distributions in nuclear spectra which show the distinguished character of intervals related to electron mass and nucleon mass splitting.

Keywords: Standard Model, radiative corrections, particle masses.
PACS: 3.40.Dk; 14.20.Dh; 21.10.-k; 12.15.Lk

Author Index

A

B

C

L

M

N

O

P

R

S

T

U

V

W

Y

Z